Biotechnology - Biosensors, Biomaterials and Tissue Engineering Annual Volume 2023

Edited by Luis Jesús Villarreal-Gómez

Published in London, United Kingdom

Biotechnology - Biosensors, Biomaterials and Tissue Engineering Annual Volume 2023
http://dx.doi.org/10.5772/intechopen.111653
Edited by Luis Jesús Villarreal-Gómez

Contributors
Babatunde Olamide Omiyale, Akeem Abiodun Rasheed, Temitope Olumide Olugbade, Robinson Omoboyode Akinnusi, Susan McDonnell, Jessica Whelan, Maycou Soares Zamprognio, Raymon Floyd Principe, Toonika Rinken, Kairi Kivirand, Eerik Jõgi, Rasmus Rohtla, Hao Yu, Zhihai Han, Cunrong Chen, Leisheng Zhang, Emilija Zdraveva, Budimir Mijovic, María Paulina Romero Obando, David Giancarlo García Vélez, Karina Janneri Lagos Álvarez, Faith H.N. Howard, Zijian Gao, Hawari Bin Mansor, Munitta Muthana, Zidi Yang, Abderrahmane Benkhalifa, Fernando Gonzales-Zubiate, Kevin Castillo-Mendieta, Jimmy Arias, Sirajudheen Anwar, Jonaid Ahmad Malik, Jeba AjgarAnsari, Sakeel Ahmed, Archana Rani, Shabana Yasmeen Ansari

First published in London, United Kingdom, 2023 by IntechOpen
IntechOpen is the global imprint of INTECHOPEN LIMITED, registered in England and Wales, registration number: 11086078, 5 Princes Gate Court, London, SW7 2QJ, United Kingdom
Printed in Croatia

British Library Cataloguing-in-Publication Data
A catalogue record for this book is available from the British Library

Additional hard and PDF copies can be obtained from orders@intechopen.com

Biotechnology - Biosensors, Biomaterials and Tissue Engineering Annual Volume 2023
Edited by Luis Jesús Villarreal-Gómez
p. cm.

This title is part of the Biomedical Engineering Book Series, Volume 20
Series Editor: Robert Koprowski

Print ISBN 978-0-85014-090-3
Online ISBN 978-0-85014-091-0
eBook (PDF) ISBN 978-0-85014-092-7
ISSN 2631-5343

IntechOpen Book Series

Biomedical Engineering

Volume 20

Aims and Scope of the Series

Biomedical Engineering is one of the fastest-growing interdisciplinary branches of science and industry. The combination of electronics and computer science with biology and medicine has improved patient diagnosis, reduced rehabilitation time, and helped to facilitate a better quality of life. Nowadays, all medical imaging devices, medical instruments, or new laboratory techniques result from the cooperation of specialists in various fields. The series of Biomedical Engineering books covers such areas of knowledge as chemistry, physics, electronics, medicine, and biology. This series is intended for doctors, engineers, and scientists involved in biomedical engineering or those wanting to start working in this field.

Meet the Series Editor

Robert Koprowski, MD (1997), Ph.D. (2003), Habilitation (2015), is an employee of the University of Silesia, Poland, Institute of Computer Science, Department of Biomedical Computer Systems. For 20 years, he has studied the analysis and processing of biomedical images, emphasizing the full automation of measurement for a large inter-individual variability of patients. Dr. Koprowski has authored more than a hundred research papers with dozens in impact factor (IF) journals and has authored or co-authored six books. Additionally, he is the author of several national and international patents in the field of biomedical devices and imaging. Since 2011, he has been a reviewer of grants and projects (including EU projects) in biomedical engineering.

Meet the Topic Editor

Dr. Luis Jesús Villarreal-Gómez is a research professor from the Faculty of Engineering and Technology Sciences, Autonomous University of Baja California, México. Dr. Villarreal is the editor-in-chief and founder of the *Revista de Ciencias Tecnológicas (Journal of Technological Sciences)* (RECIT) and is a member of several editorial and reviewer boards for numerous international journals. He has published more than 40 international papers and reviewed more than 171 manuscripts. His research interests include biomaterials, nanomaterials, bioengineering, biosensors, drug delivery systems, and tissue engineering.

Contents

Preface

This book includes ten chapters organized into four sections on "Biotechnology", "Biomaterials", "Tissue Engineering", and "Biosensors". Each chapter describes and discusses the state of the art, perspectives, and end technology of some applications in biomaterials, tissue engineering, and biosensors.

Section 1 includes two chapters. Chapter 1, "Inventoried Yeast Species in Algeria", proposes and explains biotechnology strategies that allow for the better use of yeast to benefit agricultural, environmental, food, animal, and human health. Although the case study in this chapter is from Algeria, the work can be replicated elsewhere in the world.

Chapter 2, "Upgrading Non-conventional Yeasts into Valuable Biofactories", explores the use of synthetic yeast, which promotes the production of relevant chemical compounds from biofuels to therapeutics biomolecules. *Saccharomyces cerevisiae* is the most studied yeast and has served as a model for the study of expression systems with higher reproducible results. The chapter defines the status of the most used non-conventional yeasts in metabolite production as well as the optimization engineering technologies to regulate protein expression of Pichia pastoris, *Kluyveromyces marxianus, Kluyveromyces lactis,* and *Yarrowia lipolytica*.

Section 2 includes three chapters. Chapter 3, "Silk Fibroin Nanoparticles: A Biocompatible Multi-Functional Polymer for Drug Delivery", discusses the properties of silk fibroin and biopolymers such as polylactide (PLA), poly (lactic-co-glycolic acid) (PLGA), and collagen. It also describes the structure and synthesis of silk fibroin nanoparticles. The chapter examines the use of the discussed biopolymers as nanocarriers in drug delivery systems for cancer treatment.

Chapter 4, "Emerging Selenium Nanoparticles for CNS Intervention", discusses the potential of using selenium nanoparticles (SeNPs) as a therapeutic agent in the treatment of brain pathologies. It explains the properties of nanoparticles and how these properties can be applied as nanocarriers. The chapter also highlights new developments in the application of nanoparticles for diagnostics.

Chapter 5, "Antibacterial Strategies: Photodynamic and Photothermal Treatments Based on Carbon-Based Materials", discusses the problem of bacterial resistance to antibiotics and how conventional treatments are losing efficacy for bacterial infection. This chapter proposes new therapies using multifunctional materials. It discusses photodynamic and photothermal technologies and analyzes the physico-chemical, biological, and mechanical properties of photosensitizers and photothermal agents.

Section 3 includes two chapters. Chapter 6, "Frontier Electrospun Fibers for Nanomedical Applications", describes the electrospinning technique, which is used to produce electrospun nanofibers, which are ideal for several applications such as tissue engineering, drug delivery systems, and biosensors, among others. The chapter examines the advantages and limitations of nanofibers and proposes future perspectives.

Chapter 7, "Influence of Mechanical Properties of Biomaterials on the Reconstruction of Biomedical Parts via Additive Manufacturing Techniques: An Overview", describes the state of the art of additive manufacturing and 3D printing approaches to fabricate organs of biological scaffolds for tissue engineering. It also discusses the biomaterials that can be used in these technologies. The chapter presents the final properties of printed biomaterials, focusing on cartilage and skin tissues.

Section 4 includes three chapters. Chapter 8, "Challenges and Emerging Technologies in Biomanufacturing of Monoclonal Antibodies (mAbs)", discusses the current biological and engineering challenges for the manufacturing of antibodies, developments in cell line engineering, intensified processing, continuous manufacturing, automation of and innovations in process analytical technologies, and single-use technologies.

Chapter 9, "Nanomaterials as Novel Biomarkers for Cancer Nanotheranostics: State of the Art", highlights the role of nanotechnology and nanotheranostics in early diagnosis and treatment of cancer. Lastly, Chapter 10, "Biosensor for the Detection of Cyanobacterial Toxin Microcystin-LR", explains the challenges of cyanobacteria present in the environment, its proliferation conditions, and how a biosensor detecting microcystin-LR (MC-LR) molecule can be a good strategy for monitoring and controlling this microorganism.

Dr. Luis Jesús Villarreal-Gómez
Research Professor,
Faculty of Engineering and Technology Sciences,
Autonomous University of Baja California,
Tijuana, Baja California, México

Section 1

Biotechnology

Chapter 1

Inventoried Yeast Species in Algeria

Abderrahmane Benkhalifa

Abstract

In Algeria, the study of yeasts remained marginalized for a long time. One of the supposed causes would be the reduction of examples in the school system. In medicine, species are identified because of their pathogenicity. But in food production and other fields, yeasts are mentioned as quantification in the quality-control process as well as molds. In addition to homemade breads, traditions of fermentation involving yeasts are the process of other local products, such as dairy products, vinegars, beverages, and authentic cheeses. Yeasts affect fruits and plants but also increase yields and protect species from other pathogenic microorganisms. Some conscious researchers have looked into the prospecting of yeast showing their properties and evaluating their interest, adopting biotechnology approaches, and covering several environments. 131 taxa are listed in this first compilation with 27 species in human health, 30 in animal health, 27 from dairy products, 24 taxa isolated from soils, 41 from agricultural products, and 17 concerned environmental purposes. Biotechnologies concern 44 taxa in varied topics of biomass, ethanol, vinegar, enzymes, and esters. Sixteen selected natural products inhibit 14 yeast species. Many isolated strains are promising in agriculture, agri-food, and biotechnologies and present new economic prospects. The idea of national depository is proposed.

Keywords: yeasts, biodiversity, Algeria, new economies, zymology

1. Introduction

Inventory of yeasts went from around 500 species in the mid-1980s to over 1500 species in 2010–2011 [1]. Biosystematics approaches leave many taxa in the midst of perpetual updating of their phylogenetic affiliations. Almost all species are hectic in the panoply of appellations. Moreover, interest in yeasts has widened considerably, since their development has opened up new horizons in various biotechnological uses, including enzymes and biofuels [2, 3]. The inventory of yeasts and their outlets is enriched each year with new species and new applications. Specialists know that only a tiny part is discovered [4]. In Algeria, the yeast inventory has not yet been established. We do not pretend to do it in a such rapid trial, but we try to perfect our first issue [5]. Within the national reports implementing the Convention on Biodiversity (CBD), the inventory of mushrooms needs serious updating. That of yeasts is totally neglected because it is included within fungi and they stay much less explored. Fungi and yeast remain limited in view of the attractiveness of aromatic and edible plants or those mentioned in traditional care, which are potentially targeted for their beneficial impact on human health. But looking at cultural heritage, homemade bread, vinegar, and drinks are very rich in know-how. Local peoples conserve the use of fermented foods [6], vinegar [7], and varied dairy products [8–11]. In addition to home beard, Matmoura is an exceptional preservation mode of fermented wheat [12]. It is used to prepare a famous couscous called Hamoum, whose probiotic property has been proven [13].

We are undertaking a review here to highlight the richness and interest of yeasts in order to align them with the rest of the inventories of fauna, flora, and other micro-organisms. We cover those of human and animal infections and those identified in research work devoted to food processes and fermented products as well as environmental studies. We seek to promote identified taxa or isolated strains within national research studies. It demonstrates the links between ancestral practices, interesting local products from traditions, and their socioeconomic issues. In parallel, we focus on the educational strategy to upgrade the standardization of taxa and revitalize their management as real segments of biodiversity at the service of the economy.

2. Methodology

To enrich the list of inventoried yeast species in Algeria, we spontaneously explored scholar literatures using Research gate and Google Scholar providing references in varied disciplines. They let quick access to what Algerians publish because they are the first concerned with this awareness documentation. Keywords were; yeast, fermentation, fermented foods, enzyme production, selected yeast strains... We searched through free academic search engines; PubMed, Isidor Pascal-Archives, and Agritrop. To complete local documents, such as theses and masters, we used the Algerian theses website (https://www.theses-algerie.com/), which is an aggregator portal of national universities and research institutions. We try to constitute a common list with actual taxonomic names but also with respect to those used by authors. Thus, we mentioned in some case synonyms, or the link between anamorphic and teleomorph forms. To verify favorite appellations, we used the taxonomic browser of the NCBI or the Mycobank database web site.

3. Yeasts between anthropological practices and modern economy

In addition to bread and vinegar, fundamental work is essential to enlighten the choice between traditions and modernity. There are major stakes in terms of impact on health and economy. This leads us to look into the inventory of *taxa* or even strains linked to the development of traditional leavens or the manufacture of baker's yeast in the face of the thorny problem of its importation.

3.1 Baker's yeast import

According to the National News Agency [14], requirement of baker's yeast is around 150,000 tons and we import over 100 million tons. Available data are those reported by the World Bank [15]. Top exporters to Algeria are presented in **Table 1**. The decrease observed during 2019 and 2020 is recovered in 2021 with an increase in the total imported rate. Top six countries export to Algeria less of 50,000 tons what donot corroborate with media divulgation.

(1) Worldbank Active exports	2018		2019		2020		2021	
	Trade value (1000 US$)	**Quantity (Kg)**	**Trade value (1000 US$)**	**Quantity (Kg)**	**Trade value (1000 US$)**	**Quantity (Kg)**	**Trade value (1000 US$)**	**Quantity (Kg)**
Turkey	20,693,21	9,604,790	16,075.34	7,505,800	16,966.66	8,333,850	23,173.34	11,247,600
Russian Federation	12,110,00	6,410,430	12,478.19	6,492,630	5199.70	2,949,320	10,290.50	5,039,540
Egypt, Arab Rep.	999,24	459,486	108.07	45,000	574.17	194,095	7107.63	2,117,280
China	333,39	346,250	2451.16	1,414,300	5861.58	3,092,980	2159.84	1,100,810
Morocco	100,11	48,600	222.83	106,716	805.50	411,885	1100.87	516,760
European Union	312,99	142,240	786.37	351,831	330.21	148,602	657.16	304,777
Ukraine	—	—	251,48	124,740	—	—	—	—
Tunisia	65,88	35,000	21,13	10,000	—	—	—	—
Iran, Islamic Rep.	136,00	68,000	—	—	—	—	—	—
Canada	111,96	64,800	—	—	—	—	—	—
(2) Customs Reports	77,420,00		82,390,00		68,580,00			

Table 1.
Top yeast exporters to Algeria (2018–2021) [15] compared to trade value from national customs reports [16].

From the National Customs reports, data concerning the quantity are not available in the consulted references [16]. Trade values are higher due to the grouping under reported item yeasts with other micro-organisms and baking powders. Public opinion is surprised that national production does not meet our industrial and domestic needs. Yet, giant infrastructures are established in Algeria, including that of Bechegouf in Guelma which should produce 30% of the national need [17].

3.2 Homemade sourdough and traditional breading

Through social media, there are more initiatives to promote homemade sourdough. Thus have benefited volunteers whose make efforts in showing incredible fantasies of traditions in bread making as well as other subjects of ancestral food process [18]. Sometimes, it is accepted to improve understanding and adherence with time respect and temperature control as the use of cold storage equipment. One of the YouTube descriptions is given by Y. Sellam-Benlemaalem, an agronomist basically. Having worked as a teacher of ecology for a long time, she become a recognized chef for defending local culinary traditions while accepting modernity. In her description of sourdough, she reminds us of a rare process of preserving sourdough by drying it after flattening and cutting it into thin slices, to be preserved for several years. https://www.youtube.com/watch?v=lu_FRmzBe-M&t=256s

Rural families prepare their own sourdough. This practice has been resumed in particular during recent years of the spread of COVID-19. Vernacular names are "Khemira Beldya" homemade sourdough; "Khemirat Dar," housemade sourdough; "Khemira Mahaliyah": local sourdough; "Khemira Arbiyah" Arabic Sourdough. All indicate traditional sourdough despite the fact that this practice oscillates, jealously, between maintenance and abandonment among young people and those in urban areas, particularly [19]. Homemade sourdough can be obtained automatically by soaking and emitting whole wheat flour from wheat and barley or by adding diverse sources of inoculum (Annexe I) like dates, figs, beans, watermelon juice, vinegar, whey, and curdled milk as well as fruits, leaves, and stems of some spontaneous plants, which are generally not toxic or have harmful effects such like sorrel plants *Rumex bucephalophorus* or *Rumex acetosa*. These species are widespread in rainy regions throughout North Africa and are known for their sour taste. This is the reason why it is called in Arabic "Hammeida or Hommaydha" meaning acid taste. Another originality concerns sourdough from beans inoculation, which is used in particular in Tlemcen and Oran (Western Algeria), to prepare remarkable rich soup. Unfortunately, this recipe was replaced with the use of manufactured yeast and lemon juice [20]. Actually, this tradition collapses day after day.

Sourdough starter still simple and easy to prepare—you need water and flour according to your preference. The spontaneous bacterial and yeast present in the air or that exist in the flour, especially wholemeal, will do the rest (**Figure 1**). If not, we add to it inoculum showed before but without manufactured yeast.

Under an optical microscope, the traditional sourdough shows different kind of cellular forms, oval and ovoid, and a few crescent shapes, lenticels (**Figure 2**). Smaller cylindrical shapes with interconnected spheroids in the form of rosaries are likely bacterial species. It is delicate to examine sourdough, but with dilution, we do not consider starch granules. Wild *Saccharomyces* cells can be with different diameters ranging from single to double. Sourdough is fermented by wild strains of yeast and bacteria spontaneously. *Saccharomyces cerevisiae* constitutes the species responsible for sourdough processing. It is shown that sourdough strains, which are diploid, have

Figure 1.
Sourdoughs obtained using (a) hard wheat flour (left) (b) hard wheat, soft wheat and barley flours (right).

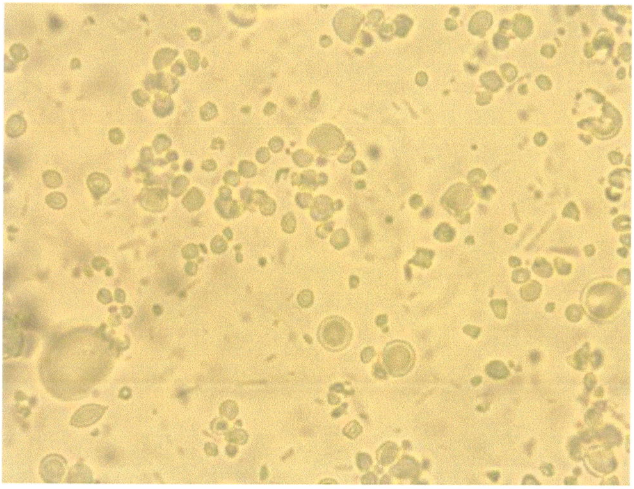

Figure 2.
Morphology of spontaneous cells observed from sourdough under a Carl Zeiss Axiostar microscope × 1000.

high copy number of genes able to use maltose, whereas industrial bakery strains are tetraploid with a rapid fermentation onset and are more efficient in CO_2 production [20]. This suggests an interest in the selection of spontaneous strains from an industry-oriented performance evaluation perspective.

In El-Oued (South-eastern Algeria) the sourdough is dried due to the favorable drying conditions in the Sahara. It is then stored in powder form like commercial yeast (**Figure 3**). This technique has become teached in recent years in the vocational training institute. The gritty brownish color is due to the use of dates in the sourdough flour mixture.

3.3 Traditional vinegar

In Algeria, vinegar is a particular case of fermentation because it is obtained from diversified sources of fruits, such as dates, apple and sometimes from pomegranate (**Figures 4** and **5**). The new one is obtained from the Indian fig (known as prickly pear), the fruit of *Opuntia ficus-indica*. Between homemade vinegars, or theses

Figure 3.
Powder sourdough obtained from local wheat flour mixed with dates (prepared by Ali Menai at the professional training Institute in El-Oued, Algeria).

Figure 4.
Left to right: Vinegar from apple (a), dates (b,c), and prickly pear (d).

proposed by companies and chemical vinegars, there are choices to be favored. In the case of the date vinegar, the fruit biomass is added with a few of other ingredients (wheat seeds, barley seeds, harmel or wild rue seeds, coriander seeds, a pinch of salt and a pinch of chili, and also two iron nails). Then, it is emerged in double quantity of water and maintained in preservation within 40–45 days [7]. During this time is done the conversion of sugar into alcohol and with the presence of acetic bacteria it transform by oxidization into vinegar, mainly with a concentration of acetic acid and distinct flavors due to the fruit parameters and the presence of other acids (malic acid in case of apple). In the case of dates, in small quantities, acetic acid, butandiol, propanone [21], amino acids, vitamins, and formic acid are also present [7]. During meals, the consumption of vinegar gives a feeling of fullness and thus limits the

Figure 5.
Vinegar manufacturing in El-Oued (Algeria) with a capacity of 3000 L.

quantities of food consumed or exaggerated. Vinegar has many properties with positive health impacts. It is reported that vinegar: 1) regulates blood sugar by improving insulin secretion, 2) curbs obesity by suppressing fats accumulation, 3) could increase HDL-cholesterol and diminishes LDL-cholesterol levels, and 4) inhibits proliferation and induces apoptosis in human cancer cells [22]. In traditional care, vinegar reduces fevers essentially. According to an ethnological survey of families in Ghardaia, more than 20 cases of date vinegar virtues are registered [23]. Like, in sourdough, we should focus also very carefully on the type of yeast that we should choose among the spontaneous yeasts, which could be observed in the traditional way or others among the strains showing better transformation of sugar into alcohol. Recently, it is demonstrated that a selected *Kasachstania unispora* strain showed performance in sourdough environments compared with commercial *S. ceverisiae,* in particular, under stress condition as acetic acid concentration, ethanol, or salinity [24].

3.4 Dairy products

The government gives specific importance to milk importation in regard to the consumption needs of more than 1500 MUS$ in 2020 (**Table 2**). This represents 15–20% of importation fees. Cheese and similar takes more than 7–8% of dairy products (**Figure 6**).

Algerian milk production is ensured by cows, sheep, goats, and a small part by camels. Cow's milk constitutes the highest share of production with more than 71% but due to the need for food, the production of cow's milk is decreasing (**Figure 5**). That of sheep and goats is maintained to ensure a little less than the third. Camel milk hardly exceeds 15,000 tons, but it remains essential to Bedouins in arid and Saharan regions. The total quantities expressed in tones are much lower than what is requested by a population that will reach 50 million inhabitants in the next few years. With 3.3 million tons per year, the maximum quota is less than 2 liters per person per week. Thus, we

	Trade Value (1000 US$)					
All milk and dairy products	2018		2019		2020	
	1,401,090		1,245,910		1,549,740	
Grated or powdered cheese and curds	107,030	7,64%	91,690	7,36%	119,850	7,73%

Table 2.
Trade values of imported milk and dairy products in Algeria (2018–2020) [16].

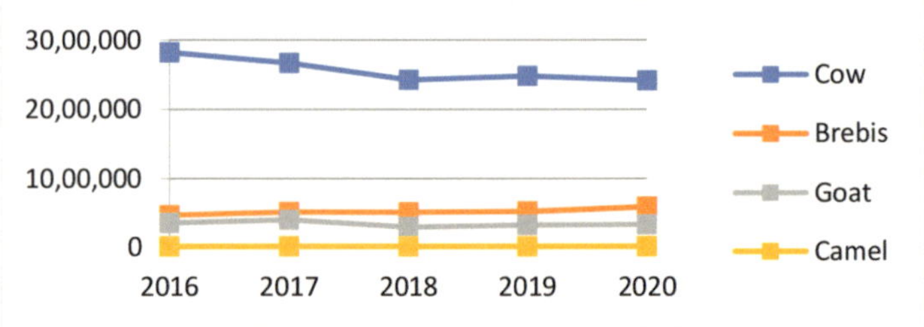

Figure 6.
Production of milk in tons in Algeria 2016–2020 (source FAO stat).

have to review the mode of consumption. Milk transformation to other products offers an economic best consumption. Fermentation is one of the success keys. Sure, there is a complexity between the import, supported by the government, and the development of local breeding. Sustained imports may not remain the ideal solution at all times. In both urban and rural areas, residents need to practice milk preservation and milk processing to prolong their nutritional components and balance their savings.

Like humans, animal health is essential. The mastitis is given as a prior problem. Dairy products are fragile and face contaminations risk. Safe row milk or fermented milk depends on respect of traditional processing like in control of modern cheeses. Cow's milk is essential to ensure the large needs but it decreases as the food load becomes more and more expensive. Algeria is mainly affected by drought and must do its utmost to fight against desertification. But, there are many types of waste that require transformation. Biomass is geared toward animal feed or other needs like energy. In addition to local knowledge, yeast should help us achieve attended goals of milk production. There are at least 20 dairy products (Annexe II). The richness of vocabulary proves how people depend on dairy products [8, 10, 25–36]. Named products have been repeatedly described in various works but they havenot been developed yet. We are proud of local products as nostalgia but we still have to deepen their social and economic characteristics to guarantee the legacy on a healthy and profitable basis. Even large-scale, family-based milk producers extract butter to be consumed or processed into cooked butter called Smen/Dhan [8, 10, 25, 26]. Those are the two products extracted from milk as lipid components, and they represent a heritage food tradition. Indication of non-identified species in the case of Smen from camel milk in western regions, Bechar, Moghrar, Ain Sefra, and Saida [26] need to make attention in analyzing dairy products more carefully. Even if it was found to contain a slight percentage of yeast (2.08–3.88 cu ft./g) despite the morphological description of the described isolates, it was not possible to identify them. This suggests to analyzed yeasts archive and products one by one in the future.

The churning of milk also allows the production of L'Ben and in a quantity that often exceeds the need for daily consumption of families. It is therefore sold if not transformed in turn into several modes of fresh or dry cheese products according to the knowledge and know-how of the populations. The consumption of curdled milk and L'Ben, the compound obtained after churning, is very famous in Algeria as dairy products. L'Ben is usually flavored with Phoenician Juniper (*Juniperus phoenicea*) dry powder (**Figure 7a**). Next are the traditional cheese-making methods even in competition with the adoption of modern cheese techniques, which constitute a regional emblem not dissociated from tradition. Jben is so popular (**Figure 7b**) but the case of Jben Al-Gafs is endemic to the Boussaada district and has a real maturity during 2 weeks (**Figure 8a** and **b**).

3.5 Fermented wheat

There is an interesting case related to ancestral food security is the technique of preserving wheat underground. Matmoura is a traditional method of preservation observed in many places in North Africa, even in Egypt. It consists in placing the wheat production crop underground, which is managed differently due to the nature

(a) (b)

Figure 7.
Ancestral traditional dairy products: Left to right; (a). L'Ben sprinkled with Phoenician Juniper, (b). Jben presented on fig leaf.

(a) (b)

Figure 8.
A, b. Jben Al-Gafs, showing ripening effect as real authentic cheese from Bousâada, Algeria (photo is imported from the website vitaminedz.com).

of the soil and the need of the families. The ground is leveled up next to the fields and the sides and bottom of the pit are covered with straw. Wheat crop is stored in the pit for several years. Because of the moisture and some water leakage, the sides ferment only. During the re-extraction of the wheat for use, the fermented dark grains are separated by themselves and used to prepare a special couscous known as Hamoum because of its dark color. This technique is still practiced in north-center and north-western Algeria, but it has become a rare product. Therefore, revealing its health secrets [12, 13, 37] and distinctive taste preoccupies some researchers and those interested in re-considering it as a cultural heritage. The subject of this fermented wheat is an authentic case of fermented food which needs deep research in microorganism identification as well as yeast. First exploration showed fourth interesting species by decreasing numeric importance on occurring culture mediums: *Saccharomyces pastorianus* (50%), *Saccharomyces boulardii* (39%), *Schizosaccharomyces pombe* (5%), and *Saccharomyces cerevisiae* (1%) [38]. This step will encourage prospecting other fermented foods and beverages and allow registering all places where Matrmoura was used as a conservation technique.

3.6 Other ethnobiological cases

We have not verified ethnological practices where yeasts are used in the care or in treatment of diseases, but in the Gourara region in the Algerian Sahara, there is a particular practice where the itching of the hands and the symptoms of fungal attacks are treated by dipping the patient's hands several times into the traditional poultry trough. This traditional material is made of pottery. It is assumed that in such a process there is probably the effect of particular yeasts but this hypothesis remains to be explored by survey in those regions where the same practice is done and try to isolate strains it should exist.

We note from some research works the use of yeast as fortifying in animal feed, in particular, dairy cows using *Saccharomyces cerevisiae* to improve milk production [39–41]. So far we cannot consider this as community adoption, but it will be in the future due to the increasing need and cost of feed. Also, because an experiment is to be done with farmers or breeders and considering the imports of many brands, it is clear that using yeast as an animal feed supplement will be a common practice. The first observation is the negligence in guiding veterinarians to adopt species other than *S. cerevisiae* the ones that should be valued as waste and be more beneficial to breeders. *Candida utilis* would be interesting to improve the yield of the degradation of cellulose of vegetable waste and to offer in addition to the energy a pre-digested animal feed. It would be a shame to not favorite the production of yeasts and choose its easier and direct use in animal feeding, knowing that it is imported. The production trials of *S. cerevisiae* are in favor of the valorization of date waste over molasses, thus recovering large quantities of date waste. In this way, we would promote the use of common dates and improve the productivity of biomass.

4. Yeasts in the school program and training strategy

Yeasts are mentioned first in the medium cycle under the theme of fermentation. The strategy requires to be reviewed in its finality because objectives are not clearly specified even less the link between the disciplinary approaches and the means of implementation. The teaching of fermentation should not be reduced to the sole case of bread or baker's yeast. It should be extended to its practical vocation linked at least to diversified and rich traditions of foods, vinegar, and dairy products. We should start with healthy examples at this stage and encourage the discovery of a very interesting living world. The current textbook should push learners to make more practical efforts. We believe, this is where the curiosity flaws are etched in the minds of future generations. At the university, the subject of yeasts is treated in several fields, including Cytology, Botany, Microbiology, and Biotechnology. In general, the star remains the baker's yeast, *Saccharomyces cerevisiae,* from school levels to university and that during more than 150 years. Ultrastructure is described only to confirm the eukaryotic cell model without worrying too much about cell's composition or physiology. In some applied microbiological evaluation, *Candida albicans* constitutes the second example. Fortunately, in medical sciences and parasitology, the clinical needs cover several taxa: *Candida albicans*, *C. tropicalis*, *Candida parapsilosis*, *C. Krusie*, *C. glabrata*, *Naganichia alba*, *Sporothrix schenckii* and sometime others. Obviously, mastering epidemiology has an evident impact on the economy regarding the risk of mortality, duration of patient hospitalization, and cost of care conditions.

We have an interest to introduce the subject of yeasts as diversified living organisms. After testing this approach in cytology, as a fundamental course of graduation, we consider this as a crucial opportunity not to be overlooked in order to initiate learners to consider the case of yeasts like any other category of cells. Since the main objective is to essentially distinguish the cellular criteria of living organisms, we have proposed the adoption of the following flowchart (**Figure 9**). With a participatory approach based on the principle of biodiversity, we encourage learners to choose for each category their own examples of the previous basic culture as a prerequisite and also of their complementary curiosity according to extra muros efforts. We gain the advantage of quickly initiating young learners and even before university to channel their school culture to recognize cellular trains and thus create the need to prepare the ground for any learning of biosystematics. For yeasts, we solve from the first step their attachment to fungi and especially their complication of sometimes overlapping unicellular eukaryotes and multi-cellular eukaryotes.

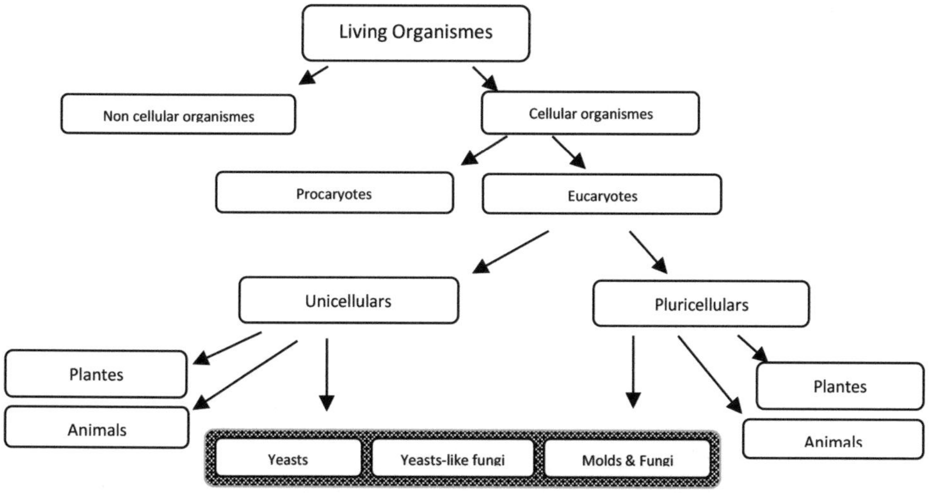

Figure 9.
Diagram of living organisms.

The curiosity of the young learners leads to favoring the choice of a few remarkable and captive criteria which distinguish yeasts from other organisms. There are three recent ones that should not be overlooked. That of the mode of division in budding or in fission without forgetting the recent discovery of the mixed mode by budding and fission at the same time and also the mode of star budding concerning the studied species [42]. The second consist of showing yeast diversity in their shape forms [4] and not focalize on the ovoid of elongated ones only. The third peculiarity is that of Ergosterol as a lipid compound in yeast membrane instead of cholesterol as in animal cells and phytosterols in plants. The rest of the sensitivity comes from the role to be discovered for each species and also from the possibilities of its appropriate culture to anticipate this or that biochemical or biotechnological exploitation.

5. Yeast inventories in Algerian research studies

Historically, Algeria is mentioned as the soil origin of type stain CBS 277 of *Pseudosaccharomyces africanus* isolated from a locality named Akbari, cited in Mycobank database and other [43]. Currently, the preferred name is *Hanseniaspora vineae* (Syn. *Kloeckera Africana*). The interest of this species concerns the increasing of flavor complexity with neutral grape varieties. Similar strain of this species was isolated in Japan and showed capacity in improving bread flavor [44]. Another strain of *Hanseniaspora osmophila* (*Kloeckera Africana*) was deposed in the UK, National Collection of Yeast Cultures in April 1920 and reported isolated from the previous habitat (Soil Akbari, Algeria) as mentioned under the number NCYC 26 [45]. Likewise, under the original epithet *Kloeckera Africana*, the strain DBVPG 6792 was deposed in 1993 in the Yeasts Collection (DBVPG) of the Dipartimento di Biologia Vegetale of the Universita' di Perugia, Italy [46].

In The Algerian Annals of Agronomy, Bremond E. tested the fermentation of mature bananas coming from Guinea in West Africa and unvaluable for the consumption to produce alcohol for the military service [47]. Regarding the purpose, active sourdough was used without the precision of yeast or any other microorganism species. At that time, they suggested that Algiers Pasteur Institute, as a professional contribution, would conserve and purify the purity of the selected strains. From that, we retain two important lessons; archive of yeast prospections will be done correctly. Any work has real importance with a clear impact on application in wine processing.

Identification of yeasts as species first concerned medical purposes to control pathogens. In parallel but more fragmented come some works confirming the role of yeasts in fermented foods or drinks. Those who focus on discovering them in environments or describing their benefits are much more recent.

5.1 Yeasts within human health

Human fungal infections including those caused by yeast and yeast-like fungi constitute a serious burden in Algeria [48–57]. At least 27 taxa are identified in care controls. Often in topic of medicine research, we observe in first *Candida* species with sometimes resistance phenomena to treatments [48, 54]. Inventoried species are: *Aureobasidium melanogenum* (*Aureobasidium pullulans* var. *melanogenum*), *C. albicans*, *Corynebacterium auris*, *C. dubliniensis*, *Candida famata*, *C. glabrata* (syn. *Torulopsis glabrata*), *C. Kefyr*, *C. Krusie*, *Candida lusitaniae* (Syn. *Clavispora lusitaniae*), *C. orthopsilosis*, *C. parapsilosis*, *C. rugosa*, *C. tropicalis, Candida zeylanoides*, *Clavispora lusitaniae* (Teleomorph of *Candida lusitaniae*) *Cr. gattii*, *Cr. neoformans* var. *grubii*, *Cr. neoformans* var. *neoformans*, *Cr. neoformans*, *Lodderomyces elongiporous* (related to sex. Form of *Candida parapsilosis*), *Meyerozyma elongisporous*, *Naganishia albida* (*Cryptococcus albidus* var. *albidus*), *Naganishia liquefaciens, Pichia kudriavzevii* (Telomorph of *Candida krusei*), *Prototheca wickerhamii, Rhodotorula mucilaginosa*, *Saccharomyces cerevisiae* and *Trichosporon* sp.

This later was observed with tenia petis infection [49, 50]. *Saccharomyces ceverivisiae* was isolated from infant fees and showed that are safe that can be considered as probiotic [54] whereas *C. parapsilosis* and *C. albicans* have potential inhibitor effects on referential strains of *Escherichia coli* and *Staphylococcus aureus*. *Cr. gattii* [55] and *C. auris* [56] are reported for the first time in Algeria. In this later work, deep identification separated the subspecies level of Cr. neoformans into its two varieties *Cr. neoformans* var. *grubii* and *Cr. neoformans* var. *neoformans*. Only one case mentioned a denture contaminated by *C. albicans* [57], the *ex vivo* decontamination of this yeast-colonized denture was realized by iodine-thiocyanate.

The urban environment is considered as a potential risk of contamination by the inhalation of fungi spores, pigeon droppings, trees, and soils are surveyed some time in urban areas and around hospitals. *Candida* sp., *C. albidus*, *C. laurantii*, *Cryptococcus neoformans*, *Rhodotorula* sp., *Saccharomyces* sp., *S. cerevisiae*, and *Trichosporon* sp. are reported from Algiers the capital [55]. Also, in Annaba and from pigeon droppings, 6 isolated species were identified; *Cryptococcus albidus* and *Cryptococcus diffluens* (for the first time) were isolated, which represents an environmental risk for humans [58]. Prospected pigeon's droppings from Constantine showed *C. albicans*, *C. glabrata*, *C. parapsilosis*, *Cr. Terreus,* and *Cr. uniguttulatus* [59].

5.2 Yeasts in animal health

From diagnosed cow cases [60–64], a total of 30 species have been identified mainly causing mastitis. Goats [7], sheep and camels are certainly affected but no isolates have been identified in the research consulted during the elaboration of this presentation. Species are; *C. albicans*, *C. colliculosa*, *C. famata*, *C. glabrata*, *Candida guilliermondii*, *C. inconspicua*, *C. kefyr*, *C. krusei*, *Candida lambica*, *C. lusitaniae* (Anamorph of *Clavispora lusitaniae*), *C. parapsilosis*, *C. pseudotropicalis, C. rugosa*, *C. tropicalis*, *C. zeylanoides, Cryptococcus albidus, Cr. laurantii, Cr. Neoformans, Cr. terreus* (Syn. *Solicoccozyma terrea*), *Cutaneotrichosporon cutaneum* (Syn. *Trichosporon cutaneum*), *Geotrichum capitatum*, *Metschnikowia pulcherrima (Syn. Torulopsis pulcherrima), Pichia kudriavzevii* (Telomorph of *Candida krusei), Rhodotorula glutinis, R. rubra, S. cerevisiae, S. fragilis, Trichosporon* sp., *T. capitatum,* and *T. fermentans* (Syn. to *Dipodascus fermentans*). With a chance that *Pichia kudriavzevii* were observed in rare cases for the moment. This species presents a most remarkable resistance to the antifungal agent fluconazole. The essential oils of *Origanum floribundum*, *Rosmarinus officinalis,* and *Thymus ciliates* were tested and shown to be favorable against *Candida albicans* and can be considered as alternatives in the control of mastitis fungi [62]. Also, essential oils of *Cinnamomum cassia* (cited as *Cinnamomum aromaticum*) and *Syzygium aromaticum* were tested against seven isolated species *C. albicans, C. lambica, C. tropicalis, C. zeylanoides, Cryptococcus albidus, Cr. Laurentii,* and *Rhodotorula glutinis* and shown strong effects to be proposed as an alternative solution to face the fungal mastitis risk [63, 64].

5.3 Yeasts from soil and agriculture products

Logically, wine and soil were the past first subject of prospecting yeast, in particular, during colonial period. That was the reason to observe today's referenced strains, *Hanseniaspora* osmophila CBS 277 and *H. vineae* (syn. *Kloeckera africana*, cited as *Pseudosaccharomyces africanus*) and in gene banks [43, 45, 46]. *Hanseniaspora opuntiae* x *pseudoguilliermondii* DBVPG 5828 was isolated from Soil close to plum tree and deposited as a reference strain in 2010 [65]. Recently, 20 other taxa were isolated from soil and agriculture products in varied agriculture areas [66–74]. Isolated taxa are *Aureobasidium pullulans*, *Candida* sp., *C. glabra, Clavispora lusitaniae*, *Cryptococcus* sp., *Cr. aerius* (Yeast state of *Solicoccozyma aeria*), *Cr. magnus* (Syn. of *Filobasidium magnum*), *Debaryomyces hansenii*, *Hanseniaspora opuntiae*, *Hanseniaspora uvarum*, *Lipomyces* sp., *Meyerozyma guilliermondii* (Teleomorphic form of *C. guilliermondii* Syn. *P. guilliermondii*), *Phyllophorus anomalia, P. kluyveri*, *Rhodotorula mucilaginosa*, *Saccharomyces* sp., *S. cerevisiae*, *Schwanniomyces* sp. *Ustilago cynodontis*, *Yarrowia lipolytica.*

Selected promising strains are; *Schwanniomyces* sp. strain LB3 [67] for amylase production, *P. kluyveri* DBVPG 5826 showing killer activity [70], and *A. pullulans* for the evaluated activity of its polygalacturonase with admissible application to reduce the cloudiness of fruit juice [72], Compared with *Saccharomyces cerevisiae*, isolated from soil strains of *Debaryomyces hansenii*, *Meyerozyma guilliermondii,* and *Rhodotorula mucilaginosa* showed the preference for alkaline pH and interesting resistance to salinity and elevated temperature and have a potential of Plant Growth Promoting (PGP) function [69]. *Ustilago cynodontis* isolated from the Sebkha of Oran (Saline soil) has multi-enzyme activities as lipolytic, proteolytic, and cellulotic [71]. *C. glabrata* has an interesting & amylase activity [75].

Fruits naturally are a suitable yeast habitat. Grapes and dates are the first prospected fruits. Twelve isolated Species from grapes [76–78] are: *Candida pseudointermedia*, *C. solani*, *Hanseniaspora uvarum, Metschnikowia pulcherrima* (Syn. *Torulopsis pulcherrima*), *Pichia deserticola, P. fermentans, S. cerevisiae, Starmerella gropengiesseri* (cited with basioname: *Candida gropengiesseri*)*, Starmerella magnoliae* (cited as homotipic basioname *Candida magnoliae*), *Torulaspora delbrueckii*, *T. microellipsoides* (cited as *Zygosaccharomyces microellipsoides*), *T. pretoriensis*. From dates, 5 identified species were isolated; *S. cerevisiae* was isolated to select several adapted strains to produce alcohol or biomass [79, 80], *Clavispora lusitaniae*, *Hanseniaspora uvarum*, *Kodamaea ohmeri* were selected to test the production of Alcohol, flavor and amino acid [80, 81] and *Candida apicola* was isolated to be targeted as temperature resistant [82]. The natural fermented green Olives contain a least 4 taxa; *Candida* sp., *C. parapsilosis* and *Saccharomyces* sp. and *S. cerevisiae* [83]. *Candida boidinii* G5 (KF156789), isolated from spent olive (Chemlal variety), showed lipase activity [84].

Like dates must sugarcane molasses and fig were explored to isolate S. cerevisiae [85]. From other fruits and products, *Candida parapsilosis* was isolated from melon, *Zygosaccharomyces bailii* from Gherkin, and *Zygosaccharomyces rouxii* from honey [81]. To produce &-amylase, *Candida guilliermondii* (Syn. *Pichia guilliermondii*) and *C. tropicalis* were isolated from potatoes [86]. Recently, *Aureobasidium pullulans*, *Rhodotorula diobovata, Vishniacozyma tephrensis* were isolated from fruits and beet peels and showed extracellular enzyme synthesis [87].

Wheat seeds were explored to isolate 8 species; *Lipomyces kononenkoae*, *Rhodotorula mucilaginosa*, *Schwanniomyces occidentalis* (Syn. *Debaryomyces occidentalis*) [73], *Clavispora lusitaniae* ABS7 [88], *Meyerozyma caribbica*, *M. guilliermondii*, *Pichia guilliermondii* (Anamophic of *M. guilliermondii*), *Rhodotorula rubra* [89].

Fermented wheat conserved within Matrouma as ethnological technique permitted to silotae *Saccharomyces boulardii*, *S. cerevisiae*, *C. pastorianus*, *Schizosacchoromyces pombe* [38]. The latter is isolated for the first time in Algeria while it could exist in many products as well as those made with seeds or powder millet.

Research works are limited to the examination of efficiency in biomass production in particular to produce *S. cerevisiae* or oriented to valorize that in bioethanol. Those oriented to biotechnological was stay at the exploration stage and need more significant interest to go up the level of applied fields in industries.

In phytopathology, the case of fungi must be extended to yeasts because they are often associated with molds and yeasts. Yeasts are considered as sources for stopping fungal development. Two new strains, isolated from the Red Sea and identified belonging to the species *Candida orthopsilosis* and *Rhodotorula mucilaginosa*, showed improvement in wheat growth parameters and its resistance against *Fusarium oxysporum*, with a complete inhibition of zearalenone production in roots and ears [90]. This indicates an interesting and amazing perspective in testing yeasts against other fusariosis cases. Basically, *Metschnikowia pulcherrima* is an epiphytic species as others. Nectar and sweet fruits are welcoming to it. This species is must be appreciated for its strong antagonistic activities against pathogenic microorganisms without producing toxic components. Thus, it is used against plant pathogens like in breaking fruit invasion by *Botrytis cinerea* [91].

5.4 Yeasts in dairy products

Raw milk evaluated in North Western Algeria showed that yeast and fungi were detected in all ewes' raw milk samples [92]. The investigation of bovine mycotic mastitis in two departments of the northeast showed that 10.17% of the samples were positive [62]. Milk handling could be prior factor causing high yeast and mold loads. Other risk factors are due to the effect of the season and the distance between farm and dairy unit. When septic conditions are ensured we have to look for those spontaneous benefit species, in particular, if dairy products are conserved or transformed. 27 identified species were isolated from dairy products. *Y. lipolytica* is isolated from all king of milk [81, 93, 94]. From fermented bovine milk 7 identified species are *C. tropicalis, Issatchenkia orientalis* (Syn. *P. Kudriavezii*), *K. marxianus*, *Saprochaete suaveolens, Trichosporon coremiiforme*, *Wickerhamomyces anomalus, Yarrowia lipolytica* [93]. The strain L2 (ACKF156787) of the last one species is isolated to evaluate the lipase enzyme in traitment of olive waste [84]. *Schizosaccharomyces* sp. with similar strain to *S. octosporus* was isolated from whey [95]. From yogurt, *K. fragilis, K. marxianus* and a similar strain to *Schizosaccharomyces malidevorans* were isolated [95]. Cheese contains *C. lactis* [95] and *Geotrichum candidum* [94]. From camel milk 16 species were isolated; *Issatchenkia orientalis* (Syn. *Pichia Kudriavezii*), *Trichosporon coremiiforme, Yarrowia lipolytica* [81], *C. maris, C. parapsilosis, C. tartarivorans, C. tropicalis, C. zylanoides, C. lusitaniae, K. marxianus, M. Farinosa, P. fermentans, P; galeiformis, manshurica, R. mucilaginosa* [96] *Candida kefyr* was isolated from cow milk, whey [95] and fermented camel milk [97]. Cooked Butter (S'men) from Camel milk showed also the contamination of Saccharomyces cerevisiae and other *Saccharomyces* sp. [26].

To conserve high volatile value in fermented cow milk (Rayeb), while this term can mean also coagulated milk, 7 species have been isolated [93]. Four selected species *S. suaveolens, I. orientalis, K. marxianus,* and *W. Anomalus* produce esters that influence the taste and ensure organoleptic parameters of the dairy product. The double-drying process of interesting yeast such as *S. suaveolens, W. Anomalus,* deserves to be popularized and taught as a way to innovate or improve traditional processes. Modern cheese ripening is considered an exogenous process but traditional or local as soft or dried can be improved by understanding of the physiological strategy of the species involved. [98] leads with the commercial strain of *Geotrichum candidum* associated with *Penicillium camembertii* the choice of the amino acid that would be consumed most efficiently and the short peptides will be targeted. This experience showed clearly the interest to select from local spontaneous *G. candidum,* which is apparently very rarest. Stay with domesticated strains and remember that we are in the fourth group described previously that of species, basically beneficial but potentially harmful. Special attention is registered to valorise the famous case of *Candida kefyr* isolated from camel milk or yogurt [95, 97]. Another targeted strain of this species has highest enzyme activity (up to 5000 EU/ml) and highest level of single-cell protein [95]. Other identified *taxa* in this study were *Candida* sp. (similar to *C. pseudotropicalis*), *Kluyveromyces fragilis, K. lactis, K. marxianus*, *Schizosaccharomyces* sp. (similar to *S. octosporus*) and *Schizosaccharomyces* sp. (similar to *S. Malidevorans*, syn. of *S. pombe*). *Yarrowia lipolytica* is isolated from sheep, goat, and camel milk [93, 96, 97]. *Candida kefyr* was isolated from Camel fermented milk in an isolated oasis; Sebseb far from Metlili [97] as an interesting safe case. Local conditions and know-how of fermentation of camel milk are to be documented from this region.

The presence of yeasts is notified in the traditional cheese Bouhezza [31, 32], as well as the local product Michouna [33, 34] but not identified. In those studies like others [27–35] the orientation is given to evaluate bacteria. Yeasts are sometimes only quantified as Yeasts and molds even in extensive studies to evaluate the sensitive properties or the chemical characteristics of the famous Bouhezza cheese [99, 100]. Like Dried cheese named Klila, the famous cheese Jbeen Al-Gafs is little studied and was analyzed only for bacteria [11]. Authors certify that yeast and molds are abundant in the final stage after 14 days of maturation.

5.5 Yeast for environmental purposes

First attention to yeasts in the environment is commented previously under human health. 16 isolated species from soil, trees, and particularly drooping pigeons in and around hospitals in urban areas [55, 58, 59] are *Candida* sp., *C. albicans, C. glabrata, C. parapsilosis, C. tropicalis, C. zeylanoides, Cr. albidus, Cr. diffluens* (Syn. of *Naganishia diffluens*), *Cr. laurentii, Cr. neoformans*, *Cr. terreus* (Syn. *Solicoccozyma terrea*), *Cr. uniguttulatus*, *Rhodotorula* sp., *S. cerevisiae,* and others *Saccharomyces* sp. and *Trichosporon* sp. We also consider all species of yeast counted as food and dairy contaminants as environmental agents. From beaches, sand affected or under risk to be affected by polluted water or charge of waste were prospected in the western region near Oran [101]. Six registered taxa are; *C. albicans, C. zeylanoides, Cr. albidus, Geotrichum* sp., *Rhodotorula* sp., and *S. cerevisiae.* What remains to be done for this purpose? is to compare this with a clean beach. The wastewater facilities of Mascara in western Algeria were examined to isolate four potentially pathogenic yeasts, *C. albicans, C. glabrata*, *Cr. Neoformans,* and *Trichosporon* sp. [102]. At least 18 species are isolated from particular environments. Recently, *Geotrichum candidum* was isolated from soils that have been used as a depository of electronic and electrical waste for more than 20 years, around Annaba in the eastern part, and was described as able to degrade *in vitro* waste of batteries and circuit boards [103]. From lakes, Chott or sebkha (saline soils) previously cited studies were reported in the section of agriculture soils. Within Dayet Oum Ghellaz, a wetland affected by drought conditions near Oran and known as highly lead and cadmium polluted, three yeast species were isolated, *Clavispora lusitaniae*, *Rhodotorula mucilaginosa* and *Wickerhamomyces anomalus* [104] and shown high tolerance to NaCl and growth with heavy metal concentrations. Isolated strains W02 (*W. anomalus*) and R07 (*R. Mucilaginosa*) gives a removal of 12.68 ± 0.91 and 15.55 ± 0.72 mg of lead/g of biomass respectively. This is a promising and interesting first result of bioremediation with should be perfected in order to be proposed in real applications. The percentage removal of heavy metals of *R. Mucilaginosa* was measured in Serbia after 48 h for Cd2+, Zn2+, and Ni2+ and showed 2.11%; 4.99%, and 29.25%, respectively [105]. In Algeria and from an oilfield soil near Hassi Messaoud, the isolated strain YBR of *Rhodotorula* sp. was tested using wastewater of olive mills as low coast substrates produce biosurfactant [106] and can be used to remove hydrocarbons from polluted soils. Microorganisms produce a large specter of surfactants as extracellular components which have potentially wide properties due to their different chemical structure. They have also antibacterial and antifungal activities as well as they offer a potential use in food processing. Bioconversion of olive mills wastewater was conducted on an experimental scale using to referenced yeast species *Yarrowia lipolytica* MUCL 28849 and *Cryptococcus curvatus* ATCC 20509 in order to produce biomass [107]. Even when Y. lipolytica decreases in biomass, the decrease in polyphenol concentration is more than 4 times lower compared to the baseline.

5.6 Research in biotechnology

5.6.1 In applied genetics

From the wild collection, two haploid strains of S. cerevisiae K10 and MYC5 were selected to produce a protoplast fusion with two other commercial strains VDH2 and FXX118 provided by the Swiss company PEC and two other strains S. cerevisiae LGI2 and *Kluyveromyces lactis* LGK1 selected by the Genetics Laboratory at the Faculty of Biological Sciences (FSB) of the University of Sciences and Technology Houari Boumediène (USTHB) in Algiers. The protoplastization was carried out using Algerian *Helix asperca digestive juice*. Three hybrids were obtained at three levels intra-generic, intra-specific and inter-specific and their physiological properties were examined in order to obtain strains that were viable and more tolerant to temperature and alcohol concentration [108]. This case study confirmed technology transfer in developed countries but should not be discontinued. The main reason is the lack of interconnection between research and yeast manufactures, which prefer or stay under the effect of imports. Yeast production units are called upon to be competitive and must have their own research laboratory which values research results instead of only focusing on overseas purchases. The university must also generalize this type of innovative research attempts so that it is taught in graduation and to all courses in the genetics module instead of increasingly theoretical or virtual teaching.

5.6.2 Yeast biomass production

The example of selection of local strains like *K. fragilis* applied to valorize whey has now been more than 20 years. That of wild *S. cerevisiae* isolated to be compared with referenced strains. Some selected local strains are sometimes more efficient in biomass production than commercial strains [79, 109–113]. While, from the case of date palm, obtained yeast biomass is different due to the date varieties and the volume of the equipments [114–116]. The second advantage is to valorize the renewable low cost substrates and finding other interesting species like *Cr. curvatus* or *Y. lipolytica* [107]. The innovative test consists of valorizing olive mill wastewater and also the

Opuntia ficus indica peels [117]. Date must stay the lower cost available substrates which need serious industrial orientation, first in yeast production. Obtained yield is about 10% [115]. Whey is also targeted to produce biomass 11–22 g/l. with *Kluyveromyces lactis*, and 10–13 g/l. *K. marxianus and C. versatilis* [112].

5.6.3 BioEthanol production

Actually, the effervescence of studies is due to real mutations under the constraints of COVID-19 and the accelerated need for ethanol at the national level. Two species were used; *S. uvarum* [118, 119] and *S. cerevisiae* [79, 85, 109, 110, 119–128] with varied substrates in particular dates and date wastes of several varieties. A newly selected strain showed a modest ethanol production but also phenyl ethanol, glycerol, acetates, and acetic acid [81, 129]. The ethanol yield of 30 ml/125 g from *Balanites* fruits [130, 131] is less than what we obtained with dates, which give 300–600 ml/1 kg dates [118]. Banana gives by far past 9 l/100 Kg [47]. To further increase bioconversion yield, a practical nature by trying to target other species of yeast such as other substrates, and even if dates are the most available, we must not forget the peelings of fruits and vegetables in the juice factories. The mixture of grape juice and date must improve yield at 160 g/L. [125]. Three major constraints need to be taken as strategic debates. First one concerns the varied substrates to be valorized. While common dates stay evident favorite due to their availability, other biomass, such as waste olive and fruit peels, are considered as well as whey. The second concerns the choice of yeast species and does not focalize only on *Saccharomyces* strains in particular because we know that waste dates and dry ones are rich in sucrose not only glucose. The test of mutant *K. marxiamus* on glucose substrate gives highly significant ethanol but lower on sucrose medium [132]. The third is that of fundamental cellular pathway of ethanol production. The selection of new wild strains gives a chance to tolerate ethanol concentration and thus can produce more efficiency [133, 134] but there are other ways to produce glycerol, acetate, and phenyl ethanol. We should know how we can concentrate on the way to produce ethanol only. We borrow from [133] the diagram of ethanol biosynthetic

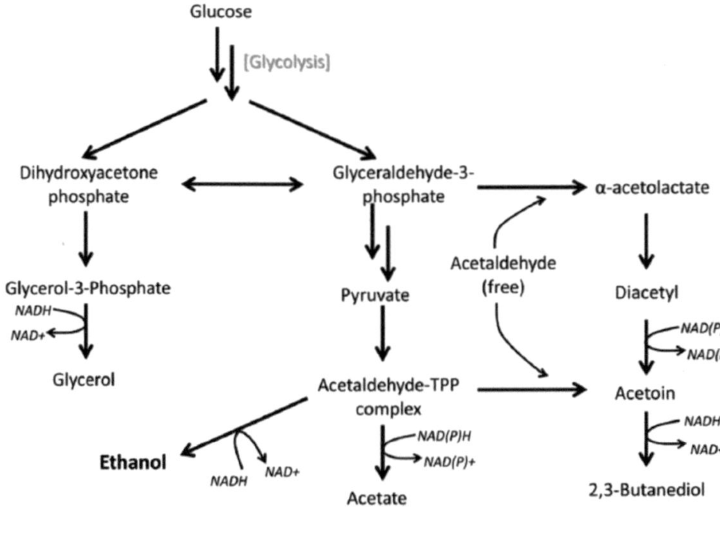

Figure 10.
Ethanol biosynthetic pathway in Saccharomyces cerevisiae [133].

pathway (**Figure 10**) in which glycerol, 2,3-butanediol, and acetate are produced, although less than ethanol. To maximize the productivity of ethanol it is necessary to minimize the flux of carbon toward glycerol and 2,3-butanediol. The reality of ethanol production route is often masked by the classic rule of sugar transformation into ethanol as being a single theoretical scheme where nothing needs to be modified.

5.6.4 Vinegar production: the best way to improve traditional process

Vinegar can be defined as a product obtained exclusively through biotechnological processes by double fermentation, alcoholic and acetic fermentation of liquids, or other substances of agricultural origin [134]. Traditional vinegar constitutes one of the natural products. It is obtained as biological vinegar from the fermentation of varied fruits. In addition to grapes and apples, dates are often presented in Algeria as ancestral ones using date varieties, and essentially those of low market values. Traditional way consists of double fermentation simultaneously and usually, has alcohol digress superior to what is accepted by standards and commercial legislation [7, 21]. Date vinegar is pronounced in flavor with a sweet taste due to esters produced by microorganisms. The initial sugar rate is not transformed totally at the end of maturation for 45 days. Thus the taste is usually sweeter [7, 21, 118, 135]. Due to the date varieties or to the process itself, sometime the total acidity of 2,5% is less than what is recommended. The spontaneous flora is showed more adequate but the resulting vinegar should be better with separating phases [118, 136].

Vinegar obtained from pomegranate, apple, and prickly pear are also made, but little research is done. Fruit vinegars have more reputation for health purposes. Experimentally biological vinegars have been shown to regulate lipid metabolism and decrease liver damage in high-fat-fed rats [137]. With apples the situation is similar and the main goal to prepare the vinegar is to valorise the huge losses of apples due to frost and convert them into producing a food health byproduct [138]. In addition to Ethanol and Acetic acid, traditional vinegar from dates contains [136]; Formic acid (to treat warts), Acetaldehyde (Ethanal), which gives ethanol by oxygenation, 1-hydroxy-2-propanone, 1–3 Butandiol, 2-Butanol, 1 Methyl Ester Formic acid (Fumigant and larvicide), 1–3 di-hydroxy propanone, 2,3-Dihydro-3,5-dihydroxy-6-methyl-4 h-pyran-4-one. The richness of those components explains or argues for the use of traditional vinegar in popular care methods and opens new perspectives for evaluating the nutritional properties of organic vinegars. The 1–3 di-hydroxy propanone is used in treatment of Vitiligo on exposed areas [98] and the 2,3-Dihydro-3,5-dihydroxy-6-methyl-4 h-pyran-4-one is an antioxidant [139].

5.6.5 Yeasts as the source of enzymes

Twelve yeast species are tested to produce varied enzymes and also to valorise different substrates as well as fruit wastes. *Aureobasidium pullulans* on tomato pomace produces polygalacturonase [68] and also Cellulase, Amylase, protease, and lipase in an experimental purpose valorising fruit and beet peels [87]. Lipase is produced by *Candida boidinii* isolated from olive pomace [84], *Ustilago cynodontis* isolated from saline soil [71], *Vishniacozyma tephrensis* [87] and *Yarrowia lipolytica* [84]. α-Amylase is produced by *C. glabrata* isolated from saline soil and valorising whey [75], *C. guilliernondii* and *C. tropicalis* [86], *Clavispora lusitaniae* [88, 140], *Schwanniomyces* sp. [67] and *Ustilago cynodontis* [71]. Cellulase is produced by *A. pullulans* [87] and *U. cynodontis* [71]. The protease is produced by *A. pullulans* [87], *Cr. neoformans*

[74], and *U. cynodontis* [71]. Maltase is produced by *C. lusitaniae* [88]. The first step of application is given by selected strains ABS7 of *C. lusitaniae* to purify the enzyme amylopullulanase in order to be used in laundry detergent [141]. A toxin from killer effect is targeted to be applied against food spoilage [142] and the third promising case came from the potential use of pectinase in clarification of fruit juices [143].

Particular attention should be given to cellulases. The case of Algerian Green energy progress plan has a potential of 0.67 Million Ton Oil Equivalent (Mtoe) from three sources of lignocellulosic; Alfa, olive pomace, and cereal straw [144]. Elsewhere, the author indicates that with the adoption of new energy crops and the strengthening of cereal technologies, Algeria would increase its bioprocesses energies up to 73.5 Mtoe and 58.9 Mtoe from these sources of biomass. We must consider more important, the lignocellulosic waste abandoned in palm groves with more than 20 Million date palms, giving at least 200.000 tons of waste, which is easy to use. This rate is suspected to increase by 4,5% by 2030 [145]. In addition to *S. cerevisiae,* selected strains concern also *K. marxianus* and *P. pastoris* [146].

5.7 Means of combating pathogenic yeasts

Honey is logically the first natural product which was applied against *C. albicans* and *Rhodotorula* sp. [147]. Traditional vinegar shows an effect on *C. albicans* [23] that is probably due to the presence of formic acid, formic acid methyl ester, and or 1–3 dihydroxy propanone. Essential oils of *Citrus sinensis* and *Citrus lemonum* give a positive impact against *C. albicans* [148]. *Cymbopogon citratis* extract has an inhibitory activity on *C. albicans* and *C. tropicalis* [149]. Essential oils of *Myrtus communis* and *Myrtus nivellei* were applied against *C. albicans*, *C. parapsilosis*, *C. tropicalis*, *Cr. neoformans* and also those isolated from vulvo-vaginal candidose patients *C. guillermondii* et *C. krusei* [150, 151]. Essential oils of *Origanum floribundum, Rosmarinus officinalis,* and *Thymus ciliates* has an effect on *C. albicans* isolated from bovine mastitis [62]. That of *Cinnomomum aromaticum* and *Syzygium aromaticum* should encourage their use against *C. tropicalis, C. albicans, Cr. neoformans, Cr. salbidus,* and *Geotrichum capitatum. Terfezia claveryi* crud extract dissolved in ethyl acetate has an effect on *C. albicans* [152]. The lichen *Xanthoria parietina* extract was tested against 4 strains isolated on patients in Thenia hospital (Algeria); *C. albicans*, *C. parapsilosis*, *Trichosporon* sp., and *Malassezia* sp. Two cases; *C. parapsilosis* and *Malassezia* sp. were mildly sensitive [153]. The isolated strain codified E96 and cited as actinomycete showed an antagonistic effect against pathogenic yeast; *Cr. albidus, Cr. diffluens,* and *Cr. Neoformans* [58]. Other cases of microorganisms were tested against referenced strains of *C. albicans* but all of those results stay at the experimental step only. From the marine ecosystem, *Cystoseira stricta* extract inhibits the referenced strains *C. albicans* [154]. *Asparagopsis armata* extract has an inhibitory activity with 0,58 mg/ml on the *C. albicans* IP 444 but with 2,34 mg/ml on *C. albicans* ATCC 1023 [155] showing the contrast within strains.

6. Conclusion

Yeast species inventoried in Algeria constitute a forgotten part of biodiversity. It aims to promote studies of yeasts oriented toward economic and cultural services. Derived from soils, plants, and animals, yeasts characterize fermented foods and drinks and can improve the quality of citizen's life. Those who refer to care in hospitals logically deserve special attention because they can be sources of complications for patients and require additional hospitalization costs. The yeasts inventoried reached 131 taxa (Annexe 1); 27 related to human health, 30 animal health, 27 from dairy products, 24 isolated from the soil, 41 taxa from agricultural products (grapes dates, olives, roots, ...), and 17 isolated from the environment (Trees, feints of pigeons, beaches, landfills). Biotechnonlogies concern 44 taxa (Biomass, Ethanol, enzymes, esters). 16 selected natural products inhibit 14 yeast species. This rich information should be used to establish a national network and help stakeholders assess and make their contributions. The inventory is not completely exhaustive because there are probably unlisted works. Anyway, the inventory will remain continuously open in view of the progress of studies which are increased by the adoption of molecular tools. Yeasts are closely linked to socio-economic purposes relating to traditions (baking, vinegars, dairy products, yeast food supplements for animals) or referred to in modern industrial processes (biomasses, ethanol, and enzymes). Their studies have socio-economic impacts linked to the benefits derived; 1) the mastery of human care with repercussions on the medical care of patients and on the strategy for the use of antibiotics, 2) in increasing milk and meat yields, 3) in environmental issues including waste treatment and depollution. The link between ancestral traditions and scientific explorations increases the chances of serving the national economy and should encourage people not to depend solely on imports. The use of biomass or ethanol is one of the most urgent needs to be supported in a national strategic plan which gives concrete form to renewable energies. Testing the production of ethanol and its mixture in the fuel is possible, but carefully choose the proprietary biomass to be transformed. Dates, waste from olive groves, and the lost share of cellulose in palm groves are most promising.

Updating the national directory of identified accessions is the first step in the collaboration between specialists to honor the standard of biodiversity inventories. A network of national skills should be set up to promote prospecting in various environments because what remains to be done undoubtedly represents 99%. We are keen to complete this initial list with a consortium of national specialists in this complex and diverse field. This can be considered an important step because it is necessary to establish a national repository of strains and listed species. The Algerian Pasteur Institute, with experience and capabilities, can fulfill this role, but it makes sense to be more specialized in pathogenic species and their antagonists. The strains prospected in other fields such as agriculture, environment, and food technology require greater attention in strengthening the technical capacities of other institutions. Whatever the applications, the repository justifies being targeted by the recently created National GenBank. The main reason is to standardize identifications, promote selected strains and establish the national register of biological resources.

Acknowledgements

The author would like to thank Mr. A. Medjahed, Director Ge-neral of INESG for his encouragement and his patience during the period of writing this synthesis. Pr Salim Mokrane and Sonia Kaci of the LBSM, at the Ecole Normale Supérieure are kindly thanked for their help in taking photos and exchanging ideas.

Conflict of interest

The author declares no conflict of interest.

Annexes

Annexes I and II referred to in this chapter are available at: https://bit.ly/3wbo74l.

Author details

Abderrahmane Benkhalifa
Higher Normal School, Bachir El-Ibrahimi, Algiers, Algeria

*Address all correspondence to: a.benkhalifa@hotmail.com;
abderrahmane.benkhalifa@g.ens-kouba.dz

References

[1] Kurtzman CP, Fell JW, Boekhout T. The Yeasts, a Taxonomic Study. Fifth ed. Vol. 1. London, UK: Elsevier; 2011. p. 537

[2] Żymańczyk-Duda E, Brzezińska-Rodak M, Klimek-Ochab M, Duda M, Zerka A. Yeast as a versatile tool in biotechnology. In: Morata A, Loira I, editors. Yeast - Industrial Applications. London, UK, UK, London: InTech; 2017. DOI: 10.5772/intechopen.70130

[3] Mota MN, Múgica P, Sá-Correia I. Exploring yeast diversity to produce lipid-based biofuels from agro-forestry and industrial organic residues. Journal of Fungi. 2022;**8**:687. DOI: 10.3390/jof8070687

[4] Boekhout T, Amend AS, El-Baidouri F, Gabald T, Geml J, Mittelbach M, et al. Trends in yeast diversity discovery. Fungal Diversity. 2022;**114**:491-537. DOI: 10.1007/s13225-021-00494-6

[5] Benkhalifa A, Toumi M. L'inventaire des levures pourrait-il booster l'économie en Algérie? In: 2ème Séminaire National d'Ethnobotanique et de Valorisation des Substances Naturelles. Algérie: Univ. Benyoucef Benkhedda, Alger I; 2021. pp. 9-12

[6] Merrad S, Kaabeche Z. État de l'art sur la conservation d'aliments par fermentation: Focus sur les aliments solides [thesis]. M'Sila, Algeria: Université M. Boudiaf; 2020. p. 50

[7] Ould El Hadj MD, Sebihi AH, Siboukeur O. Qualité hygiénique et caractéristiques physico-chimiques du vinaigre traditionnel de quelques variétés de dattes de la cuvette de Ouargla. Rev. Energ. Ren: Production et Valorisation – Biomasse. Special Issue. Alger: Centre de Développement des Energies Renouvelables (CDER); 2001. pp. 87-92

[8] Gast M. Lait. Encyclopedie berbère. 2008;**28-29**:4322-4330. DOI: 10.4000/encyclopedieberbere.294

[9] Aissaoui-Zitoun O, Pediliggieri C, Benatallah L, Lortal S, Licitra G, Zitoune MN, et al. Bouhezza, a traditional Algerian raw milk cheese, made and ripened in goatskin bags. Journal of Food, Agriculture and Environment. 2012;**10**(2):289-295

[10] Leksir C, Boudalia S, Moujahed N, Chemmam M. Traditional dairy products in Algeria: Case of Klila cheese. Journal of Ethnic Foods. 2019;**6**:7. DOI: 10.1186/s42779-019-0008-4

[11] Saidane Z, Dahou AA, Tahlaiti H, Daoudi M, Doukani K, Homrani A. Physico-chemical parameters with direct influence on the dynamism of the indigenous microflora of the traditional cheese "J'ben Elgafs". Asian Journal of Dairy and Food Research. 2021;**40**(2):157-161. DOI: 10.18805/ajdfr.DR224

[12] Mokhtari S, Taghouti M, Saidi D, Kheroua O. Traditional Algerian fermented food: First data on nutritional characteristics of wheat (*Triticum durum*) fermented in underground silos Matmora (Mascara, Algeria) compared to unfermented wheat. Advances in Biology & Earth Sciences. 2020;**5**(3):176-192

[13] Benakriche B, Bousbahi S, Gérard P. Impact nutritionnel d'un blé fermenté type Hamoum sur la translocation bactérienne intestinale chez le rat malnutri en phase de réalimentation. Nutrition Clinique et Métabolisme. 2019;**33**(1):102. DOI: 10.1016/j.nupar.2019.01.416

[14] Algérie Presse Service [Internet]. 2021. Available from: https://www.aps.dz/economie/117637-la-relance-de-l-

usine-de-production-de-levure-de-guelma-couvrira-30-du-marche-national. [Accessed: August 28, 2022]

[15] Word Integrated Trade Solution (WITS) – WorldBank. Yeasts; active exports to Algeria in 2019-2021. Available from: https://wits.worldbank.org/trade/comtrade/en/country/All/year/2021/tradeflow/Exports/partner/DZA/product/210210. [Accessed: August 28, 2022]

[16] Direction des Etudes et de la Prospective. Statistiques du commerce extérieur d'Algérie, 2018-2019, 2019-2020. Algérie: Ministère des Finances, Direction Générale des Douanes, Multigr; 2021. p. 71

[17] Hana Saada (from Radio of Algeria) Yeast production. Available from: https://www.dzbreaking.com/2021/02/18/yeast-production-guelma-plant-to-cover-30-of-national-market/. [Accessed: August 28, 2022]

[18] Vincent. Matlouh "*matloa*", an Algerian bread made out of hard wheat and semolina, leavened by either sourdough or fresh yeast. 2022. Available from: https://baking-history.com/matlouh/. [Accessed: August 28, 2022]

[19] Benkhalifa A, Kehli K, Ould Mohammed A. Levure naturelle et levain maison: Richesse et déclin des traditions en Algérie. In: 2ème Séminaire National d'Ethnobotanique et de Valorisation des Substances Naturelles. Algérie: Univ. Benyoucef Benkhedda, Alger I; 2021. pp. 68-70

[20] Bigey F, Segond D, Friedrich A, Guezenec S, Bourgais A, Huyghe L, et al. Evidence for two main domestication trajectories in Saccharomyces cerevisiae linked to distinct bread-making processes. Current Biology. 2021;**31**(4):722-732.e5. DOI: 10.1016/j.cub.2020.11.016

[21] Boukhiar A. Analyse du processus traditionnel d'obtention du vinaigre de dattes tel qu'appliqué au sud algérien: Essai d'optimisation [thesis]. Algérie: Université M'hammed Bouguera, Boumerdès; 2009

[22] Samad A, Azlan A, Ismail A. Therapeutic effects of vinegar: A review. Current Opinion in Food Science. 2016;**8**:56-61. DOI: 10.1016/j.cofs.2016.03.001

[23] Cherif B, Bouras N, Oumouna M, Ould El Hadj MD, Holtz MD, Sabaou N. Ethno-pharmacological use and antimicrobial activity of traditional date vinegar of Ghardaïa. Algerian. Journal of Arid Environment. 2014;**4**(1):83-93

[24] Korcari D, Ricci G, Capusoni C, et al. Physiological performance of *Kazachstania unispora* in sourdough environments. World Journal of Microbiology and Biotechnology. 2021;**37**:88. DOI: 10.1007/s11274-021-03027-0

[25] Bendimerad N. Caractérisation phénotypique technologique et moléculaire d'isolats de bactéries lactiques de laits crus recueillis dans les régions de l'Ouest Algérien. Essai de fabrication de fromage frais type Jben [thesis]. Tlemcen, Algeria: Université Abou Bekr Belkaid; 2013. p. 149

[26] Kacem M, Karem NE. Physicochemical and microbiological study of "shmen", a traditional butter made from camel milk in the Sahara (Algeria): Isolation and identification of lactic acid bacteria and yeasts. Grasas y Aceites. 2006;**57**(2):198-204. DOI: 10.3989/gya.2006.v57.i2.37

[27] Boulkroune A, Debbah A. Valorisation du lactosérum pour la production d'une enzyme coagulante du lait [thesis]. Constantine, Algeria: Université Frère Mentouri; 2019

[28] Guetouache M, Guessas B. Characterization and identification of lactic acid bacteria isolated from traditional cheese (Klila) prepared from cow's milk. African Journal of Microbiology Research. 2015;**9**(2):71-77. DOI: 10.5897/AJMR2014.7279

[29] Meribai A, Djenidi R, Hammouche Y, Bensoltane A. Caracterisation physicochimique et qualité microbiologique du Klila: un fromage traditionnel sec des régions arides d'Algérie: Etude préliminaire. Journal of new sciences, Agriculture and Biotechnology. 2017;**40**(4):2169-2174

[30] Benamara RN, Benahmed M, Ibri K, Boudjemaa BM, Demarigny Y. Algerian extra hard cheese of Klila: A review on the production method, and microbial, organoleptic, and nutritional properties. Journal of Ethnic Foods. 2022;**9**:41. DOI: 10.1186/s42779-022-00157-0

[31] Aissaoui-Zitoun O, Benatallah L, Ghennam EL, Zidoune MN. Manufacture and characteristics of the traditional Algerian ripened Bouhezza cheese. Journal of Food, Agriculture & Environment. 2011;9(2): 96-100

[32] Boudalia S, Boudebbouz A, Gueroui Y, Bousbia A, Benada M, Leksir CH, et al. Characterization of traditional Algerian cheese Bouhezza prepared with raw cow, goat and sheep milks. Food Science and Technology. 2020;**40**(Suppl. 2):528-537

[33] Derouiche M, Zidoune MN. Caractérisation d'un fromage traditionnel, le *Michouna* de la région de Tébessa, Algérie. Livestock Research for Rural Development. 2015;**27**(11):229

[34] Bouaguel R, Bouguedah L. Caractérisation microbiologique des fromages traditionnels Michouna et Adghess préparés à partir du lait de chèvre [thesis]. Oum El-Bouaghi, Algeria: Univ. Larbi Ben M'Hidi; 2020

[35] Khoualdia G. Caractérisation du Fromage Traditionnel algérien Medeghissa [Thesis]. Algeria: Institut de la Nutrition, de l'Alimentation et des Technologies Agro-Alimentaires (I.N.A.T.A.A.); 2017

[36] Ben Mohamed C, Bouamer N. Caractérisation Physico-Chimique du Colostrum Camelin (*Camelus dromedarius*) [Thesis]. Ouargla, Algeria: Univ. Kasdi Merbah; 2013

[37] Yssaad D, Benakriche B, Gérard G, Pochart P, Kheroua O. Fermented wheat Hamoum improves the recovery of intestinal mucosal and the short-chain fatty acids profile of colonic bacterial flora in malnourished rats. Bioscience Research. 2020;**17**(2):1166-1175

[38] Benstaali M, Benmamar KA. Approche microbiologique de la flore indigène d'un Blé fermenté type Hamoum: Les Levures comme caractère probiotique [thesis]. Mostaganem, Algeria: Univ. Abdelhamid Ibn Badis; 2021

[39] Boudjenah A. Effet d'une supplémentation de l'aliment en levure *Saccharomyces cerevisiae* sur les paramètres zootechniques de la vache laitière en Peripartum [thesis]. Alger: Ecole National Supérieure Vétérinaire; 2008

[40] Ayad MA, Benallou B, Saim MS, Smadi MA, Meziane T. Impact of feeding yeast culture on milk yield, milk components, and blood components in Algerian dairy herds. Journal of Veterinary Science and Technology. 2013;**4**:2. DOI: 10.4172/2157-7579.1000135

[41] Besseboua O, Benbarek H, Hornick JN, Ayad A. Effect of yeast *Saccharomyces cerevisiae* feed supplement on milk production and its composition

of lactating Holstein Friesian cow from northern Algeria. Journal of Veterinary Science. 2020, 2020;**4**(2):64-74

[42] Mitchison-Field LMY, Vargas-Muñiz JM, Stormo BM, Vogt EJD, Van Dierdonck S, Pelletier JF, et al. Unconventional cell division cycles from marine-derived yeasts. Current Biology. 2019;**29**(20):3439-3456. DOI: 10.1016/j.cub.2019.08.050

[43] Cadez N, Raspor P, de Cock AWAM, Boekhout T, Smith M. Th, molecular identification and genetic diversity within species of the genera *Hanseniaspora* and *Kloeckera*. FEMS Yeast Research. 2002;**1**(4):279-289. DOI: 10.1111/j.1567-1364.2002.tb00046.x

[44] Takaya M, Ohwada T, Oda T. Characterization of the yeast *Hanseniaspora vineae* isolated from the wine grape 'Yamasachi' and its use for bread making. Food Science and Technology Research. 2019;**25**(6):835-842. DOI: 10.3136/fstr.25.835

[45] C.R. Lab. Carlsberg, 1911-1, 10, p. 285. Isono M, et al. Process for producing unsaturated steroids. US Patent 3,616,225. 1971. Available from: https://www.ncyc.co.uk/catalogue/hanseniaspora-osmophila-26 [Accessed: September 30, 2022]

[46] Vaughan-Martini A, Angelini P, Cardinali G. Use of conventional taxonomy, electrophoretic karyotyping and DNA–DNA hybridization for the classification of fermentative apiculate yeasts. International Journal of Systematic and Evolutionary Microbiology. 2000;**50**:1665-1672

[47] Bremond E. L'Alcool de Bananes. Annales de l'Institut National Agronomique. 1941;**1**(2):126-134

[48] Chekiri-Talbi M, Denning DW. Burden of fungal infections in Algeria. European Journal of Clinical Microbiology & Infectious Diseases. 2017;**36**(6):999-1004. DOI: 10.1007/s10096-017-2917-8

[49] Djeridane A, Djeridane Y, Ammar-Khodja A. Epidemiological and aetiological study on tinea pedis and onychomycosis in Algeria. Mycoses. 2006;**49**:190-196

[50] Djeridane A, Djeridane Y, Ammar-Khodja A. A clinicomycological study of fungal foot infections among Algerian military personnel. Clinical and Experimental Dermatology. 2006;**32**(1):60-63. DOI: 10.1111/j.1365-2230.2006.02265.x

[51] Ait-Seddik H, Ceugniez A, Bendali F, Cudennec B, Drider D. Yeasts isolated from Algerian infants's feces revealed a burden of *Candida albicans* species, non-*albicans Candida* species and *Saccharomyces cerevisiae*. Archives of Microbiology. 2016;**198**(1):71-81. DOI: 10.1007/s00203-015-1152-x

[52] Arrache D, Madani K, Zait H, Achir I, Younsi N, Zebdi A, et al. Fongémies diagnostiquées au laboratoire de parasitologie mycologie du CHU Mustapha d'Alger. Algérie Journal of Mycology Medicine. 2016;**2016**:237-238. DOI: 10.1016/j.mycmed.2015.06.049

[53] Bendjelloul M, Boucherit-Otmani Z, Boucherit K. Study of strains of *Candida* spp. isolated from catheters in UHC of Oran (Algeria): Identification and antifungal susceptibility. Journal of Médical Mycology. 2016;**26**(3):212-216. DOI: 10.1016/j.mycmed.2016.02.022

[54] Megri Y, Arastehfar A, Boekhout T, Daneshnia F, Hörtnagl C, Sartori B, et al. *Candida tropicalis* is the most prevalent yeast species causing candidemia in Algeria: The urgent need for antifungal stewardship. Antimicrobial Resistance

and Infection Control. 2020;**9**:50. DOI: 10.1186/s13756-020-00710-z

[55] Hamroune Z. Epidémiologie de la cryptococcose en Algérie [thesis]. Algeria: Université Benyoucef Benkhedda Alger I; 2020

[56] Zerrouki H, Ibrahim A, Rebiahi SA, Elhabiri Y, Benhaddouche DE, de Groot T, et al. Emergence of *Candida auris* in intensive care units in Algeria. Mycoses. 2022;**65**(7):753-759. DOI: 10.1111/myc.13470

[57] Sebaa S, Faltot M, De Breucker S, Boucherit-Otmani Z, Bafort F, Perraudin JP, et al. Ex vivo decontamination of yeast-colonized dentures by iodine-thiocyanate complexes. Clinical, Cosmetic and Investigational Dentistry. 2018;**10**:149-158. DOI: 10.2147/CCIDE.S165377

[58] Ouargli M, Gacemi-Kirane D, Mansouri R, Al-Yasiri MS, Ranque S, RouxV. Antifungal activity of Streptomyces sp. against environmental and clinical *Cryptococcus* spp. isolates. Journal of Chemical and Pharmaceutical Research. 2015;7(10):1019-1027

[59] Touati S, Touati M. *Cryptococcus*: Recherche de la levure dans l'Est Algérien [thesis]. Algeria: Univ. Frères Mantouri; 2021

[60] Akdouche L, Aissi M, Zenia S, Saadi A. Importance of yeasts in the mammary infection of the cattle in the region of Sidi M'Hamed ben Ali, Wilaya of Relizane, Algeria. Journal of Veterinary Science & Technology. 2014;**5**(2):172. DOI: 10.4172/2157-7579.1000172

[61] Ksouri S, Djebir S, Hadef Y, Benakhla A. Survey of bovine mycotic mastitis in different mammary gland statuses in two north-eastern regions of Algeria. Mycopathologia. 2015;**179**:327-331. DOI: 10.1007/s11046-014-9845-2

[62] Ksouri S, Djebir S, Bentorki AA, Gouri A, Hadef Y, Benakhla A. Antifungal activity of essential oils extract from *Origanum floribundum* Munby, *Rosmarinus officinalis* L. and *Thymus ciliatus* Desf. Against *Candida albicans* isolated from bovine clinical mastitis. Journal of Medical Mycology. 2017;**27**(2):245-249. DOI: 10.1016/j.mycmed.2017.03.004

[63] Benbelkacem I. Suivi de quelques élevages bovins laitiers à l'ouest du pays: Etude des facteurs influençant l'apparition et/ou le maintien des mammites fongiques [thesis]. Alger: Ecole national Supérieur Vétérinaire; 2019

[64] Benbelkacem I, Selles SMA, Aissi M, Khaldi F, Ghaz K. In vitro assessment of antifungal and antistaphylococcal activities of *Cinnamomum aromaticum* essential oil against subclinical mastitis pathogens. Veterinária. 2019;**68**(1):31-37

[65] Saubin M, Devillers H, Proust L, Brier C, Grondin C, Pradal M, et al. Investigation of genetic relationships between *Hanseniaspora* species found in grape musts revealed interspecific hybrids with dynamic genome structures. Frontiers in Microbiology. 2020;**10**:2960. DOI: 10.3389/fmicb.2019.02960

[66] Labbani FZ. Activité Killer chez des levures isolées des sols du Nord-Est Algérien: Purification, caractérisation et effet sur les souches de levures indésirables [thesis]. Vol. 1. Algeria: Département de Biochimie et Biologie Cellulaire et Moléculaire, Faculté des Sciences de la Nature et de la Vie Univ. Frères Mentouri Constantine; 2015

[67] Toumi S, Tifrit A, Hadjazi D, Chama Z, Abbouni B. Production of a thermostable amylase by yeast strain

isolated from saharian soils cultivated in soft cheese whey. Der Pharmacia Lettre. 2016;**8**(17):32-41

[68] Bennamoun L. Isolement, sélection de souches levuriennes de sols arides sahariens (El-M'gheir) productrices de polygalacturonase: Purification et caractérisation enzymatique [thesis]. Algeria: Université Frères Mentouri Constantine I; 2017

[69] Bilek FN, Rezki MA, Grondin C, Yahia N, Bekki A. Plant growth promoting characteristics and stress tolerance of yeasts isolated from Algerian agricultural soils. South Asian Journal of Exprimental Biology. 2020;**10**(6):413-426

[70] Labbani F-Z, Turchetti B, Bennamoun L, Dakhmouche S, Roberti R, Corazzi L, et al. A novel killer protein from *Pichia kluyveri* isolated from an Algerian soil: Purification and characterization of its in vitro activity against food and beverage spoilage yeasts. Antonie Van Leeuwenhoek. 2015;**107**(4):961-970. DOI: 10.1007/s10482-015-0388-4

[71] Chamekh R, Deniel F, Donot C, Jany JL, Nodet P, Belabida L. Isolation, identification and enzymatic activity of halotolerant and halophilic fungi from the great Sebkha of Oran in northwestern of Algeria. Mycobiology. 2019;**47**(2):230-241. DOI: 10.1080/12298093.2019.1623979

[72] Bennamoun L, Hiligsmann S, Dakhmouche S, Ait-Kaki A, Labbani F-ZK, Nouadri T, et al. Production and properties of a thermostable, pH-stable Exo-Polygalacturonase using *Aureobasidium pullulans* isolated from Saharan soil of Algeria grown on tomato pomace. Foods. 2016;**5**(4):72. DOI: 10.3390/foods5040072

[73] Laiche AT. Production d'enzymes amylolytiques chez des souches levuriennes isolées à partir de deux origines (sol saharien et blé) et cultivées sur milieu à base de lactosérum [thesis]. Ouargla: Univ. K. Merbah; 2019. p. 181 (+Annexes)

[74] Maamra F, Maissa N. Caractérisation des enzymes protéolytiques des souches fongiques isolées à partir du sol saharien [thesis]. El-Oued: Univ. H. Lakhdar; 2017

[75] Laiche AT, Siboukeur O. Characterization of partially purified extracellular α-amylase from *Candida glabrata* grown in whey based medium. Ponte Academic Journal. 2018;**74**(4):163-173

[76] Boukhennoufa A, Berber N, Aissaoui R, Bekada AMB. Microbiological, biochemical and molecular identification (PCR-RFLP-ITS) of the yeast from sultana grape cultivated in Ain Merane (Wilaya of Chlef). Current Research on Biological Sciences. 2016;**1**(1):7-12

[77] Berber N, Aissaoui R, Bekada AMA, Coarer M. Isolation and molecular identification (PCR-Delta and PCR-RFLP-ITS) of yeasts from the black Muscat grape cultivated in El-Malah (Wilaya of Ain Temouchent, Algeria). Advances in Environmental Biology. 2016;**10**(12):55-61

[78] Berber N. Caractérisation biomoléculaire et biotechnologique des souches de *Saccharomyces cerevisiae* issues des cépages Algériens [thesis]. Mostaganem: Univ. Abdelhamid Ibn-Badis; 2018

[79] Açourene S, Ammouche A, Djaafri K. Valorisation des rebuts de dattes par la production de la levure boulangère, de l'alcool et du vinaigre. Sciences & Technologie. C, Biotechnologies. 2008;**28**:38-45

[80] Rezki-Bekki MA. Production de métabolites par les levures: Caractérisation et identification des

arômes et des alcools, Faculté des Sciences, de la Nature et de la Vie [thesis]. Algeria: Univ. d'Oran; 2014

[81] Rezki MA, Benbadis L, DeBillerbeck G, Benbayer Z, François JM. Isolation and physiological characterization of indigenous yeasts from some Algerian agricultural and dairy products. Journal of Yeast & Fungal Research. 2013;**4**(6):75-83. DOI: 10.5897/JYFR2013.0117

[82] Belbahi A, Bohuon P, Leguérine I, Meot JM, Loiseau G, Madani K. Heat resistance of *Candida apicola* and *aspergillus Niger* spores isolated from date fruit surface. Journal of Food, Processing &. Engineering. 2015;**40**(1):e12272. DOI: 10.1111/jfpe.12272

[83] Kacem M, Karam N-E. Microbiological study of naturally fermented Algerian green olives: Isolation and identification of lactic acid bacteria and yeasts along with the effects of brine solutions obtained at the end of olive fermentation on *lactobacillus plantarum* growth. Grasas y Aceites. 2006;**57**(3):292-300

[84] Bataiche I. Recherche de nouvelles potentialités de *Yarrowia lipolytica*, isolé de différents milieux naturels pour des applications biologiques [thesis]. Algeria: Université Frères Mentouri; 2014

[85] Kechkar M, Sayed W, Cabrol A, Aziza M, Ahmed ZT, Amrane A, et al. Isolation and identification of yeast strains from sugarcane molasses, dates and figs for ethanol production under conditions simulating algal hydrolysate. Brazilian Journal of Chemical Engineering. 2019;**36**(1):157-169. DOI: 10.1590/0104-6632.20190361s20180114

[86] Bessalem S, Laraba-Djebari F, Bellal M. Isolement, Purification et Caractérisation de l'a-amylase de levures isolées à partir de Pomme de terre. Recherche Agronomique. 2001;**5**(9):75-90

[87] Labbani F-ZK, Dakhmouche S, Bennamoun L, Ait-Kaki A, Nouadri T. Extracellular hydrolytic enzymes of yeasts isolated from fruit and beet peels in Algeria. International Journal of Ecosystems and Ecology Science (IJEES). 2022;**12**(1):7-16. DOI: 10.31407/ijees12.1 Abstract

[88] Dakhmouche-Djekrif S, Gillmann L, Bennamoun L, Ait-Kaki A, Labbani K, Nouadri T, et al. Amylolytic yeasts: Producers of α-amylase and Pullulanase. International Journal of Life Science & Scientific Research. 2016;**2**(4):339-354

[89] Dakhmouche-Djekrif S. Production et caractérisation de l'amylopullulanase de la levure *Clavispora lusitaniae* ABS7 isolée de blé cultivé et stocké en zones arides [thesis]. Constantine, Algeria: Université des Frères Mentouri; 2016

[90] Abdel-Kareem MM, Zohri AA, Nasr SA. Novel marine yeast strains as plant growth-promoting agents improve defense in wheat (*Triticum aestivum*) against *Fusarium oxysporum*. Journal of Plant Disease Protocol. 2021;**128**:973-988. DOI: 10.1007/s41348-021-00461-y

[91] Sipiczki M. *Metschnikowia pulcherrima* and related Pulcherrimin-producing yeasts: Fuzzy species boundaries and complex antimicrobial antagonism. Microorganisms. 2020;**8**(7):1029. DOI: 10.3390/microorganisms8071029

[92] Beldjilali AF, Belgroun F, Benlahcen K, Bettache G, Aggad H, Kihal M. Evaluation of microbiological and sanitary quality of ewe's raw milk in Western of Algeria and detection of antibiotic residue by Delvotest. Advances in Environmental Biology. 2013;**7**(6):1027-1033

[93] Hamoudi-Belarbi L, Nouri L, Belkacemi K. Effectiveness of convective drying to conserve indigenous yeasts with high volatile profile isolated from Algerian fermented raw bovine milk

(Rayeb). Food Science & Technology, Campinas. 2016;**36**(3):476-484. DOI: 10.1590/1678-457X.00416

[94] Aziza M, Amrane A. Diauxic growth of *Geotrichum candidum* and *Penicillium camembertii* on amino acids and glucose. Brazilian Journal of Chemical Engineering. 2012;**29**(2):203-210

[95] Kebbouche-Gana S, Gana ML. Algerian yeast strains: Isolation, identification and production of single cell protein from whey with strain *Candida kefyr*. International Journal of Bioscience, Biochemistry and Bioinformatics (IJBB). 2014;**4**(3):162-165. DOI: 10.7763/IJBBB.2014.V4.331

[96] Ider S, Belguesmia Y, Coucheney F, Kihal M, Drider D. Impact of seasonality and environmental conditions on yeast diversity from camel's milk collected in Algeria. Archives of Microbiology. 2019;**201**(3):399-407. DOI: 10.1007/s00203-019-01626-y Epub 2019 Feb 4

[97] Hamza H, Hinana R. Étude du potentiel biotechnologique des levures isolées du lait de chamelle dans la région de Ghardaïa [thesis]. Algeria: Université de Ghardaïa; 2020

[98] Fesq H, Brockow K, Strom K, Mempel M, Ring J, Abeck D. Dihydroxyacetone in a new formulation a powerful therapeutic option in vitiligo. Dermatology. 2001;**203**(3):241-243. DOI: 10.1159/000051757

[99] Derouiche N, Medjouj H, Aissaoui-Zitoune W, Zidoune MN. Some traditional cheeses manufactured in Algeria. In: Henriques MHF, Pereira CJD, editors. Cheese Production, Consumption and Health Benefits. New York: Food Science and Technology, Nova Science Publisher; 2017. pp. 225-242

[100] Medjoudj H, Aouar L, Zidoune MN, Hayaloglu AA. Proteolysis, microbiology, volatiles and sensory evaluation of Algerian traditional cheese Bouhezza made using goat's raw milk. International Journal of Food Properties. 2017;**20**(3):S3246-S3265. DOI: 10.1080/10942912.2017.1375515

[101] Matallah-Boutiba A, Benmessaoud N, Messaoui N, Boutiba Z. Microbiological Sandy Beach quality in Western Algeria. Journal of Marine Biology & Oceanography. 2016;**6**:1. DOI: 10.4172/2324-8661.1000171

[102] Benfreha-Benyelles M, Mokrani S, Belgharbi AA, Meddah A, Moussa-Boudjemaa B. Enquête microbiologique et possibilité de réutilisation des eaux usées pour la production végétale. Un cas de la station d'épuration d'El-Kouwaer pendant la période de dysfonctionnement. Algérie, Revue Agrobiologia. 2021;**11**(2):2566-2578

[103] Bourzama G, Iratni N, Ennaghra N, Ouled-Haddar H, Soumati B, Amour K, et al. *In vitro* removal of electronic and electrical wastes by fungi isolated from soil at Annaba area in northeast of Algeria. Environment and Natural Resources Journal. 2021;**19**(4):302-309

[104] Aibeche C, Selami N, Zitouni-Haouar FE, Oeunzar K, Addou A, Kaid-Harche M, et al. Bioremediation potential and lead removal capacity of heavy metal-tolerant yeasts isolated from Dayet Oum Ghellaz Lake water (northwest of Algeria). International Microbiology. 2022;**25**(1):61-73. DOI: 10.1007/s10123-021-00191-z

[105] Grujić SM, Radojević ID, Vasić SM, Čomić LR, Ostojić AM. Heavy metal tolerance and removal efficiency of the *Rhodotorula mucilaginosa* and *saccharomyces boulardii* planktonic cells and biofilm. Kragujevac Journal of Science. 2018;**40**:217-226

[106] Derguine-Mecheri L, Kebbouche-Gana S, Djenane D.

Biosurfactant production from newly isolated *Rhodotorula* sp. YBR and its great potential in enhanced removal of hydrocarbons from contaminated soils. World Journal of Microbiology and Biotechnology. 2021;**37**:18. DOI: 10.1007/s11274-020-02983-3

[107] Dermeche S, Moulti-Mati F. Assessment of olive mill wastewaters bioconversion potential into biotechnological and health interest microbial biomass, Algerian. Journal of Environmental Science and Technology. 2019, 2019;**5**(1):898-908

[108] Belhocine H. Contribution à l'amélioration génétique des levures d'intérêt industriel par la technique de fusion des protoplastes [thesis]. Vol. 2011. Alger: Université des sciences et de la technologie Houari Boumediène; 2011. p. 89

[109] Chibi S, Rabet S, El-hadi D. Etude des paramètres environnementaux sur la croissance de *Saccharomyces cerevisiae* isolée de rebuts de dattes. Algerian Journal of Environmental Science and Technology. 2016;**2**(3):62-69

[110] Chibi S, El-Hadi D. The isolation and characterization of yeast strains *Saccharomyces cerevisiae* cultivated on the musts of rebuts of dates: Physiological studies of the adaptation and the resistance to ethanol. Algerian Journal of Environmental Science & Technology. 2019;**5**(2):937-946

[111] Gana S. et Touzi A. Valorisation du lactosérum par la production de levures lactiques avec les procédés de fermentation discontinue et continue. Rev. Energ. Ren. Special Issue: Production et Valorisation–Biomasse. Alger: Centre de Développement des Energies Renouvelables (CDER); 2001. pp. 51-58

[112] Moeini H, Nahvi I, Tavassoli M. Improvement of SCP production and BOD removal of whey with mixed yeast culture. Electronic Journal of Biotechnology. 2004;**7**(3):249-255

[113] Ould El Hadj MD, Bitour Z, Siboukeur O. Etude de la production de levure boulangère (*Saccharomyces cerevisiae*) cultivée sur mout de rebuts de dates. Courrier du Savoir Scientifique et Technique. 2006;**7**(7):13-18

[114] Açourene S, Ammouche A. Optimization of culture medium for Baker's yeast, ethanol, citric acid and &-amylase production from dates syrup. Research Journal of Agriculture & Biological Sciences. 2010;**6**(6):846-860

[115] Oulad-Belkhir A. Fabrication de la levure boulangère à base de rebuts de dates Ghars [thesis]. Ouarlga: Univ. Kasdi Merbah; 2016

[116] Kara AM, Outili N, Ait KA, Cherfia R, Benhassine S, Benaissa A, et al. Optimization of Baker's yeast production on date extract using response surface methodology (RSM). Food. 2017;**6**(8):64. DOI: 10.3390/foods6080064

[117] Diboune N, Nancib A, Nancib N, Anibal J, Boudrant J. Utilization of prickly pear waste for baker's yeast production. Biotechnology and Applied Biochemistry. 2019;**66**(5):744-754. DOI: 10.1002/bab.1753

[118] Boughnou S. Essai de production du vinaigre à partir de déchets de dattes. Ann. Inst. Nat. Agro., El-Harrach. 1988;**12**(2):65-83

[119] Kaidi F, Touzi A. Production de Bioalcool à partir des déchets de dattes. Rev. Energ. Ren. Special Issue: Production et Valorisation – Biomasse. Alger: Centre de Développement des Energies Renouvelables (CDER); 2001. pp. 75-78

[120] Ould E-HMD, Cheick M, Hamdi W, Sayah Z, Bouaziz S. Etude comparative de la production d'éthanol brut à partir de trois variétés de dattes communes (Degla Beida, Tacherwit et Hamraya) reparties dans les différentes classes de dattes (molle, demi-molle et sèche) de la cuvette de Ouargla (Sahara septentrional Est algérien). Algerian. Journal of Arid Environment. 2012;**2**(2):78-87

[121] Mehani I, Boulal A, Bouchekima B. Biofuel production from waste of starting dates in South Algeria. International Journal of Environmental and Ecological Engineering. 2013;**7**(8):572-574

[122] Oucif-Khaled MTO, Segni L. Production of bioethanol from dates of poor quality. African Journal of Agriculture Research. 2014;**9**(37):2814-2818

[123] El-Hadi D, Korteby S, Chibi S. Production de bioethanol à partir de deux variétés de dates (Deglet Nour et Hamraya). Revue Agrobiologia. 2016;**6**(1):111-120

[124] Boulal A, Kihal M, Khelifi C. Fermentative strength of yeasts strain, naturally isolated using common date in south-west of Algeria. International Journal of ChemTech Research. 2017;**10**(1):180-188

[125] Mansour A, Rihani R, Bentahar F. Etude de la production de bioethanol-biocarburant à partir de sous produits agricoles: Effet de l'aération. Algerian Journal of Environmental Science ALJEST. 2019;**5**(3):1087-1093

[126] Laouar A. Caractérisation physicochimique et microbiologique de deux variétés de dattes "Hmira, Feggous" et production de bioéthanol à partir de rebuts de dattes "Hmira" [thesis]. Béchar: Univ. Mohamed Tahri; 2020

[127] Edjekouane M, Lansari F, Khelifi O, Boukheteche I, Laksaci H. Production of Bioéthanol from a local natural resource. Algerian Journal of Renewable Energy and Sustainable Development. 2020;**2**(1):56-59. DOI: 10.46657/ajresd.2020.2.1.8

[128] El-Bey S, Ouakil A, Zoubiri F-Z, Rihani R, Bentahar F. Production de bioéthanol à partir de jus d'orange de Rouïba. Revue Nature et Technologie. 2022;**14**(2):1-7

[129] Rezki MA, Aouad L, Bekki A. Production of ethanol and polyethanol by yeasts isolated from date (*Phoenix dactylifera* L.) wastes. African Journal of Biotechnology. 2015, 2015;**14**(50):3288-3294. DOI: 10.5897/AJB2015.14979

[130] Hadri K, Boulal A, Cheikh N. Ethanol production by *Balanites aegyptiaca* fruits valorization in the Adrar region of Algeria. Journal of Renewable and Sustainable Energies. 2021;**1**(1):109-115

[131] Hadri K, Cheikh N, Ammar M, Messaoudi F-Z, Boulal A. Valorization of *Balanites aegyptiaca* fruits by production of bioethanol: Study and optimization. Current Trends in Natural Sciences. 2022;**11**(21):290-303. DOI: 10.47068/ctns.2022.v11i21.032

[132] Aziz S. Production of Ethanol from Molasses Using Thermotolerant *Kluyveromyces Marxianus* [Thesis]. Jamshoro: Mehran University of Engineering & Technology; 2010

[133] Macedo N, Brigham CJ. From beverages to biofuels: The journeys of ethanol-producing microorganisms. International Journal of Biotechnology for Wellness Industries. 2014;**3**:79-87

[134] Dabija A, Hatnean CA. Study concerning the quality of apple vinegar obtained through classical method. Journal of Agroalimentary Processes and Technologies. 2014;**20**(4):304-310

[135] Bouaziz S., Ould El-hadj M. D., Contribution a l'étude des caractéristiques physico-chimiques et biochimiques de quelques types de vinaigres traditionnels de dattes obtenues a partir de quelques variétés de la région de Ouargla, Annales des Sciences et Technologie. 2010;**2**(1):80-86

[136] Benahmed-Djilali A, Benrachedi K, Benamara S. La qualité du vinaigre de dattes obtenu par le procède traditionnel du sud algérien; composés volatils. In: X colloque international, Sécurité alimentaire: Réalités et perspectives. Adrar: Univ. Ahmed Draia; 2007. pp. 443-460

[137] Bouazza A, Bitam A, Amiali M, Bounihi A, Yargui L, Koceir E. Ahmed, effect of fruit vinegars on liver damage and oxidative stress in high-fat-fed rats. Pharmaceutical Biology. 2016;**54**(2):260-265. DOI: 10.3109/13880209.2015.1031910

[138] Ameur D, Heleili N. Comparative study on physico-chemical and sensory properties of vinegar produced from apple varieties. Agricultural Science Digest. 2022;**42**:534-540. DOI: 10.18805/ag.DF-401

[139] Yu X, Zhao M, Liu F, Zeng S, Hu J. Identification of 2,3-dihydro-3,5-dihydroxy-6-methyl-4H-pyran-4-one as a strong antioxidant in glucose–histidine Maillard reaction products. Food Research International. 2013;**51**(1):397-403. DOI: 10.1016/j.foodres.2012.12.044

[140] Dali NS, Hamame A. Recherche de levures productrices d'enzymes glycolytiques exocellulaires thermostables: Production (sur boite de Pétri et en batch) et Caractérisation des enzymes produites. Constantine, Algeria: Univ. F. Mantouri; 2016

[141] Dakhmouche-Djekrif S, Bennamoun L, Labbani FZK, Ait-Kaki A, Naouadri T, Pauss A, et al. An alkalothermophilic amylopullulanase from the yeast *Clavispora lusitaniae* ABS7: Purification, characterization and potential application in laundry detergent. Catalysts. 2021;**11**:1438. DOI: 10.3390/catal11121438

[142] Labbani F-ZK, Bennamoun L, Dakhmouche S, Ait-Kaki A, Nouadri T. Production, characterization and in vitro evaluation of a yeast killer toxin against some food and beverage spoilage yeasts, in the abstracts of the 1st international webinar on innovative applications of biotechnologies in the food industry: From the laboratory to the enterprise. Algerian Journal of Nutrition and Food Sciences (AJNFS). 2021;**1**(4):26-27

[143] Bennamoun L, Dakhmouche S, Ait-Kaki A, Labbani F-ZK, Nouadri T, Meraihi Z. Production et caractérisation de pectinase levurienne: Essais d'application dans la clarification de jus de citron in the abstracts of the 1st international webinar on innovative applications of biotechnologies in the food industry: From the laboratory to the enterprise. Algerian Journal of Nutrition and Food Sciences (AJNFS). 2021;**1**(4):3

[144] Gares M, Hiligsmann S, Chaouche NK. Lignocellulosic biomass and industrial bioprocesses for the production of second generation bioethanol, does it have a future in Algeria? SN Applied Sciences. 2020;**2**:1680. DOI: 10.1007/s42452-020-03442-2

[145] Mihi A. Dynamic simulation of future date palm plantation (Phoenix dactylifera L.) growth using CA–Markov model and FAO-LCCS data in Algerian dryland oases desert. Modeling Earth Systems and Environment. 2022;**8**:3215-3230. DOI: 10.1007/s40808-021-01289-z

[146] Šuchová K, Fehér C, Ravn JL, Bedő S, Biely P, Geijer C. Cellulose- and

xylan-degrading yeasts: Enzymes, applications and biotechnological potential. Biotechnology Advances. 2022;**59**:107981. DOI: 10.1016/j.biotechadv.2022.107981

[147] Moussa A, Djebli N, Aissat S, Meslem A, Benhalima A. Antifungal activity of four honeys of different types from Algeria against pathogenic yeast: *Candida albicans* and *Rhodotorula* sp. Asian Pacific Journal of Tropical Biomedicine. 2012;**2**(7):554-557. DOI: 10.1016/S2221-1691(12)60096-3

[148] Saadoune Z, Boutoumi Y. Impact of the composition of the essential oils of *Citrus sinensis* (Orange) and *Citrus limonum* (lemon) on the microbiological activity. Algerian Journal of Engineering Research, AJER. 2018;**3**:1-8

[149] Boukhatem MN, Ferhat MA, Kameli A, Saidi F, Kebir HT. Lemon grass (*Cymbopogon citratus*) essential oil as a potent anti-inflammatory and antifungal drugs. Libyan Journal of Medicine. 2014;**19**(9):25431. DOI: 10.3402/ljm.v9.25431

[150] Bouzabata A, Cabral C, Gonçalves MJ, Cruz MT, Bighelli A, Cavaleiro C, et al. *Myrtus communis* L. as source of a bioactive and safe essential oil. Food and Chemical Toxicology. 2015;**75**:166-172. DOI: 10.1016/j.fct.2014.11.009

[151] Bouzabata A, Bazzali O, Cabral C, Gonçalves MJ, Cruz MT, Bighelli A, et al. New compounds, chemical composition, antifungal activity and cytotoxicity of the essential oil from *Myrtus nivellei* batt. & Trab., an endemic species of Central Sahara. Journal of Ethnopharmacology. 2013;**149**(3):613-620. DOI: 10.1016/j.jep.2013.06.054

[152] Neggaz S, Fortas Z. Tests of antibiotic properties of Algerian Desert truffle against bacteria and fungi. Journal of Life Sciences. 2013;**7**(3):259-266

[153] Bouchenak O, Boumaza S, Yahiaoui K, Benhabyles N, Laoufi R, Toubal S, et al. Evaluation de l'activité antioxydante et l'effet antimicrobien des composes phénoliques extraits du lichen *Xanthoria parietina* de la région de Boumerdes. Revue des BioRessources. 2020;**10**(2):93-103

[154] Boukhatem T, Chadli R, Bouchama A, Hamed D. Cytotoxicity test on *Artemia salina*, antibacterial and antifungal activities of *Cystoseira stricta* extracts from the coast of Mostaganem, western Algeria. Journal of Materials and Environmental Sciences. 2018;**9**(12):3190-3196

[155] Messahli I, Gouzi H, Sifi I, Chaibi R, Rezzoug A, Rouari L. Anticandidal activity of dichloromethane extract obtained from the red algae a. armata of the Algerian coast. Acta Ecologica Sinica. 2022;**42**(5):461-466. DOI: 10.1016/j.chnaes.2021.08.005

Chapter 2

Upgrading Non-Conventional Yeasts into Valuable Biofactories

Kevin Castillo-Mendieta, Jimmy Arias and Fernando Gonzales-Zubiate

Abstract

The use of synthetic biology on yeasts has enhanced the production of commercially relevant chemicals, from biofuels to recombinant therapeutic proteins, to name just a few. Despite most of these advances had already been studied and described in *Saccharomyces cerevisiae*, during the last years the attention has turned to the use of alternative expression systems with a higher yield and quality such as non-conventional yeasts. Recently, there has been an increase in studies about non-conventional yeasts due to advantages based on their natural capacity to tolerate harsh conditions or the wide range of carbon sources they need during the generation of specific products. This chapter, therefore, aims to describe the current status of the most used non-conventional yeasts in metabolite production as well as the engineering behind them in order to optimize or regulate protein expression: *Pichia pastoris*, *Kluyveromyces marxianus*, *Kluyveromyces lactis* and *Yarrowia lipolytica*.

Keywords: non-conventional yeasts, bioengineering, synthetic biology, CRISPR-Cas9, Golden Gate cloning, TALENs

1. Introduction

Yeast is probably one of the oldest domesticated organisms, since it was used for beer brewing already in Sumer and Babylonia around 6000 BC [1]. Not surprisingly, yeast cells were among the first microorganisms seen after the invention of the microscope in the seventeenth century, but their recognition as a living organisms did not come until two centuries later [2]. Yeasts, as such, do not form a single taxonomic or phylogenetic group in the kingdom fungi, rather, they occur in different subdivisions belonging to *Ascomycota and Basidiomycota*. Moreover, they are unicellular with budding and binary fission as the main asexual reproduction, and sexual spore production in stress conditions [3, 4].

Interestingly, yeast cells can exhibit a variety of cell sizes, shapes and colors. Even individual cells from a pure strain of a single species can display morphological heterogeneity. Moreover, yeast cell size varies widely from 2 up to 50 μm in length. Many yeast species are ellipsoidal or ovoid, but other cell shapes can be also observed as in *Candida albicans* and *Yarrowia lipolytica* which are mostly filamentous [1].

With respect to their diversity, there are around 2000 accepted yeast species included in the Yeast Trust Database (theyeasts.org) [4]. They have been isolated from highly diverse environments such as insect guts, food products, soil, oceans and even ancient ice fields [1]. However, it seems that we have just scratched the tip of the iceberg. According to Fell's estimation, what we have found represents only 1% of the species that might exist in nature [4].

Furthermore, yeasts as a whole are interesting because they are capable of metabolizing a wide variety of carbon sources including glucose, fructose, lactose, xylose and arabinose [3]. Besides, the metabolic activity of some yeasts can be dependent on the sugar concentration present in the medium: fermentation in high sugar concentration and aerobic respiration in low sugar concentrations (Crabtree effect) which can be advantageous in some industrial processes [3, 5].

2. Industrial applications of yeasts

Industrially, yeasts possess many attractive features that confer them some benefits in relation to bacteria such as *Escherichia coli*. For instance, yeasts have the capacity to grow on a wide variety of carbon sources, perform post-translational modifications, and compartmentalize reactions in organelles, they also present high secretion capacity, and are less susceptible to infectious agents like bacteriophages [6].

For these reasons, natural yeasts have been used in a lot of industrial processes. For example, in the food industry, the alcoholic fermentation of *S. cerevisiae* is used for the production of bread, beer and wine [7]. Furthermore, other yeasts species take different roles in the elaboration of food products such as yoghurts, in which *Torulopsis candida*, and *Kluyveromyces fragilis* are used for the improvement of aroma, texture and addition of nutrients by fermenting lactose with hydrolysis of milk casein [8].

Another important application of *S. cerevisiae* is the production of biofuels such as bioethanol, which is a result of sugar fermentation under anaerobic conditions. *S. cerevisiae* catabolizes sugars by glycolysis until it produces pyruvate that is then converted to acetaldehyde and carbon dioxide, which is reduced to ethanol by an alcohol dehydrogenase [9]. In addition, other yeast species also have the capacity to produce bioethanol. In fact, *Kluyveromyces marxianus*, *Dekkera bruxellensis* and *Scheffersomyces stipitis* are capable of producing bioethanol by fermentation of polyfructan substrates, hexoses and lignocelluloses substrates respectively [9]. Some yeasts are able to naturally produce bioethanol using lignocellulose resources (cheap, abundant and renewable) making them of great interest in second-generation biofuels, thus providing a clear advantage over first-generation biofuels that employ large cultivated areas [9–11].

On the other hand, despite *S. cerevisiae* has been widely studied and its industrial applications being countless, other yeast species, known as non-conventional yeasts (NCYs[1]) are becoming more popular in industrial applications. Several NCYs have diverse advantages compared to *S. cerevisiae*, mainly they are more suitable for a big number of biotechnological processes since they present natural tolerance to stresses like extreme pH, temperatures and osmolarity conditions [12]. Some of the most studied NCY species that are capable to withstand harsh conditions are *Yarrowia lipolytica*, *Hanensula polymorpha*, *Pichia pastoris* and *Kluyveromyces lactis* [13]. In fact, *K. lactis* is widely used in the cheese industry, replacing the conventional rennet, due

[1] Currently, there is not an accepted definition of NCYs, but many scientists consider NCYs as "non-*Saccharomyces*" yeasts [7].

to its ability to produce lactic acid from lactose. On the other hand, *Y. lipolytica* is used for the production of biosurfactants, carotenoids and lipids and *K. marxianus* for the production of bioethanol, aroma compounds and biosurfactants [6, 12].

Furthermore, some of the most interesting features of NCYs are their capacity to accumulate metabolites, synthesize and secrete recombinant proteins and enzymes [12]. For instance, yeasts like *Y. lipolytica* and *K. lactis* are able to secrete high titers of proteins extracellularly better than *S. cerevisiae* [13]. Moreover, some useful industrial enzymes like amylases, cellulases, proteases and lipases, have been reported to be produced by several strains of the NCY *Aureobasidium pullulans* [14]. These interesting properties have led to a proliferation of studies in NCYs aiming to improve their performance in the production of important metabolites and proteins. Hence, this review focuses on providing a clear description and analysis of the use of synthetic biology tools and strategies at the expression level that helps enhance four of the most popular NCYs: *P. pastoris*, *K. marxianus*, *K. lactis* and *Y. lipolytica* into valuable biofactories.

3. Use of synthetic biology in NCYs

Synthetic biology relies on the premise that a biological system can be built using a collection of previously described parts and subsystems [6]. This is achieved by standardization and modularization of useful biological parts, mechanisms and systems, or redesign the existing ones to provide new and better qualities [15]. Therefore, it is capable to define building blocks at various levels such as expression, protein and pathway levels [15]. The ability to control the dynamics at each level is important in order to establish unique and robust expression and production platforms for biomanufacturing [6]. For example, in synthetic biology-inspired therapies, the regulation of gene expression is important to determine the amount of the therapeutic and allows for accurate control over the design of synthetic cells [16].

Furthermore, with the emergence of modern genome editing tools, the synthetic biological capabilities to rewire and engineer organisms for production purposes have enabled the application of engineering efforts in non-conventional yeast of interest to industrial biotechnology [6].

This section will focus on the description of synthetic biology tools at the expression level, covering the engineering of genetic parts which include promoter, terminator and signal peptide, as well as codon optimization. In addition, available genome editing tools like CRISPR-Cas and cloning methods such as Golden Gate are also discussed. Their applications will be described in detail, later on, in Section 4.

3.1 Engineering of genetic parts

Codon optimization. The degeneracy of the genetic code means that several amino acids can be encoded by more than one codon (e.g. Leu = CUU, CUC, CUA, CUG), thus a random codon usage would be expected for those amino acids [17]. However, the expression of the same gene is different depending on the organism, because of the availability of host's tRNA pool. This is known as codon bias and is thought to affect the translation efficiency [18]. As a result, codon optimization is an important strategy when considering the expression of heterologous proteins.

Promoters. Selecting an adequate promoter is an important step since it can affect the level of expression of the desired protein. There can be constitutive or inducible promoters, the latter more advantageous since they allow researchers to separate cell growth from the production of the desired protein. This avoids potential toxic effects due to a constitutive expression of heterologous proteins [19]. In addition, having a variety of promoters available is desired to fine-tune and optimize pathways that involve the de-expression of several proteins [20].

Terminators. It not only plays a critical role in transcription but is also able to influence mRNA stability and lifetime. This provides a new level of regulating protein expression; however, the impact of terminators is sometimes underestimated compared to promoters [21, 22]. In *S. cerevisiae* expression vectors, some endogenous terminators such as T_{CYC1} and T_{ADH1} are commonly used. Nevertheless, it has been demonstrated that *S. cerevisiae* terminators can show a high degree of transferability across other non-conventional yeasts [23].

Signal peptides. If a protein of interest is desired to be secreted, it is only required to add a secretion signal peptide at the N-terminus of the nascent polypeptide [24]. However, selecting an appropriate signal peptide is crucial since the protein quality and yield may vary widely depending on the heterologous protein being expressed [25]. Therefore, screening and characterizing many signal peptide sequences is a good approach to have adequate expression levels of different proteins.

3.2 Cloning methods and genome editing tools

Golden Gate. Golden Gate cloning method uses type II restriction enzymes (*Bsa*I and *Bpi*I) to precisely assemble multiple genetic parts by simultaneous restriction and ligation steps. These restriction enzymes cut outside their recognition sequences leaving 4-letter overhangs which can be freely designed; hence this method offers important benefits as it is cheaper than other advanced techniques, it does not require long flanking DNA and it allows scar-less cloning [26].

TALEN. Transcription-activator-like effector (TALE) nuclease is a genome editing tool based on type II effector proteins from bacterial plant pathogens of the genus *Xanthomonas* fused with the non-specific nuclease domain of the restriction endonuclease *Fok*I [27]. These TALENs are designed such that they bind separate targets in opposition to each other with an appropriate spacer between them allowing the *Fok*I nucleases to cause a double-strand break (DSB) and subsequently allowing the editing of the genome [28]. This genome editing tool has many advantages such as easy assembly, availability of powerful resources, cross-species flexibility and a high rate of success [27].

CRISPR-Cas9. This tool revolutionized the field of gene editing, since it is capable of creating a DSB in a specific DNA site by just using a single-guide RNA (sgRNA) complementary to the targeted region and an endonuclease (Cas9). This makes CRISPR-Cas9 a better genome editing tool compared to TALENs, which require laborious protein engineering steps for each new editing target. This editing tool has been successfully used for knock-out of a gene, but it can be adapted for other applications such as regulating transcription and facilitating metabolic engineering [29, 30].

4. Engineering non-conventional yeasts (NCYs)

This section will discuss the application of synthetic biology tools and strategies in four of the most popular NCYs: *P. pastoris*, *K. marxianus*, *K. lactis* and *Y. lipolytica*, yeasts that have been selected for several reasons. First, they naturally present

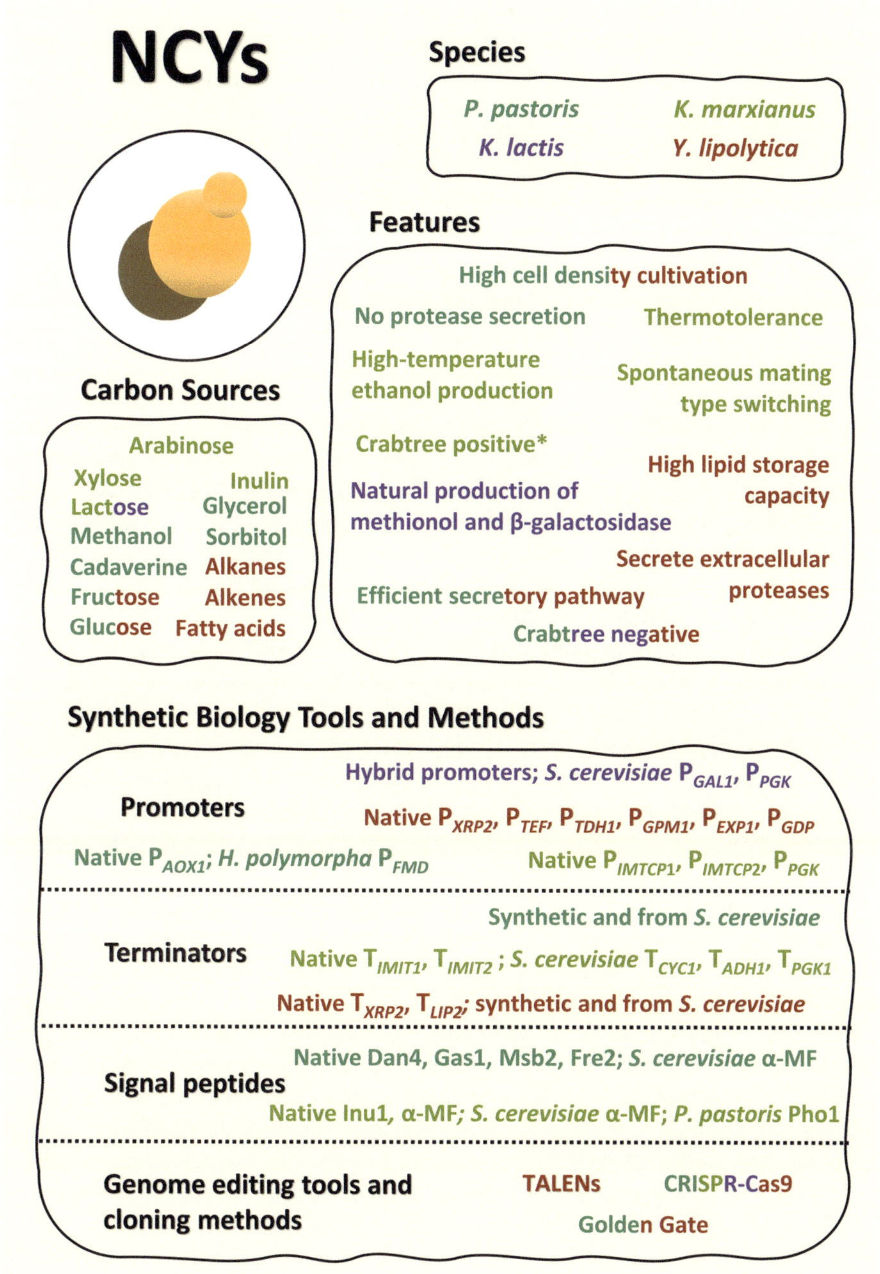

Figure 1.
Description of the non-conventional yeasts (NCYs): P. pastoris, K. marxianus, K. lactis *and* Y. lipolytica *and the synthetic biology tools and strategies applied to each of them. Descriptions of the same color represent a specific yeast. Example: All text in red belongs to* Y. lipolytica. *K. marxianus *is classified as Crabtree-negative, although some reports consider this species as Crabtree-positive due to strain variability.*

interesting characteristics with potential benefits in industrial applications such as the use of "waste products" as substrates, thermotolerance, high capacity to store lipids and the capacity to efficiently secrete desired compounds (**Figure 1**).

Second, they have their genomes sequenced and available at the National Center for Biotechnology Information (NCBI) website [31] and several metabolomic and transcriptomic studies have been reported [32–37]. This information is key when searching for new sources of BioBricks, since novel genetic parts or metabolites can be discovered under different contexts such as carbon sources or stress conditions. Third, information about regulatory associations between transcription factors (TFs) and target genes in these four species is available at the N. C. Yeastract database [38]. Last but not least, all the referred species have genome-scale metabolic models reported in the literature [39–42], allowing researchers to predict metabolic fluxes, and subsequently, optimize the production of relevant compounds in these microorganisms.

4.1 *Pichia pastoris*

Pichia pastoris[2] was initially developed by Phillips Petroleum Company for the production of single-cell protein for feedstock, but it was then repurposed as a promising expression system for the production of recombinant proteins [44] due to its many advantages such as high cell density cultivation, an efficient secretory capacity with a low background of endogenous proteins, the absence of protease secretion, the presence of alternative constitutive and inducible promoters and the ability to perform post-translational modifications to proteins [24, 45].

P. pastoris is a Crabtree-negative yeast that is able to utilize a variety of carbon sources including glucose, glycerol, fructose, sorbitol, methanol, alanine and cadaverine [43, 44]. As a matter of fact, it has the ability to use methanol as a sole carbon source (methylotrophic) due to several adaptations such as the expression of enzymes involved in methanol metabolism (e.g. alcohol oxidase) and the proliferation of peroxisomes (reaching over 80% of the cell volume) [44].

For these reasons, *P. pastoris* has become an industrially important microorganism. This is evidenced by the over 300 industrial processes that have been licensed and more than 70 commercial products that are currently on the market [46]. The use of synthetic biology tools in this yeast has led to a whole new level of potential industrial applications for this yeast. Some examples of the application of synthetic biology strategies in *P. pastoris* are provided below.

Codon Optimization. A clear application of this strategy in this yeast is the heterologous expression of enzymes such as phytases which are important enzymes that once included in animal feed can help not only increase the absorption of phosphorus in monogastric animals (e.g. pigs and horses) but at the same time reduce phosphorus levels in manure, thus finding a cheaper and more efficient way of producing this enzyme is desired. For instance, Xiong et al. compared the expression of a recombinant *Aspergillus niger* phytase in *P. pastoris* before and after codon optimization of both the phytase gene and the *S. cerevisiae* signal sequence α mating factor (α-MF). They obtained a phytase activity of 865 Units/ml, resulting in a 14.5-fold increase in the production/activity of phytase in comparison with the non-optimized gene and signal sequence [47]. The following year the same group applied this strategy to express a *Peniophora lycii* phytate enzyme achieving a phytase activity of 10,540 Units/ml and a 13.6-fold yield increase compared with the non-optimized gene and signal sequence [48]. In the last years, heterologous expression of other industrially

[2] Although initially named *Pichia pastoris* in the 1950s, it was then reclassified into the genus *Komagataella* in 1995, splitting up into the two species *K. pastoris* and *K. phaffii* [43]. Here, we still use the name *P. pastoris* for simplicity.

relevant proteins has been enhanced using this method, including but not limited to keratinases, endoinulinases, α-amylases, lipases, xylanases, fibases, pectinases, IFN-ω and hydroxynitrile lyases [49–57].

Promoters. Since all gene promoters of the methanol utilization (MUT) pathway are strongly repressed by carbon sources such as glucose, they can be a useful tool for the induced expression of heterologous proteins [58]. In fact, most of the heterologous expression of proteins in *P. pastoris* is carried out using one of those promoters, P*AOX*$_1$ [59]. In addition, other orthologous MUT promoters from related species have been evaluated in this species such as *Hansenula polymorpha* P_{FMD} which showed a 3.5-fold higher expression compared to the strongest endogenous MUT promoters [58].

On the other hand, using methanol (a flammable and toxic compound) in a large-scale fermentation process can be potentially dangerous [60]. For this reason, researchers have tried to find alternatives to overcome this limitation using different approaches such as employing orthologous promoters from related methylotrophic yeasts [58] or even engineering cis- and trans-acting elements in the P_{AOX1} [60]. For instance, using a trans-acting approach, Wang et al. developed a methanol-free method for protein expression using the P_{AOX1} promoter. They developed a strain that overexpressed the transcription activator Mit1 and repressed glucose- or glycerol-dependent transcriptional repressors Mig1, Mig2 and Nrg1. Then they evaluated the performance of both the methanol-free system and the wild-type system under their respective optimal culture conditions. Although the expression level of the recombinant insulin precursor protein in the methanol-free system was only 58.6% of the wild-type, they claim that it can be further improved by overexpressing unfolded protein response activators, protein foldases or chaperons [60].

Terminators. Knowledge about terminators is rather limited in *P. pastoris*, although in recent years it has been gaining more attention. In 2020 Ito et al. developed a terminator catalog of 72 sequences including synthetic, endogenous and heterogenous (*S. cerevisiae*) terminators with a 17-fold range of expression when using red fluorescent protein and *Aspergillus aculeatus* β-glucosidase as reporter proteins. Moreover, interesting results were found from this study including independence of terminator activity from the upstream gene and high degree of transferability of *S. cerevisiae* terminators to *P. pastoris* [61].

Signal Peptides. The *S. cerevisiae* signal sequence α-MF is widely used for secreted expression of recombinant proteins in *P. pastoris* [62], where a peptidic pre-region leads the translocation from the cytoplasm to the ER lumen. Additionally, it has a peptidic pro-region which facilitates the proper transit of target polypeptides from the ER to the Golgi apparatus; unfortunately, this pro-region tends to aggregate in the ER, impeding a proper secretion of the target protein [63]. Several strategies can be applied to overcome this issue such as codon optimization or directed evolution of the signal sequence α-MF [63]. Lately, new endogenous signal peptides could outperform signal sequence α-MF. For instance, in 2019 Duan and collaborators were able to identify four endogenous signal peptides (Dan4, Gas1, Msb2 and Fre2) in *P. pastoris*. All of them showed more than fourfold enhancement of total β-galactosidase activity over the signal sequence α-MF, with being Gas1 the one that showed the best results (230-fold increase). Moreover, Msb2 signal peptide had a better performance in the expression of β-galactosidase, yEGFP and cephalosporin C acylase; therefore, it could be considered as a more effective signal for heterologous protein secretion in *P. pastoris* [62].

Golden Gate. Prielhofer et al. developed GoldenPiCS a modular cloning system that facilitates the engineering of *P. pastoris* by generating episomal plasmids with up to 8 expression units. The main feature of this method is that researchers can easily exchange genetic parts and quickly create and test new variants. The feasibility of this method was demonstrated with the optimization of a CRISPR-Cas9 system for *P. pastoris* using different combinations of humanized Cas9 and sgRNA on one single episomal plasmid [26].

In a more recent study, Cheng et al. used the Golden Gate cloning method to develop a versatile and easy way of assembling eukaryotic gene exons into both prokaryotic and eukaryotic plasmids in a one-step reaction [64]. Thus, this new approach enables researches to rapidly identify the optimal expression host for the production of specific proteins, overcoming some disadvantages of traditional methods to obtain intron-free eukaryotic genes (e.g. whole-gene synthesis or reverse transcription methods) which are time-consuming, expensive and complicated to operate [64].

CRISPR-Cas9. In *P. pastoris* exogenous cassettes with long homology arms are integrated ectopically and homologous recombination (HR) occurs only at variable frequencies of <0.1 to 30% [29]. However, high efficiency for gene insertion by HR can be achieved by deleting the protein Ku70 which is involved in nonhomologous end-joining (NHEJ) repair [65]. Weninger et al. used this strategy to develop integration cassettes of CRISPR-Cas9 marker-free with close to 100% efficiency [66]. Yang et al. developed a high-efficiency CRISPR-Cas system in *P. pastoris* synthetizing Cas9 (codon-optimized for *Homo sapiens)* and the sgRNA on different plasmids. They validated the editing efficiency in gene deletion of six genes, reaching or exceeding 75% of efficiency for each target gene. This system eliminates the episomal sgRNA plasmid through sub-culture to allow editing of another gene with a consistent single gene editing efficiency [65]. However, it was demonstrated that this system performed poorly when editing multiple genes. Finally, even though huge progress has been made to improve CRISPR-Cas9 efficiency in *P. pastoris*, it is still lower than other yeasts such as *S. cerevisiae* and *Kluyveromyces lactis* [65].

4.2 *Kluyveromyces marxianus*

K. marxianus is phylogenetically related to *S. cerevisiae* and more closely related to *K. lactis*. It is a hemiascomycetous yeast that can exist as stable haploid or diploid cells and is able to spontaneously switch its mating type (homothallic) [67, 68]. *K. marxianus* strains have been isolated from a great variety of habitats including dairy products, soil, sugarcane bagasse, insects and fruits. Therefore, this yeast presents a high metabolic diversity and a significant degree of intraspecific polymorphism [68–71], in point of fact, several industrially relevant compounds have been found to be naturally produced by *K. marxianus* including pectinase, aroma compounds, inulinase, lipase and lactase [69, 72]. Furthermore, some strains have been described to exhibit multi-stress resistance [68].

K. marxianus is classified as facultative fermentative and Crabtree negative, although some reports consider this species as Crabtree positive due to strain variability [69, 73]. *K. marxianus* cannot naturally grow under strictly anaerobic conditions [69], but it can be genetically modified to grow under such conditions. However, growth rates are still lower than anaerobic growth in *S. cerevisiae* [74]. *K. marxianus* is capable of using non-conventional sugars such as xylose, arabinose and inulin as carbon sources [40]. In addition, this yeast has the ability to use lactose which cannot be accomplished by *S. cerevisiae* and can grow at higher temperatures with a wider range of substrates than *K. lactis* [67, 69].

All these features make *K. marxianus* a promising biofactory, having a wide range of applications such as host for the production of heterologous proteins;

alternative to baker's yeast; bioremediation of textile dyes, cheese whey and copper; biomass for animal feeding; probiotics and high-temperature bioethanol production [67, 69, 72, 73, 75]. For instance, Nonklang et al. found remarkable differences in high-temperature ethanol production between *S. cerevisiae* and *K. marxianus* as *K. marxianus* DMKU3-1042 was the fastest to convert glucose to ethanol at 45°C whereas none of the *S. cerevisiae* strains were able to grow at this temperature [67].

In the last years, some progress in synthetic biology has been accomplished in this species; however, it is still limited compared to other NCYs such as *P. pastoris* since *K. marxianus* still lacks efficient genetic tools, there are also limited auxotrophic markers and very few constitutive and inducible promotes have been described [21]. Despite that, some examples and applications are discussed below.

Codon Optimization. This strategy is currently used to improve the expression of recombinant proteins in *K. marxianus*, especially vaccines. For instance, codon optimization has been used for the heterologous expression of the porcine circovirus type 2 (PCV2) Cap protein in *K. marxianus* as an alternative to produce PCV2 virus-like particle vaccines to treat porcine circovirus disease and help reduce economical losses in the swine industry. Duan et al. reported in their experiment higher yields compared with *E. coli* and *P. pastoris* as host vectors [76]. Other examples where codon optimization has been applied to *K. marxianus* include heterologous expression of single-chain antibodies, overproduction of inulinase, expression of the dengue virus type 1 nonstructural protein 1 and porcine parvovirus-like particles [77–80].

Promoters. In the last years, several promoters have been identified for this species which can be induced by heat, xylose, lactose or inulin [22]. In addition, several strong endogenous promoters of genes such as purine-cytosine permease, inulinase, enolase and glyceraldehyde 3-phosphate have been characterized [21, 69]. More recently, Kumar et al. identified two new strong promoters (P_{IMTCP1} and P_{IMTCP2}) which are more efficient at different temperatures and carbon sources than previously known promoters in this species [21].

Interestingly, the relative strength of promoters can change depending on the carbon source provided [81]. For instance, Kumar et al. showed that the *K. marxianus* inulinase promoter has relatively higher activity in the presence of xylose than dextrose [21]. In addition to finding new promoters, already described promoters have been also engineered to improve their features. For example, Zhou et al. improved the expression of lignocellulolytic enzymes in *K. marxianus* by a mutation inside the inulinase promoter and a deletion of an A-T-rich region inside the 5'UTR [82].

Notably, if thermotolerance of this yeast is to be exploited when expressing heterologous proteins, it is not enough to only focus on utilizing thermostable proteins but also identifying thermotolerant promoters because promoter activity tends to decrease with elevated temperature as it was demonstrated by Yang et al. [81]. Despite the fact that lower promoter activity is observed when increasing the temperature, *K. marxianus* promoters have been found to be stronger than their corresponding promoters from *S. cerevisiae* at such temperatures. In fact, the *K. marxianus* constitutive promoter P_{PGK} has been shown to retain relatively strong activity with an increase in temperature [81].

Terminators. Only a limited number of terminators have been examined in *K. marxianus* including terminators from *S. cerevisiae* such as T_{CYC1}, T_{PGK1} and T_{ADH1} [21, 22]. Additionally, new recently described *K. marxianus* terminators have widened the range of regulation of protein expression in this species. For instance, in a recent study researchers found an increase in EGFP expression (fourfold increase of mRNA level) in *K. marxianus* when using the endogenous terminators T_{IMTT1} or T_{IMTT2} instead of *S. cerevisiae* T_{CYC1} [21].

Signal Peptides. Research about signal peptides in this species is still limited since only a few signal sequences have been characterized and employed in heterologous protein expression on *K. marxianus* including signal sequences from *K. marxianus* Inu1, *P. pastoris* Pho1, *S. cerevisiae* α-MF and *K. marxianus* α-MF [21]. Moreover, some experiments of signal sequence engineering conducted to improve its activity have been reported, for instance, Yarimizu et al. developed a synthetic signal sequence in the yeast *K. marxianus* by redesigning the hydrophobic core of *Gaussia princeps* secretory luciferase signal sequence. The hydrophobic sequence was replaced by a repeat of 16 methionine residues, resulting in 20-fold higher activity than that from the wild type [83].

CRISPR-Cas9. Cernak et al. first used a CRISPR-Cas9 system to inactivate genes responsible for spontaneous mating-type switching (common phenomenon in *K. marxianus*), enabling the production of stable heterothallic haploid strains which can mate. As a result, they combined three complex traits found in different strains (ability to take up exogenous DNA, thermotolerance capacity and higher lipid production) into single *K. marxianus* clones [71].

Li et al. developed a one-step multigene integration system based on CRISPR-Cas9, which is capable of integrating up to three cassettes in a single, targeted genomic locus in *K. marxianus.* It consists of the CRISPR plasmid (expression of the sgRNA and Cas9) and the homology donor plasmid (700 bp up- and down-stream homology to the targeted site). This system has been proven to have an efficiency comparable to single-gene integration and it can be performed within 4 days from transformation to confirmation [84].

In 2022 Bever et al. developed a highly efficient CRISPR-Cas9 system in *K. marxianus* that allows editing of multiple genes which can be used in both NHEJ-functional and -deficient strains showing nearly 100% efficiency of gene disruption in those two strains, whereas 100% efficiency of donor integration was observed only in NHEJ-deficient strains. In addition, this system achieved a dual integration efficiency of 25.5% in an NHEJ-deficient strain [85].

4.3 *Kluyveromyces lactis*

Kluyveromyces lactis is an NCY known for its capacity to assimilate lactose and convert it into lactic acid. *K. lactis* is a respiratory Crabtree-negative yeast highly used in industries due to its ability to secrete the protein β-galactosidase, used for making lactose-free products [23]. Moreover, this yeast is also capable to produce methionol, which is a flavor-active compound important in the overall aroma of soy sauce and cheese [7].

Some advantages of working with *K. lactis* yeasts are the capacity of producing heterologous proteins in simple growth medium, complete knowledge of their genome and more importantly, they can be easily genetically manipulated [75]. Due to its similarity in biosynthetic capacities to *S. cerevisiae*, *K. lactis* toolkits for heterologous gene expression are mostly the same. However, *K. lactis* presents many attributes that make it more suitable for protein expression and extracellular secretion than *S. cerevisiae* [13]. In fact, *K. lactis* uses an inducible promoter P_{LAC4} which is commercially available due to its capacity to secrete recombinant proteins in the culture fluid under lactose presence, a very useful feature for protein purification [13].

Promoters. In general, the promoters used for heterologous protein production strategies in *K. lactis* are the same as *S. cerevisiae*, P_{GAL1} or P_{PGK}, which, due to their high level of transferability, have shown the potential of promoter engineering in *S. cerevisiae* to be applied in the *K. lactis* expression system [6]. However, other engineering strategies involving *K. lactis* promoters have also been developed. For example, Sakhtah et al. have recently developed a novel auto-inducible promoter system in *K. lactis*. For this, portions of two promoters, the constitutive P_{GAP1} and the carbon source-sensitive P_{ICL1}, were combined to form a hybrid promoter called P_{350} [86]. This novel promoter is induced by the depletion of glucose or glycerol in the medium, making it auto-inducible as the carbon sources are consumed by the growing cells. The development of this hybrid promoter promises to be useful for the implementation of one-step protein expression methods for small- and large-scale bioprocesses [86]. Moreover, another hybrid promoter approach used in K. *lactis* involves the combination of core promoter elements of *Trichoderma reesei* P_{CBH1} and *K. lactis* P_{LAC4}, which showed an increase in protein production in this yeast [13].

CRISPR-Cas9. For the implementation of CRISPR-Cas9 in *K. lactis*, Horwitz et al. adapted an *S. cerevisiae* system by exchanging the 2 μ element with the *K. lactis* specific pKD1 vector-stabilizing element and the constitutive promoter P_{FBA1} at a *GAL80* site [87–89]. Moreover, sgRNA expression was driven by the typical P_{SNR52} pol III promoter and a T*SUP*$_4$ terminator, and the deletion of the *KU80* gene was performed to reduce NHEJ [6, 29, 87–89]. The implementation of this system resulted in the successful integration of three donor six-gene-DNA parts into three separate loci (*DIT1, ADH1 and NDT80*) with a triple integration efficiency of 2.1% [6, 29, 87, 88]. Despite this low efficiency, the speed and ability to screen strains reduced the design-built-test cycle for this non-conventional yeast, and further improvements in targeting this efficiency could enhance genome editing for wild-type or industrial strains [6].

On the other hand, CRISPR-Cas9 genome editing was used by Burghardt et al. in order to increase the enzymatic production of the prebiotic fructo-oligosaccharides (FOS) in *K. lactis*. For this, the fructosyltransferase gene (*FFT*), needed for forward reactions, from *Aspergillus terreus* NIH2624 was integrated with a *K. lactis* GG799 production host. Furthermore, a CRISPR-Cas9 system was used to delete a native invertase gene, involved in reverse reactions. The results showed an increase in transferase activity by 66.9% when grown in a fed-batch process [90].

4.4 *Yarrowia lipolytica*

Yarrowia lipolytica belongs to the *Ascomycota*, *Dipodascaceae* family. It is naturally found in lipid and protein-rich substrates such as soil, sewage and oil-polluted environments due to their capacity to hydrolyse lipids, assimilate hydrocarbons and fatty acids and secrete extracellular proteases [91]. *Y. lipolytica* is a Crabtree-negative haploid, heterothallic yeast with mating types Mat A and Mat B, and low mating frequency in nature [92]. In a laboratory setting, cells appear spherical, ellipsoidal or elongated and arranged singly, in pairs or clustered in groups. Furthermore, colonies present a creamy texture and a convoluted pale white matte surface [91]. About the carbon sources, *Y. lipolytica* is capable to assimilate hydrophobic substrates like alkanes, alkenes, fatty acids, fatty acid methyl esters, triglycerides and hydrophilic substrates like glucose, fructose, some alcohols, many polyols and many organic acids [91]. Some important characteristics of this yeast are its efficient secretion pathway and lipid storage capacity, two qualities that have made it a research model for protein secretion and lipid metabolism [92]. Moreover, due to its lipogenic metabolism, *Y. lipolytica* has been studied for the biosynthesis of acetyl-CoA-derived molecules such as terpenes [93]. In the last years, because of its production capacity of industrial interest compounds and the ability to grow at high cell densities, different synthetic biology tools have been developed and applied in *Y. lipolytica* [94]. As matter of fact, Wong and colleagues designed a collection of BioBricks for *Y. lipolytica* called YaliBricks, which contains compatible restriction enzyme sites that allows modular genetic engineering [95].

Promoters. In *Yarrowia lipolytica*, two important promoters have been isolated and characterized; the promoter from the *XRP2* gene, which codes for an alkaline extracellular protease and the constitutive promoter from the *TEF* gene, which codes for translational elongation factor-1 [96]. Furthermore, recent studies are focusing on the development of hybrid promoters that could increase the strength of the available ones. Early approaches to promoter hybridization led to the characterization of upstream activating sequences (UASs), native to *XPR2*, which resulted in an increase on promoter activity when hybridized in several tandem repeats [97]. Madzak et al. engineered four hybrid promoters (named hp1d, hp2d, hp3d and hp4d) containing up to four copies of one of its upstream activation sequences ($UAS1_{XPR2}$) fused upstream from a P_{LEU2} promoter [98]. The resulting promoters showed an increase in their strength depending on the number of tandem $UAS1_{XPR2}$ elements, with hp4d being the strongest hybrid promoter and therefore used widely for heterologous gene expression in *Y. lipolytica* [94, 99]. On the other hand, Schwarts et al. constructed a synthetic hybrid promoter, using *GAL1* UAS from *S. cerevisiae* and the *TEF* core promoter from *Y. lipolytica*, that achieved a slightly higher expression than P*UAS1B8*-*TEF*, hybrid promoter that has been widely used [93]. Moreover, other native promoters like P_{TDH1}, P_{GPM1}, P_{EXP1}, P_{FBAINm}, P_{GPAT}, P_{GPD} and P_{YAT} have been characterized and used in expression of heterologous genes with promising results [94].

Terminators. The most commonly used terminators for expression of heterologous genes in *Y. lipolytica* are derived from the native *XPR2* and *LIP2* genes [99]. Moreover, some *S. cerevisiae* terminators have shown a high degree of transferability in *Y. lipolytica* [100]. Indeed, synthetic terminators designed for *S. cerevisiae* have been used in *Y. lipolytica* with an increase of 60% in expression level over some wild-type terminators [56]. Additionally, these synthetic terminators are commonly smaller than the natural ones, conferring them an advantage for transcription units (TU) and vector design since they show low risk of undesired HR between TU or with the genome, contributing high stability to genetically modified strains [50]. However, despite these advances, the number of studies of terminators in *Y. lipolytica*, in comparison to promoters, is still scarce [96].

Golden Gate. Larroude and collaborators have developed a modular toolkit based on the Golden Gate strategy that allows assembly in one step of three transcript units together with integration into *Y. lipolytica* genome. This approach comes with a collection of six selective markers and sequences for random or specific integration, nine promoters of variable strength and five terminators [100]. In such manner, the heterologous production of β-carotene is possible with the expression of three genes involved in the carotenoid pathway after a single transformation [101], making *Y. lipolytica* a competitive biotechnological producer of β-carotene.

CRISPR-Cas9. The use of CRISPR-Cas9 in *Y. lipolytica* has been widely studied. In recent years, a system from *Streptococcus pyogenes* has been adapted, with a synthetic RNA polymerase III promoter and an optimized Cas9 to perform a marker-free gene disruption and integration in *Y. lipolytica*. In fact, five loci have been recently identified that could serve as hotspots for targeting marker-free gene integration [97]. This system resulted in a single-gene disruption and HR with a 90 and 70% of efficiency respectively when Cas9 and the sgRNA were co-transformed using donor DNA [94]. In addition, systems like CRISPR interference (CRISPRi) and CRISPR activation (CRISPRa) have been developed for controlling gene expression in *Y. lipolytica* [99]. Here, a deactivated Cas9 (dCas9) is fused to transcriptional repressors or activators, allowing binding to sgRNA-complementary DNA without cleavage that could result in DSBs [91]. An implementation of the CRISPRa system was performed by Schwarts et al. in order to upregulate *BGL1* and *BGL2* (β-glucosidase genes that are transcriptionally silent), and allow *Y. lipolytica* strains to use cellobiose as a carbon source. For this, a VPR activator was identified and fused to dCas9 to enable gene activation [93, 96].

Other CRISPR-Cas9 strategies have been developed, for example, a paired sgRNA, consisting of two vectors, each containing Cas9 gene and a sgRNA cassette, was used in order to target areas upstream and downstream the start and stop codon, respectively, and allow a complete gene knockout via gene excision with a 20% of efficiency [94].

TALEN. Used to direct DNA DSBs to occur at a specific target site, was applied in *Y. lipolytica* to produce structure-based mutagenesis of a fatty acid synthase (FAS) domain, and allow the synthesis of fatty acids with shorter chain lengths [99]. Moreover, site-directed mutagenesis improved in efficiency when homologous exogenous DNA was added to the targeted site, resulting in HR-mediated repair in 40% of clones [94].

5. Future perspectives

To date, only about 1% of the yeast species found in nature have had their genomes fully characterized. Thus, it is not surprising, with the accessibility of new sequencing technologies, the complete genome analysis of many newly discovered yeast species with unique characteristics will be available. This will greatly expand the catalog of genetic parts allowing a more sensitive fine-tuning of desired economically relevant compounds and the discovery of new genes of interest. In addition, new yeast host vectors with desirable characteristics such as faster growth rates, stress-tolerance, efficient secretion systems and desired metabolic pathways will be engineered and domesticated to facilitate their use in industrial applications.

Moreover, in silico simulations will play a crucial role in the efficient design of new synthetic yeast biofactories since more accurate predictions will be made. Nevertheless, there are still limitations that have to be overcome such as the absence of gene regulatory information, lack of accurate metabolic models at genomic scale [102], or missing experimental design and testing of potential NCY biofactories. For instance, CRISPR-Cas9 still performs poorly regarding the adequate sgRNA production in NCYs. sgRNA expression is normally accomplished using RNA polymerase III (RNAP III) promoters (not well characterized in NCYs), implying more studies are needed for effective genetic engineering. Wagner and Alper suggest two approaches to overcome this issue: optimization of heterologous RNAP III promoters or the screening of native RNAP III promoters [6, 13] which are currently being tested in some NCYs such as *Y. lipolytica* [97].

Finally, a quite positive outcome of the use of NCYs as biofactories is the production of industrially relevant compounds in an economical manner. Synthetic biology helps to search and engineer strains capable of utilizing cheaper substrates, including "waste products" (e.g. whey and molasses), supporting a sustainable circular economy which in the future will certainly have a more relevant role.

6. Conclusions

For several decades yeasts have proven to be of great importance for the development of modern society, contributing to industrial processes including food and pharmaceutics. In addition, the current application of synthetic biology techniques in these organisms has given them a greater potential to be used as substitutes for organisms commonly used in the industry, such as bacteria, given the benefits they present. The use of these techniques in unconventional yeasts such as *P. pastoris*, *K. marxianus*, *K. lactis* and *Y. lipolytica* has increased very rapidly in recent years. For example, the genome editing technique CRISPR-Cas9 has been developed in these four species to improve the production of compounds, such as the prebiotic fructooligosaccharides in the case of *K. lactis*. Other techniques, despite being recently applied, have shown promising results in improving the expression of genes and the production of compounds of interest. This is the case of the use of TALENs in *Y. lipolytica* and Golden Gate in *P. pastoris*.

On the other hand, the engineering of genetic parts has also been developed in these unconventional yeasts. Codon optimization in *P. pastoris* and *K. marxianus* has allowed the production of heterologous enzymes such as phytases, important in animal feed, and recombinant proteins such as PCV2 Cap protein, which can be used in vaccines against porcine circovirus disease. Modifications of promoters and terminators have also been investigated, with promoter studies being the most common, as shown by the recent literature existing on the four species described in this chapter. In sum, the advances shown here demonstrate the potential of non-conventional yeasts as alternatives to traditionally used organisms, or even for the discovery of new systems with potential industrial use, capable of improving the quality of people's life.

Author details

Kevin Castillo-Mendieta[1], Jimmy Arias[1] and Fernando Gonzales-Zubiate[1,2]*

1 School of Biological Sciences and Engineering, Yachay Tech University, Urcuquí, Ecuador

2 MIND Research Group, Model Intelligent Networks Development, Urcuquí, Ecuador

*Address all correspondence to: fgonzales@yachaytech.edu.ec

References

[1] Feldmann H. Yeast Molecular and Cell Biology. Weinheim, Germany: Wiley-Blackwell; 2012

[2] Nicholas P. Money. The Rise of Yeast: Oxford University Press; 2018

[3] Satyanarayana T, Kunze G. Yeast Diversity in Human Welfare. Singapore: Springer; 2017. DOI: 10.1007/978-981-10-2621-8

[4] Boekhout T, Amend AS, el Baidouri F, Gabaldón T, Geml J, Mittelbach M, et al. Trends in yeast diversity discovery. Fungal Diversity. 2021;**114**:491-537. DOI: 10.1007/S13225-021-00494-6

[5] de Deken RH. The Crabtree effect: A regulatory system in yeast. Journal of General Microbiology. 1966;**44**:149-156. DOI: 10.1099/00221287-44-2-149

[6] Wagner JM, Alper HS. Synthetic biology and molecular genetics in non-conventional yeasts: Current tools and future advances. Fungal Genetics and Biology. 2016;**89**:126-136. DOI: 10.1016/j.fgb.2015.12.001

[7] Forti L, Cramarossa MR, Filippucci S, Tasselli G, Turchetti B, Buzzini P. Nonconventional yeast-promoted biotransformation for the production of flavor compounds. Handbook of Food Bioengineering. 2018;7:165-187. DOI: 10.1016/B978-0-12-811518-3.00006-5

[8] Nandy SK, Srivastava RK. A review on sustainable yeast biotechnological processes and applications. Microbiological Research. 2018;**207**:83-90. DOI: 10.1016/j.micres.2017.11.013

[9] Parapouli M, Vasileiadis A, Afendra AS, Hatziloukas E. *Saccharomyces cerevisiae* and its industrial applications. AIMS Microbiology. 2020;**6**:1-31. DOI: 10.3934/microbiol.2020001

[10] Unrean P, Khajeeram S. Model-based optimization of *Scheffersomyces stipitis* and *Saccharomyces cerevisiae* co-culture for efficient lignocellulosic ethanol production. Bioresources and Bioprocessing. 2015;**2**:1-11. DOI: 10.1186/S40643-015-0069-1

[11] Lignocellulose SK. A chewy problem. Nature. 2011;**474**:S12-S14. DOI: 10.1038/474s012a

[12] Geijer C, Ledesma-Amaro R, Tomas-Pejo E. Unraveling the potential of non-conventional yeasts in biotechnology. FEMS Yeast Research. 2022;**22**:1-6. DOI: 10.1093/femsyr/foab071

[13] Madhavan A, Jose AA, Binod P, Sindhu R, Sukumaran RK, Pandey A, et al. Synthetic biology and metabolic engineering approaches and its impact on non-conventional yeast and biofuel production. Frontiers in Energy Research. 2017;**5**:1-12. DOI: 10.3389/fenrg.2017.00008

[14] Prasongsuk S, Lotrakul P, Ali I, Bankeeree W, Punnapayak H. The current status of Aureobasidium pullulans in biotechnology. Folia Microbiologia (Praha). 2018;**63**:129-140. DOI: 10.1007/s12223-017-0561-4

[15] Liu Z, Zhang Y, Nielsen J. Synthetic biology of yeast. Biochemistry. 2019;**58**:1511-1520. DOI: 10.1021/acs.biochem.8b01236

[16] Bojar D, Scheller L, El HGC, Xie M, Fussenegger M. Caffeine-inducible gene switches controlling experimental

diabetes. Nature Communications. 2018;**9**:1-10. DOI: 10.1038/s41467-018-04744-1

[17] Kurland CG. Codon bias and gene expression. FEBS Letters. 1991;**285**:165-169. DOI: 10.1016/0014-5793(91)80797-7

[18] Quax TEF, Claassens NJ, Söll D, van der Oost J. Codon bias as a means to fine-tune gene expression. Molecular Cell. 2015;**59**:149-161. DOI: 10.1016/J.MOLCEL.2015.05.035

[19] Türkanoğlu Özçelik A, Yılmaz S, Inan M. *Pichia pastoris* promoters. Methods in Molecular Biology. 2019;**1923**:97-112. DOI: 10.1007/978-1-4939-9024-5_3

[20] Vogl T, Sturmberger L, Kickenweiz T, Wasmayer R, Schmid C, Hatzl AM, et al. A toolbox of diverse promoters related to methanol utilization: Functionally verified parts for heterologous pathway expression in *Pichia pastoris*. ACS Synthetic Biology. 2016;**5**:172-186. DOI: 10.1021/acssynbio.5b00199

[21] Kumar P, Sahoo DK, Sharma D. The identification of novel promoters and terminators for protein expression and metabolic engineering applications in *Kluyveromyces marxianus*. Metabolic Engineering Communications. 2021;**12**:1-13. DOI: 10.1016/J.MEC.2020.E00160

[22] Rajkumar AS, Varela JA, Juergens H, Daran JMG, Morrissey JP. Biological parts for *Kluyveromyces marxianus* synthetic biology. Frontiers in Bioengineering and Biotechnology. 2019;**7**:97. DOI: 10.3389/fbioe.2019.00097

[23] Gomes AMV, Carmo TS, Carvalho LS, Bahia FM, Parachin NS. Comparison of yeasts as hosts for recombinant protein production. Microorganisms. 2018;**6**:38. DOI: 10.3390/microorganisms6020038

[24] Obst U, Lu TK, Sieber V. A modular toolkit for generating *Pichia pastoris* secretion libraries. ACS Synthetic Biology. 2017;**6**:1016-1025. DOI: 10.1021/acssynbio.6b00337

[25] Dou W, Zhu Q, Zhang M, Jia Z, Guan W. Screening and evaluation of the strong endogenous promoters in *Pichia pastoris*. Microbial Cell Factories. 2021;**20**:1-12. DOI: 10.1186/S12934-021-01648-6

[26] Prielhofer R, Barrero JJ, Steuer S, Gassler T, Zahrl R, Baumann K, et al. GoldenPiCS: A golden gate-derived modular cloning system for applied synthetic biology in the yeast *Pichia pastoris*. BMC Systems Biology. 2017;**11**:1-14. DOI: 10.1186/S12918-017-0492-3

[27] Wright DA, Li T, Yang B, Spalding MH. TALEN-mediated genome editing: Prospects and perspectives. Biochemical Journal. 2014;**462**:15-24. DOI: 10.1042/BJ20140295

[28] Li T, Wright DA, Spalding MH, Yang B. TALEN-based Genome Editing in Yeast. Cham, Switzerland: Springer 2015. pp. 289-307. DOI: 10.1007/978-3-319-10142-2_27

[29] Raschmanová H, Weninger A, Glieder A, Kovar K, Vogl T. Implementing CRISPR-Cas technologies in conventional and non-conventional yeasts: Current state and future prospects. Biotechnology Advances. 2018;**36**:641-665. DOI: 10.1016/J.BIOTECHADV.2018.01.006

[30] Yip BH. Recent advances in CRISPR/Cas9 delivery strategies. Biomolecules. 2020;**10**:839. DOI: 10.3390/BIOM10060839

[31] Sayers EW, Bolton EE, Brister JR, Canese K, Chan J, Comeau DC, et al. Database resources of the national center

for biotechnology information. Nucleic Acids Research. 2022;**50**:D20-D26. DOI: 10.1093/NAR/GKAB1112

[32] Love KR, Shah KA, Whittaker CA, Wu J, Bartlett MC, Ma D, et al. Comparativegenomicsandtranscriptomics of *Pichia pastoris*. BMC Genomics. 2016;**17**:1-17. DOI: 10.1186/S12864-016-2876-Y

[33] Nurcholis M, Lertwattanasakul N, Rodrussamee N, Kosaka T, Murata M, Yamada M. Integration of comprehensive data and biotechnological tools for industrial applications of *Kluyveromyces marxianus*. Applied Microbiology and Biotechnology. 2020;**104**:475-488. DOI: 10.1007/S00253-019-10224-3

[34] Becerra M, González-Siso MI, Cerdán ME. A transcriptome analysis of *Kluyveromyces lactis* growing in cheese whey. International Dairy Journal. 2006;**16**:207-214. DOI: 10.1016/J.IDAIRYJ.2005.03.005

[35] Morin N, Cescut J, Beopoulos A, Lelandais G, le Berre V, Uribelarrea JL, et al. Transcriptomic analyses during the transition from biomass production to lipid accumulation in the oleaginous yeast yarrowia lipolytica. PLoS One. 2011;**6**:e27966. DOI: 10.1371/JOURNAL.PONE.0027966

[36] Pomraning KR, Wei S, Karagiosis SA, Kim YM, Dohnalkova AC, Arey BW, et al. Comprehensive metabolomic, lipidomic and microscopic profiling of yarrowia lipolytica during lipid accumulation identifies targets for increased lipogenesis. PLoS One. 2015;**10**:1-17. DOI: 10.1371/JOURNAL.PONE.0123188

[37] Alvim MCT, Vital CE, Barros E, Vieira NM, da Silveira FA, Balbino TR, et al. Ethanol stress responses of *Kluyveromyces marxianus* CCT 7735 revealed by proteomic and metabolomic analyses. Antonie van Leeuwenhoek, International Journal of General and Molecular Microbiology. 2019;**112**:827-845. DOI: 10.1007/S10482-018-01214-Y

[38] Godinho CP, Palma M, Oliveira J, Mota MN, Antunes M, Teixeira MC, et al. The N.C.yeastract and communityyeastract databases to study gene and genomic transcription regulation in non-conventional yeasts. FEMS Yeast Research. 2021;**21**:1-10. DOI: 10.1093/FEMSYR/FOAB045

[39] Ye R, Huang M, Lu H, Qian J, Lin W, Chu J, et al. Comprehensive reconstruction and evaluation of *Pichia pastoris* genome-scale metabolic model that accounts for 1243 ORFs. Bioresour Bioprocess. 2017;**4**:1-12. DOI: 10.1186/S40643-017-0152-X

[40] Marcišauskas S, Ji B, Nielsen J. Reconstruction and analysis of a *Kluyveromyces marxianus* genome-scale metabolic model. BMC Bioinformatics. 2019;**20**:1-9. DOI: 10.1186/S12859-019-3134-5

[41] Dias O, Pereira R, Gombert AK, Ferreira EC, Rocha I. iOD907, the first genome-scale metabolic model for the milk yeast *Kluyveromyces lactis*. Biotechnology Journal. 2014;**9**:776-790. DOI: 10.1002/BIOT.201300242

[42] Xu Y, Holic R, Hua Q. Comparison and analysis of published genome-scale metabolic models of yarrowia lipolytica. Biotechnology and Bioprocess Engineering. 2020;**25**:53-61. DOI: 10.1007/S12257-019-0208-1

[43] Peña DA, Gasser B, Zanghellini J, Steiger MG, Mattanovich D. Metabolic engineering of *Pichia pastoris*. Metabolic Engineering. 2018;**50**:2-15. DOI: 10.1016/J.YMBEN.2018.04.017

[44] Sreekrishna K, Kropp KE. *Pichia pastoris*. In: Nonconventional Yeasts in Biotechnology. 1996. pp. 203-253. DOI: 10.1007/978-3-642-79856-6_6

[45] Kang Z, Huang H, Zhang Y, Du G, Chen J. Recent advances of molecular toolbox construction expand *Pichia pastoris* in synthetic biology applications. World Journal of Microbiology and Biotechnology. 2016;**33**:1-8. DOI: 10.1007/S11274-016-2185-2

[46] García-Ortega X, Cámara E, Ferrer P, Albiol J, Montesinos-Seguí JL, Valero F. Rational development of bioprocess engineering strategies for recombinant protein production in *Pichia pastoris* (Komagataella phaffii) using the methanol-free GAP promoter. Where do we stand? New Biotechnology. 2019;**53**:24-34. DOI: 10.1016/J.NBT.2019.06.002

[47] Xiong AS, Yao QH, Peng RH, Han PL, Cheng ZM, Li Y. High level expression of a recombinant acid phytase gene in *Pichia pastoris*. Journal of Applied Microbiology. 2005;**98**:418-428. DOI: 10.1111/J.1365-2672.2004.02476.X

[48] Xiong AS, Yao QH, Peng RH, Zhang Z, Xu F, Liu JG, et al. High level expression of a synthetic gene encoding Peniophora lycii phytase in methylotrophic yeast *Pichia pastoris*. Applied Microbiology and Biotechnology. 2006;**72**:1039-1047. DOI: 10.1007/S00253-006-0384-8

[49] Hu H, Gao J, He J, Yu B, Zheng P, Huang Z, et al. Codon optimization significantly improves the expression level of a keratinase gene in *Pichia pastoris*. PLoS One. 2013;**8**:1-8. DOI: 10.1371/JOURNAL.PONE.0058393

[50] Wang Y, Jiang S, Jiang X, Sun X, Guan X, Han Y, et al. Cloning and codon optimization of a novel feline interferon omega gene for production by *Pichia pastoris* and its antiviral efficacy in polyethylene glycol-modified form. Virulence. 2022;**13**:297-309. DOI: 10.1080/21505594.2022.2029330

[51] He M, Wu D, Wu J, Chen J. Enhanced expression of endoinulinase from Aspergillus niger by codon optimization in *Pichia pastoris* and its application in inulooligosaccharide production. Journal of Industrial Microbiology & Biotechnology. 2014;**41**:105-114. DOI: 10.1007/S10295-013-1341-Z

[52] Wang JR, Li YY, Liu DN, Liu JS, Li P, Chen LZ, et al. Codon optimization significantly improves the expression level of α-Amylase gene from Bacillus licheniformis in *Pichia pastoris*. BioMed Research International. 2015;**2015**:1-9. DOI: 10.1155/2015/248680

[53] Ahn J, Jang MJ, Ang KS, Lee H, Choi ES, Lee DY. Codon optimization of Saccharomyces cerevisiae mating factor alpha prepro-leader to improve recombinant protein production in *Pichia pastoris*. Biotechnology Letters. 2016;**38**:2137-2143. DOI: 10.1007/S10529-016-2203-3

[54] Hu H, Dai S, Wen A, Bai X. Efficient expression of xylanase by codon optimization and its effects on the growth performance and carcass characteristics of broiler. Animals. 2019;**9**:65. DOI: 10.3390/ANI9020065

[55] Che Z, Cao X, Chen G, Liang Z. An effective combination of codon optimization, gene dosage, and process optimization for high-level production of fibrinolytic enzyme in Komagataella phaffii (*Pichia pastoris*). BMC Biotechnology. 2020;**20**:1-13. DOI: 10.1186/S12896-020-00654-7

[56] Karaoğlan M, Erden-Karaoğlan F. Effect of codon optimization and

promoter choice on recombinant endo-polygalacturonase production in *Pichia pastoris*. Enzyme and Microbial Technology. 2020;**139**:1-9. DOI: 10.1016/J.ENZMICTEC.2020.109589

[57] Zhai Z, Nuylert A, Isobe K, Asano Y. Effects of codon optimization and glycosylation on the high-level production of hydroxynitrile lyase from Chamberlinius hualienensis in *Pichia pastoris*. Journal of Industrial Microbiology & Biotechnology. 2019;**46**:887-898. DOI: 10.1007/S10295-019-02162-W

[58] Vogl T, Fischer JE, Hyden P, Wasmayer R, Sturmberger L, Glieder A. Orthologous promoters from related methylotrophic yeasts surpass expression of endogenous promoters of *Pichia pastoris*. AMB Express. 2020;**10**:1-9. DOI: 10.1186/S13568-020-00972-1

[59] Nong L, Zhang Y, Duan Y, Hu S, Lin Y, Liang S. Engineering the regulatory site of the catalase promoter for improved heterologous protein production in *Pichia pastoris*. Biotechnology Letters. 2020;**42**:2703-2709. DOI: 10.1007/S10529-020-02979-X

[60] Wang J, Wang X, Shi L, Qi F, Zhang P, Zhang Y, et al. Methanol-independent protein expression by AOX1 promoter with trans-acting elements engineering and glucose-glycerol-shift induction in *Pichia pastoris*. Scientific Reports. 2017;7:1-12. DOI: 10.1038/srep41850

[61] Ito Y, Terai G, Ishigami M, Hashiba N, Nakamura Y, Bamba T, et al. Exchange of endogenous and heterogeneous yeast terminators in *Pichia pastoris* to tune mRNA stability and gene expression. Nucleic Acids Research. 2020;**48**:13000-13012. DOI: 10.1093/NAR/GKAA1066

[62] Duan G, Ding L, Wei D, Zhou H, Chu J, Zhang S, et al. Screening endogenous signal peptides and protein folding factors to promote the secretory expression of heterologous proteins in *Pichia pastoris*. Journal of Biotechnology. 2019;**306**:193-202. DOI: 10.1016/J.JBIOTEC.2019.06.297

[63] Ito Y, Ishigami M, Hashiba N, Nakamura Y, Terai G, Hasunuma T, et al. Avoiding entry into intracellular protein degradation pathways by signal mutations increases protein secretion in *Pichia pastoris*. Microbial Biotechnology. 2022;**15**:2364-2378. DOI: 10.1111/1751-7915.14061

[64] Cheng J, Wu M, Zhong R, Si D, Meng G, Zhang R, et al. Rapid golden gate assembly of exons from genomic DNA for protein expression in Escherichia coli and *Pichia pastoris*. BioTechniques. 2021;**71**:445-450. DOI: 10.2144/BTN-2021-0039

[65] Yang Y, Liu G, Chen X, Liu M, Zhan C, Liu X, et al. High efficiency CRISPR/Cas9 genome editing system with an eliminable episomal sgRNA plasmid in *Pichia pastoris*. Enzyme and Microbial Technology. 2020;**138**:109556. DOI: 10.1016/J.ENZMICTEC.2020.109556

[66] Weninger A, Fischer JE, Raschmanová H, Kniely C, Vogl T, Glieder A. Expanding the CRISPR/Cas9 toolkit for *Pichia pastoris* with efficient donor integration and alternative resistance markers. Journal of Cellular Biochemistry. 2018;**119**:3183-3198. DOI: 10.1002/JCB.26474

[67] Nonklang S, Abdel-Banat BMA, Cha-aim K, Moonjai N, Hoshida H, Limtong S, et al. High-temperature ethanol fermentation and transformation with linear DNA in the thermotolerant yeast *Kluyveromyces marxianus* DMKU3-1042. Applied and Environmental Microbiology. 2008;**74**:7514-7521. DOI: 10.1128/AEM.01854-08

[68] Lane MM, Burke N, Karreman R, Wolfe KH, O'Byrne CP, Morrissey JP. Physiological and metabolic diversity in the yeast *Kluyveromyces marxianus*. Antonie Van Leeuwenhoek. 2011;**100**:507-519. DOI: 10.1007/S10482-011-9606-X

[69] Fonseca GG, Heinzle E, Wittmann C, Gombert AK. The yeast *Kluyveromyces marxianus* and its biotechnological potential. Applied Microbiology and Biotechnology. 2008;**79**:339-354. DOI: 10.1007/S00253-008-1458-6

[70] Morrissey JP, Etschmann MMW, Schrader J, de Billerbeck GM. Cell factory applications of the yeast *Kluyveromyces marxianus* for the biotechnological production of natural flavour and fragrance molecules. Yeast. 2015;**32**:3-16. DOI: 10.1002/YEA.3054

[71] Cernak P, Estrela R, Poddar S, Skerker JM, Cheng YF, Carlson AK, et al. Engineering *Kluyveromyces marxianus* as a robust synthetic biology platform host. MBio. 2018;**9**:1-16. DOI: 10.1128/MBIO.01410-18

[72] Karim A, Gerliani N, Aïder M. *Kluyveromyces marxianus*: An emerging yeast cell factory for applications in food and biotechnology. International Journal of Food Microbiology. 2020;**333**:108818. DOI: 10.1016/J.IJFOODMICRO.2020.108818

[73] Lane MM, Morrissey JP. *Kluyveromyces marxianus*: A yeast emerging from its sister's shadow. Fungal Biology Reviews. 2010;**24**:17-26. DOI: 10.1016/J.FBR.2010.01.001

[74] Dekker WJC, Ortiz-Merino RA, Kaljouw A, Battjes J, Wiering FW, Mooiman C, et al. Engineering the thermotolerant industrial yeast *Kluyveromyces marxianus* for anaerobic growth. Metabolic Engineering. 2021;**67**:347-364. DOI: 10.1016/J.YMBEN.2021.07.006

[75] Bilal M, Ji L, Xu Y, Xu S, Lin Y, Iqbal HMN, et al. Bioprospecting *Kluyveromyces marxianus* as a Robust Host for Industrial Biotechnology. Frontiers in Bioengineering and Biotechnology. 2022;**10**:562. DOI: 10.3389/FBIOE.2022.851768

[76] Duan J, Yang D, Chen L, Yu Y, Zhou J, Lu H. Efficient production of porcine circovirus virus-like particles using the nonconventional yeast *Kluyveromyces marxianus*. Applied Microbiology and Biotechnology. 2018;**103**:833-842. DOI: 10.1007/S00253-018-9487-2

[77] Yang D, Chen L, Duan J, Yu Y, Zhou J, Lu H. Investigation of *Kluyveromyces marxianus* as a novel host for large-scale production of porcine parvovirus virus-like particles. Microbial Cell Factories. 2021;**20**:1-13. DOI: 10.1186/S12934-021-01514-5

[78] Bragança CRS, Colombo LT, Roberti AS, Alvim MCT, Cardoso SA, Reis KCP, et al. Construction of recombinant *Kluyveromyces marxianus* UFV-3 to express dengue virus type 1 nonstructural protein 1 (NS1). Applied Microbiology and Biotechnology. 2015;**99**:1191-1203. DOI: 10.1007/S00253-014-5963-5

[79] Nambu-Nishida Y, Nishida K, Hasunuma T, Kondo A. Genetic and physiological basis for antibody production by *Kluyveromyces marxianus*. AMB Express. 2018;**8**:1-9. DOI: 10.1186/S13568-018-0588-1

[80] Zhang Y, Li YF, Chi Z, Liu GL, Jiang H, Hu Z, et al. Inulinase hyperproduction by *Kluyveromyces marxianus* through codon optimization, selection of the promoter, and high-cell-density fermentation for efficient inulin

hydrolysis. Annales de Microbiologie. 2019;**69**:647-657. DOI: 10.1007/S13213-019-01457-8

[81] Yang C, Hu S, Zhu S, Wang D, Gao X, Hong J. Characterizing yeast promoters used in *Kluyveromyces marxianus*. World Journal of Microbiology and Biotechnology. 2015;**31**:1641-1646. DOI: 10.1007/S11274-015-1899-X

[82] Zhou J, Zhu P, Hu X, Lu H, Yu Y. Improved secretory expression of lignocellulolytic enzymes in *Kluyveromyces marxianus* by promoter and signal sequence engineering. Biotechnology for Biofuels. 2018;**11**:1-14. DOI: 10.1186/S13068-018-1232-7

[83] Yarimizu T, Nakamura M, Hoshida H, Akada R. Synthetic signal sequences that enable efficient secretory protein production in the yeast *Kluyveromyces marxianus*. Microbial Cell Factories. 2015;**14**:1-14. DOI: 10.1186/S12934-015-0203-Y

[84] Li M, Lang X, Moran Cabrera M, de Keyser S, Sun X, da Silva N, et al. CRISPR-mediated multigene integration enables Shikimate pathway refactoring for enhanced 2-phenylethanol biosynthesis in *Kluyveromyces marxianus*. Biotechnology for Biofuels. 2021;**14**:1-15. DOI: 10.1186/S13068-020-01852-3

[85] Bever D, Wheeldon I, da Silva N. RNA polymerase II-driven CRISPR-Cas9 system for efficient non-growth-biased metabolic engineering of *Kluyveromyces marxianus*. Metabolic Engineering Communications. 2022:1-10. DOI: 10.1016/J.MEC.2022.E00208

[86] Sakhtah H, Behler J, Ali-Reynolds A, Causey TB, Vainauskas S, Taron CH. A novel regulated hybrid promoter that permits autoinduction of heterologous protein expression in *Kluyveromyces lactis*. Applied and Environmental Microbiology. 2019;**85**:1-12. DOI: 10.1128/AEM.00542-19

[87] Horwitz AA, Walter JM, Schubert MG, Kung SH, Hawkins K, Platt DM, et al. Efficient multiplexed integration of synergistic alleles and metabolic pathways in yeasts via CRISPR-Cas. Cell Systems. 2015;**1**:88-96. DOI: 10.1016/j.cels.2015.02.001

[88] Cai P, Gao J, Zhou Y. CRISPR-mediated genome editing in non-conventional yeasts for biotechnological applications. Microbial Cell Factories. 2019;**18**:1-12. DOI: 10.1186/s12934-019-1112-2

[89] Bisschoff E. The Development of a Wide Range CRISPR-Cas9 Gene Editing System. Bloemfontein, South Africa: University of the Free State; 2019

[90] Burghardt JP, Fan R, Baas M, Eckhardt D, Gerlach D, Czermak P. Enhancing the heterologous fructosyltransferase activity of *Kluyveromyces lactis*: Developing a scaled-up process and abolishing invertase by CRISPR/Cas9 genome editing. Frontiers in Bioengineering and Biotechnology. 2020;**8**:1-15. DOI: 10.3389/fbioe.2020.607507

[91] Abdel-Mawgoud AM, Markham KA, Palmer CM, Liu N, Stephanopoulos G, Alper HS. Metabolic engineering in the host Yarrowia lipolytica. Metabolic Engineering. 2018;**50**:192-208. DOI: 10.1016/J.YMBEN.2018.07.016

[92] Madzak C. Yarrowia lipolytica strains and their biotechnological applications: How natural biodiversity and metabolic engineering could contribute to cell factories improvement. Journal of Fungi. 2021;**7**:1-67. DOI: 10.3390/jof7070548

[93] Schwartz C, Curtis N, Löbs AK, Wheeldon I. Multiplexed CRISPR

activation of cryptic sugar metabolism enables Yarrowia Lipolytica growth on cellobiose. Biotechnology Journal. 2018;**13**:1-7. DOI: 10.1002/biot.201700584

[94] Larroude M, Rossignol T, Nicaud JM, Ledesma-Amaro R. Synthetic biology tools for engineering Yarrowia lipolytica. Biotechnology Advances. 2018;**36**:2150-2164. DOI: 10.1016/j.biotechadv.2018.10.004

[95] Wong L, Engel J, Jin E, Holdridge B, Xu P. YaliBricks, a versatile genetic toolkit for streamlined and rapid pathway engineering in Yarrowia lipolytica. Metabolic Engineering Communications. 2017;**5**:68-77. DOI: 10.1016/J.METENO.2017.09.001

[96] Larroude M, Celinska E, Back A, Thomas S, Nicaud JM, Ledesma-Amaro R. A synthetic biology approach to transform Yarrowia lipolytica into a competitive biotechnological producer of β-carotene. Biotechnology and Bioengineering. 2018;**115**:464-472. DOI: 10.1002/bit.26473

[97] Markham KA, Alper HS. Synthetic biology expands the industrial potential of Yarrowia lipolytica. Trends in Biotechnology. 2018;**36**:1085-1095. DOI: 10.1016/j.tibtech.2018.05.004

[98] Madzak C, Tréton B, Blanchin-Roland S. Strong hybrid promoters and integrative expression/secretion vectors for quasi-constitutive expression of heterologous proteins in the yeast Yarrowia lipolytica. Journal of Molecular Microbiology and Biotechnology. 2000;**2**:207-216

[99] Ma J, Gu Y, Marsafari M, Xu P. Synthetic biology, systems biology, and metabolic engineering of Yarrowia lipolytica toward a sustainable biorefinery platform. Journal of Industrial Microbiology & Biotechnology. 2020;**47**:845-862. DOI: 10.1007/s10295-020-02290-8

[100] Larroude M, Park YK, Soudier P, Kubiak M, Nicaud JM, Rossignol T. A modular golden gate toolkit for Yarrowia lipolytica synthetic biology. Microbial Biotechnology. 2019;**12**:1249-1259. DOI: 10.1111/1751-7915.13427

[101] Larroude M, Nicaud JM, Rossignol T. Golden gate multigene assembly method for Yarrowia lipolytica. Methods in Molecular Biology. 2022;**2513**:205-220. DOI: 10.1007/978-1-0716-2399-2_12

[102] Patra P, Das M, Kundu P, Ghosh A. Recent advances in systems and synthetic biology approaches for developing novel cell-factories in non-conventional yeasts. Biotechnology Advances. 2021;**47**:1-29. DOI: 10.1016/j.biotechadv.2021.107695

Section 2

Biomaterials

Chapter 3

Silk Fibroin Nanoparticles: A Biocompatible Multi-Functional Polymer for Drug Delivery

Faith H.N. Howard, Zijian Gao, Hawari Bin Mansor, Zidi Yang and Munitta Muthana

Abstract

The versatility of nanomedicines allows for various modifications of material type, size, charge and functionalization, offering a promising platform for biomedical applications including tumor targeting. One such material, silk fibroin (SF) has emerged, displaying an excellent combination of mechanical and biological properties characterized by its high tensile and breaking strength, elongation, stiffness and ductility. High stability allows SF to maintain its chemical structure even at high temperatures (around 250°C) and compared with other biological polymers like polylactide (PLA), poly(lactic-co-glycolic acid) (PLGA), and collagen, SF shows excellent biocompatibility and lower immunogenic response making it a very suitable material for drug delivery and tissue engineering. Here we describe the structure, synthesis and properties of SF nanoparticles. We evaluate its emergence as a multi-functional polymer for its utility as a nanocarrier to deliver cancer therapies directly to tumors together with considerations for its clinical use.

Keywords: silk fibroin, polymer, nanomedicine, nanocarrier, drug delivery

1. Introduction

As stated by the World Health Organization, cancer is the second leading cause of death globally, accounting for an estimated 9.6 million deaths, or 1 in 6 deaths, in 2018 [1]. GLOBOCAN also predicts an increase in cancer rates, with more than 20 million new cancer cases expected annually by 2025 [2]. Over time, as our understanding and knowledge of molecular and tumor biology has increased, the cancer treatment paradigms have notably changed, particularly during the past 20 years. Previously, cancer was listed and treated based on its origin or its unique histomorphologic characteristics. However, in 2002, Schiller et al. reported that third generation chemotherapy administered to non-small-cell lung cancer showed almost similar survival curves [3]. Although the results are only limited to lung cancer, this indicates that cancer treatment using general (non-specific) cytotoxic chemotherapies have reached a therapeutic plateau. Since then, this research area has evolved based on two focus areas: tumor molecular profiling and molecular targets. Together, these efforts have further realized two recent revolutions in cancer research. Firstly, genotype-directed precision oncology which focuses on personalized therapies to treat specific genomic abnormalities regardless of the cancer type. Secondly, the targeting of particular components in the tumor microenvironment. As a result, the discovery of an abundance of anticancer drugs have showed a promising early step for cancer eradication. However, most of these therapeutics agents have undesirable characteristics, limiting their clinical usage and invalidating further drug research [4].

Recent studies have suggested that the use of nanoparticles as a drug carrier may be one of the best alternatives to improve the therapeutic effect of anticancer drugs. Nanotechnology, or in this case, nanomedicine, is the use of materials usually in nanometer scale in the fields of medicine and health [5]. Nanoparticle-based drug delivery systems have shown remarkable progress in overcoming the limitations of conventional drug delivery or drug therapy method. The unique characteristics that usually accompany these potential carriers include nanoscale size, high surface-to-volume ratio, auspicious physical and chemical properties and most importantly, endless possibilities for modifications that support cell targeting, gene delivery etc. [5].

Nanoparticles fall into two different categories, namely soft/organic and hard/inorganic nanoparticles. Soft nanoparticles are based on organic material, typically prepared from polymers or molecules that can self-assemble (coacervation) into large particles. The materials can range from full synthetic polymer to natural materials such as silk proteins [6]. Hard nanoparticles, on the other hand, are inorganic and usually insoluble, e.g. silver, gold nanoparticles, and carbon nanotubes [7, 8]. The benefits of organic nanoparticles as drug nanocarriers have been reported numerously in recent years, citing desirable characteristics including, biodegradability, non-antigenic and superior biocompatibility [9]. Thus, the delivery system's advancement holds the promise of future precision medicine, which would greatly improve cancer survival rate by treating each cancer patient with the most effective drugs in the most efficient ways [10].

One such natural polymer which promises great potential as a drug delivery system is silk fibroin (SF). Silk has been recognized as a valuable natural material for the fabric industry for centuries, but during the last few decades, it has attracted immense attention as a promising biopolymer for biomedical and pharmaceutical applications [11]. Silk from the domesticated silkworm *Bombyx mori (B. mori)* is well characterized and has been approved as a safe biomaterial by the US Food and Drug Administration (FDA) [12]. This chapter focuses on the use of *B. mori* derived SF as a functional material for cancer drug delivery.

2. Silk fibroin

2.1 Structure

Most SF utilized for biomedical and commercial applications are derived from cocoons of *B. mori* domestic silk moths. Constructed from a continuous fiber strand comprised of two cores of fibroin protein held together by sericin protein [13, 14]. The primary structure of *B. mori* SF mainly consists of glycine (Gly) (43%), alanine (Ala) (30%) and serine (Ser) (12%) [15]. The secondary structure of SF exists in three different structural forms including silk I, silk II and silk III. Silk I consists of α-helix domains which is in a water-soluble state and easy to convert to silk II

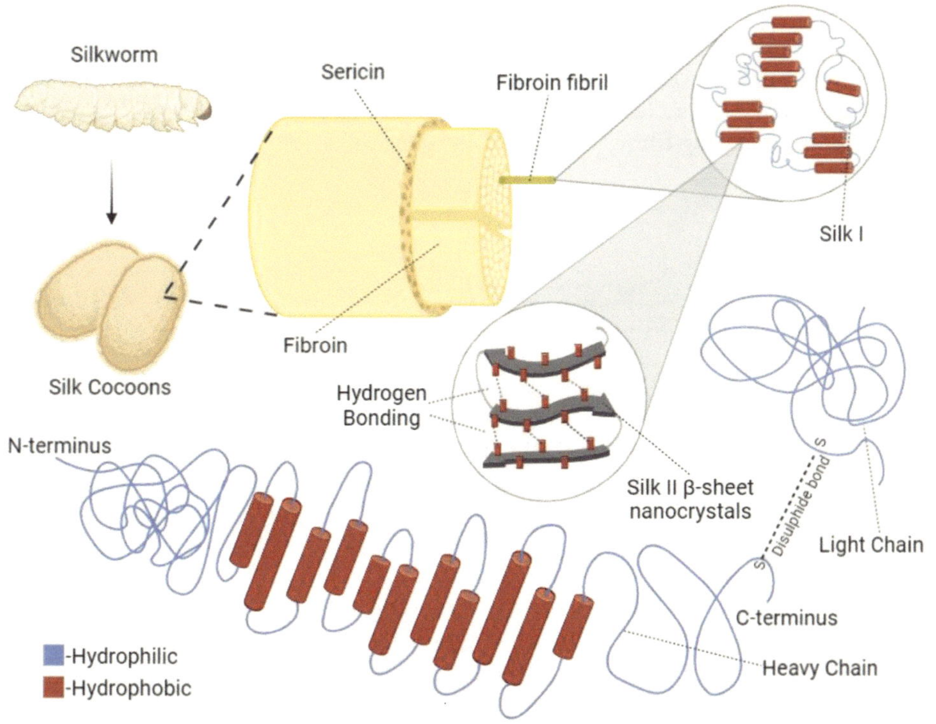

Figure 1.
Cross-sectional and structural composition of silk fibers. Created with BioRender.

structure once treated with organic solvents, electromagnetic fields, or physical spinning environments. Unlike silk I, silk II contains an antiparallel β-sheet/crystal molecular model which has hydrogen side chains from glycine and methyl side chains from the alanines resulting in higher stability and both water and solvent insolubility. Silk III prevails at the water/air interface (**Figure 1**) [16–18].

2.2 Properties

In the past centuries, silk has been used as a natural material for the fabric industry, but recently SF nanoparticles have been considered as a potential alternative carrier for anticancer drug delivery because of its physicochemical, mechanical, and biological properties [11]. Compared with other materials, SF is characterized by its high tensile and breaking strength, elongation, stiffness and ductility [19]. High stability allows maintenance of SF chemical structure even at high temperatures (around 250°C) [20]. In addition, compared with other biological polymers like polylactide (PLA), poly(lactic-co-glycolic acid) (PLGA), and collagen, silk fibroin shows excellent biocompatibility and lower immunogenic response [21–23] making it a very suitable material for drug delivery and tissue engineering. It is commonly used as a suture material, demonstrating comparable immunogenicity and biocompatibility to other natural and synthetic suture materials [24]. Moreover, it has consistently shown to be non-toxic and fully resorbable, this includes any degradation products [25]. Surprisingly, SF has been proved to have an intrinsic anti-inflammatory ability which could be used in the treatment of inflammatory bowel disease [26].

Degradability is another important property of silk fibroin. The degradation rate of SF is related to the molecular weight, the degree of crystallinity, morphological features and crosslinking [27]. This affords a tunability from seconds to years, another unique feature of SF. As a result of these properties, SF has been used in different nanosystems, including films, sponges, hydrogels, tubes [28]. Here, we focus on silk nanoparticles.

3. Silk nanoparticles

3.1 Synthesis of silk nanoparticles

Silk nanoparticles are an excellent candidate as a carrier for drugs, targeted therapies and contrast agents. They are synthesized from regenerated SF through a variety of methods based on their self-assembly behavior. Among all of the approaches, desolvation (**Figure 2**) is the most common method for SF nanoparticle synthesis, in which dissolving agents including ethanol, acetone, dimethyl sulfoxide (DMSO) and methanol could be used to dehydrate and package the silk chain, leading to the change from silk I to silk II structure and forming SF nanoparticles [29]. During the desolvation method, lipophilic active drugs (e.g. curcumin and 5-fluorouracil) can be easily dissolved in the organic solvents allowing nanoencapsulation of anticancer drugs [30]. The main challenge of the desolvation method is to find the ideal SF/dissolving agent ratio and ensure adequate mixing during SF nanoparticle formation which plays an important role in the control of the nanoparticle size.

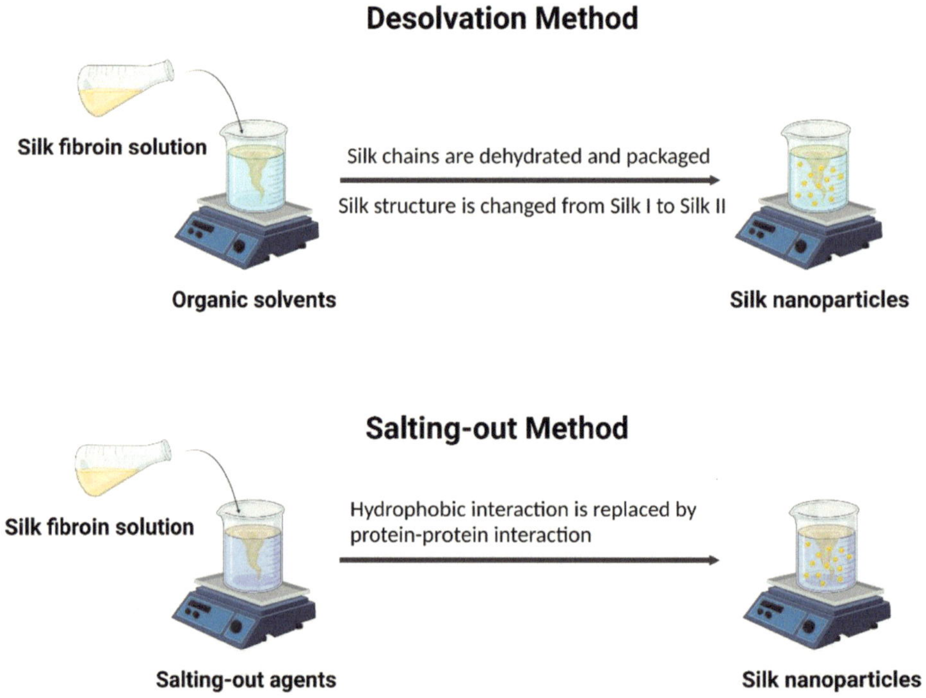

Figure 2.
Representation of the desolvation and salting-out method for the production of silk fibroin nanoparticles. Created with BioRender.

Salting-out (**Figure 2**) is another widely used method in the preparation of protein-based nanoparticles by removing the water barriers between protein molecules and increasing the interactions between proteins, leading to aggregation and precipitation from the solution [31]. The salt, pH, and ionic strength are major factors influencing the yield, particle morphology, zeta potential, and nanoparticle stability [30]. The salting-out method has advantages in avoiding the usage of toxic solvents, therefore maintaining the activity of the protein. However, the synthesized particle size is relatively large (500–2000 nm), and the high amounts of salts are difficult to remove [29].

The electrospraying method uses electrical forces to obtain liquid atomization where the liquid flowing out of a capillary nozzle is dispersed into small droplets by an electric field [32, 33]. Similar to the desolvation method, the main limitation of electrospraying is the organic solvents used during the synthesis that may damage the bioactivity of their cargo (e.g. enzymes, genes, and cell vitality) [33]. Other silk nanoparticle synthesis methods which are not used widely used include supercritical fluid technologies [34], mechanical comminution [35], capillary-microdot technique [36], and microemulsion [37].

3.2 Functionalization

Many drug delivery systems lack a targeting mechanism, resulting in poor accumulation at tumor targets. To achieve efficacious concentrations at the tumor, saturating drug doses are often administered which can lead to non-specific or toxic effects. Nanoparticles have been designed to overcome this problem by delivering the drug to specific tissue instead of a more generalized treatment. Currently, four different targeting mechanism have been explored; passive targeting, targeted recognition, triggered release and guided delivery (**Figure 3**). Historically, the enhanced permeability and retention (EPR) effect (a unique phenomenon of solid tumors) has been utilized whereby nanoparticles can extravasate through the leaky blood vessels of the tumor tissue without the need for surface modification [5, 38, 39]. Some surface modification that affects circulation time may also indirectly affect passive targeting such as PEGylation of the nanoparticles. This is due to EPR effect increases proportional to the circulation time [40]. However, some tissues may also contain fenestrated blood vessels resulting in a similar effect on nanoparticles, causing them to accumulate there [5]. Additionally, the tumor microenvironment varies depending on the tumor type and passive targeting may not be as efficient in that particular condition.

Targeted recognition involves the use of targeting molecules as an attachment to the drug-loaded nanoparticles. Targeting molecules such as ligands have high specificity to receptors and other cancer-specific target molecules available on the surface of cancerous tissue such as glycans [4]. Conjugation of nanoparticles with ligands such as transferrin, folic acid, enzymes, antibodies and other macromolecules has demonstrated enhanced uptake of nanoparticles by cancer cells [5, 39]. However, they still rely on passive accumulation at the tumor site.

Traditional cancer treatment methods suffer from a lack of specific regional and temporal activation leading to off-target effects. The concept of smart nanosystems uses the intrinsic and environmental differences between normal cells and cancer cells to trigger activation or release of drugs at a tumor site [41]. Intrinsic activation strategies include using the differences in pH, enzyme expression level, the concentration of membrane proteins and soluble molecules between healthy cells and cancer cells to trigger drug release at a specific location in the

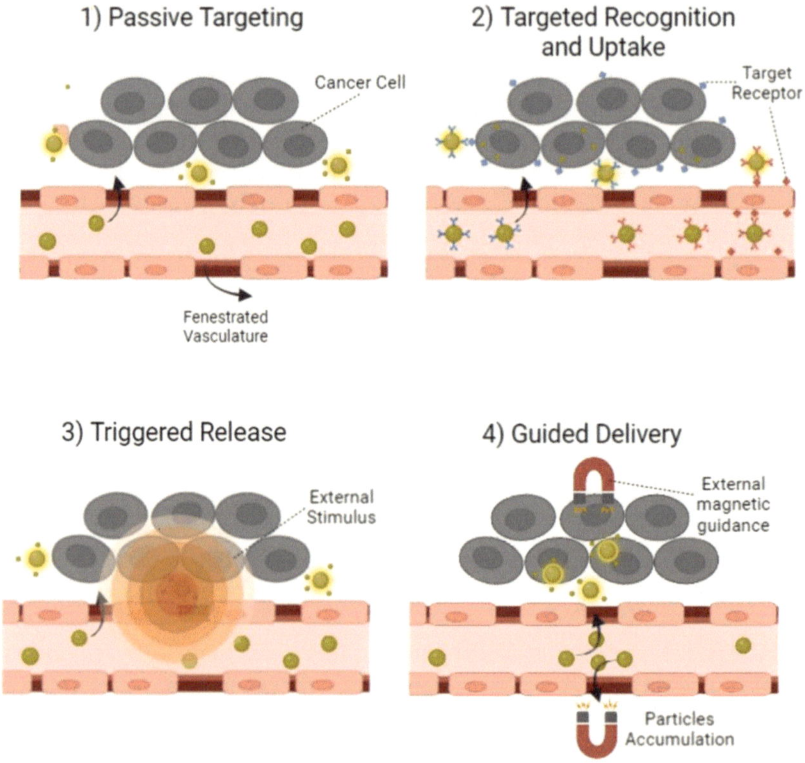

Figure 3.
Schematic of different drug targeting approaches for nanoparticles. Created with BioRender.

body [5, 41–43]. Extrinsic activation includes using ultrasound, magnetic field, light and photodynamic therapy to provide an activatable system with less toxic, safe, and minimal adverse effects [44–46]. At the moment, the development of nanoparticles activation strategy is not only satisfied with a single treatment but also presume a multimodal trigger system. Therefore, the synthesis and analysis of multifunctional nanomaterials have been extensively researched *in vitro* and *in vivo* for cancer [47, 48].

The above strategies, however, lack a navigational force to the desired target as well as the ability to penetrate tumors beyond diffusion limits. Magnetic tumor targeting, using magnetic carriers and an external magnetic field is demonstrating promise for enhanced tumor accumulation of chemotherapy [49] and virotherapy [50] following their systemic administration. In the presence of a magnetic field, peptide-functionalized magnetic silk nanoparticles demonstrated increased cellular uptake of an anticancer agent (ASC-J9) by HCT116 colorectal cancer spheroids [51]. Additionally, in an orthotopic model of breast cancer, magnetic targeting enriched Doxorubicin-loaded magnetic SF nanoparticles at the tumor site with a concomitant suppression of uptake by the liver, resulting in a significant decrease in tumor volume and survival [52]. The provision of an external driving force expands therapeutic use to a wide variety of tumors, independent of specific receptor expression.

3.3 Nanocarrier for cancer

3.3.1 Chemotherapy delivery

Due to excellent stability in the change of temperature [53], humidity [54], and pH [55], silk-based nanocarriers have been widely studied for the delivery of numerous chemotherapeutic substances such as doxorubicin, cisplatin, paclitaxel, 5′-fluorouracil, and floxuridine for cancer treatment (**Table 1**). In addition, silk nanoparticles can also be used in the delivery of natural plant-derived therapeutics including curcumin [65], celastrol and triptolide [70], which are limited by their poor water solubility. To reduce systemic toxicity and adverse side effects of chemotherapeutics, targeted delivery can be achieved by conjugation of various targeting ligands to the silk material (as described above), recognizing overexpression of particular epitopes on the surface of target cells [71]. For example, Lei Huang et al. [63] designed a folate (FA) conjugated silk nanoparticle double loaded with doxorubicin (FA-SFPs-DOX-DOX), which provided a pH-dependent targeted drug release, lasting for over 30 hours. This

Therapeutic agent	Silk source	Functionalization	Cancer Type	*In vitro* model	*In vivo* model	Reference
Doxorubicin	Silk Protein	RGD peptide	Breast	4 T1	4 T1 (F)	[56]
	Silk sericin	—	Breast	4 T1	—	[57]
	B. mori	—	Breast	4 T1	4 T1 (F)	[58]
	B. mori	—	Breast, colon	4 T1, MCF7, Caco2-BBE	4 T1 (OT)	[59]
	Spider silk	Herceptin	Breast	D2F2, D2F2E2	D2F2, D2F2E2 (OT)	[60]
	RSF	Hyaluronic acid	Lung	A459	A459 (X)	[61]
	B. mori	—	Breast	4 T1	4 T1 (F)	[62]
	B. mori	Folate	—	KB, C2C12	—	[63]
Cisplatin	*B. mori*	—	Lung	A549	—	[64]
Curcumin	*B. mori*	—	Breast	MDA-MB-231	—	[65]
Curcumin/ 5-FU	*B. mori*	Hyaluronic acid	Breast	4 T1	4 T1 (F)	[66]
	B. mori	—	Breast	4 T1	4 T1 (OT)	[30]
5-FU	RSF	—	Colorectal	HT-29	—	[67]
Paclitaxel	*B. mori*	—	Liver	H22	H22 (F)	[68]
	B. mori	—	Gastric	BGC-823, SGC-7901	BGC-823 (X, F)	[69]

RSF = regenerated silk fibroin, 5-FU = 5-fluorouracil, F = flank, OT = orthotopic, and X = xenograft. All in vivo models performed in mice.

Table 1.
Silk nanoparticles as nanocarriers for chemotherapy.

study demonstrated the importance of the target ligand, with FA-SFPs-DOX-DOX inducing greater cytotoxicity against HeLa cells compared with SFPs-DOX-DOX.

3.3.2 Peptide and protein delivery

Silk-based nanocarriers can also bind with peptides and proteins improving their *in vivo* stability. Lactoferrin is one such protein showing anti-cancer properties, whereby apo-bovine lactoferrin loaded silk nanoparticles induces significantly higher internalization and cytotoxicity towards the MDA-MB-231 and MCF-7 breast cancer cell lines [72]. Peptide-based cancer vaccines are another important therapeutic agent in cancer treatment. However, peptide vaccines suffer from short *in vivo* stability caused by proteolytic degradation and rapid clearance from the bloodstream [73]. A silk nanoparticle delivery system is an effective way to improve the bioavailability and stability of peptide tumor vaccines [74]. Using engineered spider silk nanoparticles a peptide-based vaccination resulted in successful activation of cytotoxic T-cells, without unspecific immune responses [75]. How these antigens are delivered can also influence the developing vaccination response. It is thought that controlled, persistent antigenic signals elicit stronger responses than transient bolus vaccine exposure [76–78], such as that seen with microneedle skin patches. Microneedle vaccines exploit the skin's accessibility, both in terms of ease of administration as well as access to densely populated areas of antigen presenting cells. Silk microneedles therefore represent an attractive prospect due to their tunable release kinetics of encapsulated cargo as well as their overall biodegradability. This system demonstrated a > 10-fold increase in ovalbumin (OVA)-specific T cell and humoral responses in C57/Bl6 mice when compared with parenteral immunization [79], warranting further investigation.

3.3.3 Gene delivery

Viral vectors are traditional carriers for gene delivery, however, their drawbacks in inducing high systemic toxicity and immune responses limit their application in cancer treatment [22]. Thus, non-viral vectors have emerged to address challenges surrounding improving transfection efficiency, target specificity and cytotoxicity [80]. Among various materials, silk-based nanocarriers have been reported to provide biodegradability, biocompatibility, high transfection efficiency, and DNase resistance in gene delivery [9]. Through genetic engineering, the transfection efficiency of silk nanoparticles could be further improved. Numata et al. [81] combined silk protein-based nanocarriers with poly(L-lysine) (PLL) for gene delivery, resulting in improved transfection efficiency of pDNA in human embryonic kidney (HEK) cells. Additionally, to further enhance target specificity of the silk-based gene delivery system, they included tumor homing peptides (THP) [80]; F3 peptide (specifically targeted towards nucleolin expressing tumor and endothelial cells) and Lyp1 peptide (shows target specificity towards the p32 receptor overexpressed in tumor cells) [82, 83]. The use of cationic polymers with silk-based nanocarriers is another popular strategy due to their high cellular uptake efficiency, good water solubility, excellent transferability and easy synthesis [84]. Polyethyleneimine (PEI) is one of the commonly used cationic polymers which easily assembles with gene therapies and demonstrates improved cellular uptake due to their positive charge [85]. Song et al. [65] designed magnetic-silk/PEI core-shell nanoparticles for targeted delivery of c-myc antisense oligodeoxynucleotides (ODNs), which had high uptake efficiencies and significantly inhibit the growth of MDA-MB-231 cells.

3.3.4 Diagnostics and theranostics

In addition to therapeutic delivery, silk nanoparticles are also promising as non-invasive imaging components and provide an opportunity to augment existing imaging modalities for diagnostic purposes. These modalities are often limited by inadequate contrast between healthy and diseased tissue, contributing to failure to detect signs of illness, particularly early signs. Silk nanoparticles can be used as a vehicle for loading magnetic resonance contrast agents, overcoming agglomeration limitations of magnesium oxide nanospheres [86]. Alternatively the use of fluorescent dyes and carbon dots for modification of the silk fibroin itself has applications for live cell imaging or visualizing degradation of silk-fibroin implants [87]. The production of fluorescent silk nanoparticles can be created using simple dyes, chemical modification of the fibroin, conjugation or entrapment of fluorescent proteins and even doping the silkworm larvae's diet with fluorescent dyes such as rhodamine and fluorescein [88]. Additionally, carbon quantum dots (CQDs) generated from SF are strongly fluorescent, resist photobleaching, can be further functionalized [89–91] and in comparison to other colloidal materials, avoid the need for toxic heavy metals. However, this process does require controlled and pressurized heating of the fibroin for carbonization into CQDs.

By combining these imaging modalities with their role as a drug carrier, silk nanoparticles are becoming an important theranostic device. Theranostics is an approach that combines cancer treatment and diagnosis, in which efficient imaging guidance of therapy is necessary for detecting the drug loading, targeted delivery, and release[123]. For example, Levodopa (a PTT agent) and manganese dioxide particles (a contrast agent) were formulated with silk sericin from *B. mori* cocoons to create a one-step method for MRI-guided photothermal therapy [92]. The composition of spheres made of spider silk and iron oxide nanoparticles have also demonstrated drug loading and release capacity with potential to be used in both hyperthermia and magnetic resonance imaging (MRI) applications combined with drug delivery against tumor cells [93].

4. Conclusion

Nanoparticles used for drug delivery require desired physicochemical properties including size [94], shape [95], structure [96], rigidity [97], and surface modification [98]. Translation and application of nanoparticles, including silk, to the clinic must first overcome a number of challenges including their heterogeneity, reproducibility and upscale production. Silk derived from different sources will possess different amino acid sequences and morphology, whilst LPS contamination of recombinant silk is a major obstacle for progression to clinic, requiring careful characterization of its toxicity and immunogenicity. Additionally, traditional nanoparticle preparation methods involving breaking down of large particles, nanoprecipitation, or self-assembly of monomers, suffer from wide size distribution and large batch-to-batch variability [99]. To obtain more stable and controllable nanoparticles, microfluidics has emerged for manipulating tiny fluids (1×10^{-9} L–1×10^{-18} L) in micro-channels with dimensions of tens of micrometers [100]. Several flow patterns including laminar flow, turbulent flow and droplet flow could be achieved under microfluidic control with potential to enhance fluid mixing, reduce reagent consumption and batch-to-batch variations [101, 102]. Interestingly, the introduction of superparamagnetic magnetic nanoparticles (used to provide magnetic targeting capabilities) during the SF formation process provided artificial regulation of this process as well as drug entrapment, preventing agglomeration of SF and resulting in uniform, spherical nanoparticles [52]. Ultimately, SF nanoparticles provide many attractive properties for multi-functional drug delivery strategies but future use relies on reliable, reproducible manufacture to ensure appropriate comparisons can be made for their translation.

Acknowledgements

Financial support provided by Cancer Research UK (CRUK grant reference: C25574/A24321) and Public Service Department Malaysia (JPA).

Author details

Faith H.N. Howard*†, Zijian Gao†, Hawari Bin Mansor†, Zidi Yang and Munitta Muthana
University of Sheffield, Sheffield, UK

*Address all correspondence to: f.howard@sheffield.ac.uk

† Joint first authors.

References

[1] WHO. Cancer: Overview. 3 Feb 2022. Available from: https://www.who.int/health-topics/cancer

[2] GLOBOCAN. Global Cancer Observatory. 2022. Available from: https://gco.iarc.fr/

[3] Schiller JH, Harrington D, Belani CP,Langer C, Sandler A, Krook J, et al. Comparison of four chemotherapy regimens for advanced non-small-cell lung cancer. The New England Journal of Medicine. 2002;**346**(2):92-98

[4] Cho K, Wang X, Nie S, Chen ZG, Shin DM. Therapeutic nanoparticles for drug delivery in cancer. Clinical Cancer Research. 2008;**14**(5):1310-1316

[5] Wicki A, Witzigmann D, Balasubramanian V, Huwyler J. Nanomedicine in cancer therapy: Challenges, opportunities, and clinical applications. Journal of Controlled Release. 2015;**200**:138-157

[6] Young SWS, Stenzel MH, Yang J-L. Nanoparticle-siRNA: A potential cancer therapy? Critical Reviews in Oncology/Hematology. 2016;**98**:159-169

[7] Tomalia DA. In quest of a systematic framework for unifying and defining nanoscience. Journal of Nanoparticle Research. 2009;**11**(6):1251-1310

[8] Kim T, Hyeon T. Applications of inorganic nanoparticles as therapeutic agents. Nanotechnology. 2014;**25**(1):012001

[9] Zhao Z, Li Y, Xie MB. Silk fibroin-based nanoparticles for drug delivery. International Journal of Molecular Sciences. 2015;**16**(3):4880-4903

[10] Bernards R. A missing link in genotype-directed cancer therapy. Cell. 2012;**151**(3):465-468

[11] Wenk E, Merkle HP, Meinel L. Silk fibroin as a vehicle for drug delivery applications. Journal of Controlled Release. 2011;**150**(2):128-141

[12] Kundu B, Rajkhowa R, Kundu SC, Wang X. Silk fibroin biomaterials for tissue regenerations. Advanced Drug Delivery Reviews. 2013;**65**(4):457-470

[13] Cao Y, Wang B. Biodegradation of silk biomaterials. International Journal of Molecular Sciences. 2009;**10**(4):1514-1524

[14] Guidetti G, d'Amone L, Kim T, Matzeu G, Mogas-Soldevila L, Napier B, et al. Silk materials at the convergence of science, sustainability, healthcare, and technology. Applied Physics Reviews. 2022;**9**(1):50

[15] Kaplan DL, Mello CM, Arcidiacono S, Fossey S, Senecal K, Muller W. Silk. In: McGrath K, Kaplan D, editors. Protein-Based Materials. Boston, MA: Birkhäuser Boston; 1997. pp. 103-131

[16] Jin HJ, Kaplan DL. Mechanism of silk processing in insects and spiders. Nature. 2003;**424**(6952):1057-1061

[17] Matsumoto A, Lindsay A, Abedian B, Kaplan DL. Silk fibroin solution properties related to assembly and structure. Macromolecular Bioscience. 2008;**8**(11):1006-1018

[18] Chen X, Shao Z, Knight DP, Vollrath F. Conformation transition kinetics of Bombyx mori silk protein. Proteins: Structure, Function, and Bioinformatics. 2007;**68**(1):223-231

[19] Zhang C, Song D, Lu Q, Hu X, Kaplan DL, Zhu H. Flexibility regeneration of silk fibroin in vitro. Biomacromolecules. 2012;**13**(7):2148-2153

[20] Lu Q, Hu X, Wang X, Kluge JA, Lu S, Cebe P, et al. Water-insoluble silk films with silk I structure. Acta Biomaterialia. 2010;**6**(4):1380-1387

[21] Yucel T, Lovett ML, Kaplan DL. Silk-based biomaterials for sustained drug delivery. Journal of Controlled Release. 2014;**190**:381-397

[22] Numata K, Hamasaki J, Subramanian B, Kaplan DL. Gene delivery mediated by recombinant silk proteins containing cationic and cell binding motifs. Journal of Controlled Release. 2010;**146**(1):136-143

[23] Meinel L, Karageorgiou V, Hofmann S, Fajardo R, Snyder B, Li C, et al. Engineering bone-like tissue in vitro using human bone marrow stem cells and silk scaffolds. Journal of Biomedical Materials Research Part A. 2004;**71A**(1):25-34

[24] Patil PP, Reagan MR, Bohara RA. Silk fibroin and silk-based biomaterial derivatives for ideal wound dressings. International Journal of Biological Macromolecules. 2020;**164**:4613-4627

[25] Kim DH, Kim YS, Amsden J, Panilaitis B, Kaplan DL, Omenetto FG, et al. Silicon electronics on silk as a path to bioresorbable, implantable devices. Applied Physics Letters. 2009;**95**(13):133701

[26] Lozano-Perez AA, Rodriguez-Nogales A, Ortiz-Cullera V, Algieri F, Garrido-Mesa J, Zorrilla P, et al. Silk fibroin nanoparticles constitute a vector for controlled release of resveratrol in an experimental model of inflammatory bowel disease in rats. International Journal of Nanomedicine. 2014;**9**:4507-4520

[27] You R, Zhang Y, Liu Y, Liu G, Li M. The degradation behavior of silk fibroin derived from different ionic liquid solvents. Natural Science. 2013;**5**:10-19

[28] Crivelli B, Perteghella S, Bari E, Sorrenti M, Tripodo G, Chlapanidas T, et al. Silk nanoparticles: From inert supports to bioactive natural carriers for drug delivery. Soft Matter. 2018;**14**(4):546-557

[29] Zhang Y-Q, Shen W-D, Xiang R-L, Zhuge L-J, Gao W-J, Wang W-B. Formation of silk fibroin nanoparticles in water-miscible organic solvent and their characterization. Journal of Nanoparticle Research. 2007;**9**(5):885-900

[30] Li H, Tian J, Wu A, Wang J, Ge C, Sun Z. Self-assembled silk fibroin nanoparticles loaded with binary drugs in the treatment of breast carcinoma. International Journal of Nanomedicine. 2016;**11**:4373-4380

[31] Lammel AS, Hu X, Park SH, Kaplan DL, Scheibel TR. Controlling silk fibroin particle features for drug delivery. Biomaterials. 2010;**31**(16):4583-4591

[32] Anatol J, Arkadiusz TS. Electrospraying route to nanotechnology: An overview. Journal of Electrostatics. 2008;**66**:197-219

[33] BockN,WoodruffMA,HutmacherDW, Dargaville TR. Electrospraying, a reproducible method for production of polymeric microspheres for biomedical applications. Polymers. 2011;**3**(1):131-149

[34] Zhao Z, Li Y, Chen A-Z, Zheng Z-J, Hu J-Y, Li J-S, et al. Generation of silk fibroin nanoparticles via solution-enhanced dispersion by

supercritical CO2. Industrial & Engineering Chemistry Research. 2013;**52**(10):3752-3761

[35] Grbavčić ŽB, Arsenijević ZL, Garić-Grulović RV. Prediction of single particle settling velocities through liquid fluidized beds. Powder Technology. 2009;**190**(3):283-291

[36] Gupta V, Aseh A, Ríos CN, Aggarwal BB, Mathur AB. Fabrication and characterization of silk fibroin-derived curcumin nanoparticles for cancer therapy. International Journal of Nanomedicine. 2009;**4**:115-122

[37] Myung SJ, Kim H-S, Kim Y, Chen P, Jin H-J. Fluorescent silk fibroin nanoparticles prepared using a reverse microemulsion. Macromolecular Research. 2008;**16**(7):604-608

[38] Albanese A, Tang PS, Chan WC. The effect of nanoparticle size, shape, and surface chemistry on biological systems. Annual Review of Biomedical Engineering. 2012;**14**:1-16

[39] Bregoli L, Movia D, Gavigan-Imedio JD, Lysaght J, Reynolds J, Prina-Mello A. Nanomedicine applied to translational oncology: A future perspective on cancer treatment. Nanomedicine. 2016;**12**(1):81-103

[40] van Vlerken LE, Vyas TK, Amiji MM. Poly(ethylene glycol)-modified nanocarriers for tumor-targeted and intracellular delivery. Pharmaceutical Research. 2007;**24**(8):1405-1414

[41] White BD, Duan C, Townley HE. Nanoparticle activation methods in cancer treatment. Biomolecules. 2019;**9**(5):26

[42] Boron WF. Regulation of intracellular pH. Advances in Physiology Education. 2004;**28**(4):160-179

[43] Li X, Kim J, Yoon J, Chen X. Cancer-associated, stimuli-driven, turn on Theranostics for multimodality imaging and therapy. Advanced Materials. 2017;**29**(23):1606857

[44] Mullin LB, Phillips LC, Dayton PA. Nanoparticle delivery enhancement with acoustically activated microbubbles. IEEE Transactions on Ultrasonics, Ferroelectrics, and Frequency Control. 2013;**60**(1):65-77

[45] Pearce J, Giustini A, Stigliano R, Jack HP. Magnetic heating of nanoparticles: The importance of particle clustering to achieve therapeutic temperatures. Journal of Nanotechnology in Engineering and Medicine. 2013;**4**(1):110071-1100714

[46] Huang L, Li Z, Zhao Y, Zhang Y, Wu S, Zhao J, et al. Ultralow-power near infrared lamp light operable targeted organic nanoparticle photodynamic therapy. Journal of the American Chemical Society. 2016;**138**(44):14586-14591

[47] Chen KJ, Liang HF, Chen HL, Wang Y, Cheng PY, Liu HL, et al. A thermoresponsive bubble-generating liposomal system for triggering localized extracellular drug delivery. ACS Nano. 2013;**7**(1):438-446

[48] Guduru R, Liang P, Runowicz C, Nair M, Atluri V, Khizroev S. Magneto-electric nanoparticles to enable field-controlled high-specificity drug delivery to eradicate ovarian cancer cells. Scientific Reports. 2013;**3**:2953

[49] Lu YJ, Lin PY, Huang PH, Kuo CY, Shalumon KT, Chen MY, et al. Magnetic graphene oxide for dual targeted delivery of doxorubicin and photothermal therapy. Nanomaterials (Basel). 2018;**8**(4):20

[50] Howard FHN, Al-Janabi H, Patel P, Cox K, Smith E, Vadakekolathu J, et al.

Nanobugs as drugs: Bacterial derived Nanomagnets enhance tumor targeting and oncolytic activity of HSV-1 virus. Small. 2022;**18**(13):e2104763

[51] Tomeh MA, Hadianamrei R, Xu D, Brown S, Zhao X. Peptide-functionalised magnetic silk nanoparticles produced by a swirl mixer for enhanced anticancer activity of ASC-J9. Colloids and Surfaces B: Biointerfaces. 2022;**216**:112549

[52] Tian Y, Jiang X, Chen X, Shao Z, Yang W. Doxorubicin-loaded magnetic silk fibroin nanoparticles for targeted therapy of multidrug-resistant cancer. Advanced Materials. 2014;**26**(43):7393-7398

[53] Liu Y, Zheng Z, Gong H, Liu M, Guo S, Li G, et al. DNA preservation in silk. Biomaterials Science. 2017;**5**(7):1279-1292

[54] X-m Z, Berghe IV, Wyeth P. Heat and moisture promoted deterioration of raw silk estimated by amino acid analysis. Journal of Cultural Heritage. 2011;**12**:408-411

[55] Zong X-H, Zhou P, Shao Z-Z, Chen S-M, Chen X, Hu B-W, et al. Effect of pH and copper(II) on the conformation transitions of silk fibroin based on EPR, NMR, and Raman spectroscopy. Biochemistry. 2004;**43**(38):11932-11941

[56] Bin L, Yang Y, Wang F, Wang R, Fei H, Duan S, et al. Biodegradable silk fibroin Nanocarriers to modulate hypoxia tumor microenvironment favoring enhanced chemotherapy. Frontiers in Bioengineering and Biotechnology. 2022;**10**:960501

[57] Li X, Hou S, Chen J, He CE, Gao YE, Lu Y, et al. Engineering silk sericin decorated zeolitic imidazolate framework-8 nanoplatform to enhance chemotherapy. Colloids and Surfaces. B, Biointerfaces. 2021;**200**:111594

[58] Norouzi P, Motasadizadeh H, Atyabi F, Dinarvand R, Gholami M, Farokhi M, et al. Combination therapy of breast cancer by Codelivery of doxorubicin and Survivin siRNA using Polyethylenimine modified silk fibroin nanoparticles. ACS Biomaterials Science & Engineering. 2021;**7**(3):1074-1087

[59] Chen Q, Ma Y, Bai P, Li Q, Canup BSB, Long D, et al. Tumor microenvironment-responsive Nanococktails for synergistic enhancement of cancer treatment via Cascade reactions. ACS Applied Materials & Interfaces. 2021;**13**(4): 4861-4873

[60] Florczak A, Deptuch T, Lewandowska A, Penderecka K, Kramer E, Marszalek A, et al. Functionalized silk spheres selectively and effectively deliver a cytotoxic drug to targeted cancer cells in vivo. Journal of Nanobiotechnology. 2020;**18**(1):177

[61] Gou S, Yang J, Ma Y, Zhang X, Zu M, Kang T, et al. Multi-responsive nanococktails with programmable targeting capacity for imaging-guided mitochondrial phototherapy combined with chemotherapy. Journal of Controlled Release. 2020;**327**:371-383

[62] Tan M, Liu W, Liu F, Zhang W, Gao H, Cheng J, et al. Silk fibroin-coated Nanoagents for acidic lysosome targeting by a functional preservation strategy in cancer chemotherapy. Theranostics. 2019;**9**(4):961-973

[63] Huang L, Tao K, Liu J, Qi C, Xu L, Chang P, et al. Design and fabrication of multifunctional Sericin nanoparticles for tumor targeting and pH-responsive subcellular delivery of cancer chemotherapy drugs. ACS Applied Materials & Interfaces. 2016;**8**(10):6577-6585

[64] Kim SY, Naskar D, Kundu SC, Bishop DP, Doble PA, Boddy AV, et al.

Formulation of biologically-inspired silk-based drug carriers for pulmonary delivery targeted for lung cancer. Scientific Reports. 2015;**5**(1):11878

[65] Song W, Muthana M, Mukherjee J, Falconer RJ, Biggs CA, Zhao X. Magnetic-silk Core–Shell nanoparticles as potential carriers for targeted delivery of curcumin into human breast cancer cells. ACS Biomaterials Science & Engineering. 2017;**3**(6):1027-1038

[66] Huang Y, Xie D, Gou S, Canup BSB, Zhang G, Dai F, et al. Quadruple-responsive nanoparticle-mediated targeted combination chemotherapy for metastatic breast cancer. Nanoscale. 2021;**13**(11):5765-5779

[67] Hudita A, Radu IC, Galateanu B, Ginghina O, Herman H, Balta C, et al. Bioinspired silk fibroin nano-delivery systems protect against 5-FU induced gastrointestinal mucositis in a mouse model and display antitumor effects on HT-29 colorectal cancer cells in vitro. Nanotoxicology. 2021;**15**(7):973-994

[68] Wu P, Liu Q, Wang Q, Qian H, Yu L, Liu B, et al. Novel silk fibroin nanoparticles incorporated silk fibroin hydrogel for inhibition of cancer stem cells and tumor growth. International Journal of Nanomedicine. 2018;**13**:5405-5418

[69] Wu P, Liu Q, Li R, Wang J, Zhen X, Yue G, et al. Facile preparation of paclitaxel loaded silk fibroin nanoparticles for enhanced antitumor efficacy by Locoregional drug delivery. ACS Applied Materials & Interfaces. 2013;**5**(23):12638-12645

[70] Chen S-R, Dai Y, Zhao J, Lin L, Wang Y, Wang Y. A mechanistic overview of Triptolide and Celastrol, natural products from Tripterygium wilfordii hook F. Frontiers in Pharmacology. 2018;**9**:13

[71] Gobin AS, Butler CE, Mathur AB. Repair and regeneration of the abdominal wall musculofascial defect using silk fibroin-chitosan blend. Tissue Engineering. 2006;**12**(12):3383-3394

[72] Roy K, Patel YS, Kanwar RK, Rajkhowa R, Wang X, Kanwar JR. Biodegradable Eri silk nanoparticles as a delivery vehicle for bovine lactoferrin against MDA-MB-231 and MCF-7 breast cancer cells. International Journal of Nanomedicine. 2016;**11**:25-44

[73] Malonis RJ, Lai JR, Vergnolle O. Peptide-based vaccines: Current Progress and future challenges. Chemical Reviews. 2020;**120**(6):3210-3229

[74] Oroojalian F, Beygi M, Baradaran B, Mokhtarzadeh A, Shahbazi M-A. Immune cell membrane-coated biomimetic nanoparticles for targeted cancer therapy. Small. 2021;**17**(12):2006484

[75] Lucke M, Mottas I, Herbst T, Hotz C, Römer L, Schierling M, et al. Engineered hybrid spider silk particles as delivery system for peptide vaccines. Biomaterials. 2018;**172**:105-115

[76] Mohanan D, Slütter B, Henriksen-Lacey M, Jiskoot W, Bouwstra JA, Perrie Y, et al. Administration routes affect the quality of immune responses: A cross-sectional evaluation of particulate antigen-delivery systems. Journal of Controlled Release. 2010;**147**(3):342-349

[77] Jewell CM, López SC, Irvine DJ. In situ engineering of the lymph node microenvironment via intranodal injection of adjuvant-releasing polymer particles. Proceedings of the National Academy of Sciences of the United States of America. 2011;**108**(38):15745-15750

[78] Johansen P, Storni T, Rettig L, Qiu Z, Der-Sarkissian A, Smith KA,

et al. Antigen kinetics determines immune reactivity. Proceedings of the National Academy of Sciences. 2008;**105**:5189-5194

[79] DeMuth PC, Min Y, Irvine DJ, Hammond PT. Implantable silk composite microneedles for programmable vaccine release kinetics and enhanced immunogenicity in transcutaneous immunization. Advanced Healthcare Materials. 2014;**3**(1):47-58

[80] Numata K, Mieszawska-Czajkowska AJ, Kvenvold LA, Kaplan DL. Silk-based Nanocomplexes with tumor-homing peptides for tumor-specific gene delivery. Macromolecular Bioscience. 2012;**12**(1):75-82

[81] Numata K, Subramanian B, Currie HA, Kaplan DL. Bioengineered silk protein-based gene delivery systems. Biomaterials. 2009;**30**(29):5775-5784

[82] Christian S, Pilch J, Akerman ME, Porkka K, Laakkonen P, Ruoslahti E. Nucleolin expressed at the cell surface is a marker of endothelial cells in angiogenic blood vessels. The Journal of Cell Biology. 2003;**163**(4):871-878

[83] Song N, Zhao L, Zhu M, Zhao J. Recent progress in LyP-1-based strategies for targeted imaging and therapy. Drug Delivery. 2019;**26**(1):363-375

[84] Zakeri A, Kouhbanani MAJ, Beheshtkhoo N, Beigi V, Mousavi SM, Hashemi SAR, et al. Polyethylenimine-based nanocarriers in co-delivery of drug and gene: A developing horizon. Nano Reviews & Experiments. 2018;**9**(1):1488497

[85] Chen Z, Krishnamachary B, Bhujwalla ZM. Degradable dextran Nanopolymer as a carrier for choline kinase (ChoK) siRNA cancer therapy. Nanomaterials (Basel). 2016;**6**(2):8

[86] Li J, Khalid A, Verma R, Abraham A, Qazi F, Dong X, et al. Silk fibroin coated magnesium oxide Nanospheres: A biocompatible and biodegradable tool for noninvasive bioimaging applications. Nanomaterials (Basel). 2021;**11**(3):20

[87] Kim JH, Park CH, Lee OJ, Lee JM, Kim JW, Park YH, et al. Preparation and in vivo degradation of controlled biodegradability of electrospun silk fibroin nanofiber mats. Journal of Biomedical Materials Research. Part A. 2012;**100**(12):3287-3295

[88] Sagnella A, Chieco C, Di Virgilio N, Toffanin S, Posati T, Pistone A, et al. Bio-doping of regenerated silk fibroin solution and films: A green route for biomanufacturing. RSC Advances. 2014;**4**(64):33687-33694

[89] Colusso E, Cicerchia L, Rigon M, Gomes V, Martucci A. Photoluminescence properties of silk–carbon quantum dots composites. Journal of Sol-Gel Science and Technology. 2022:8

[90] Horo H, Saha M, Das H, Mandal B, Kundu LM. Synthesis of highly fluorescent, amine-functionalized carbon dots from biotin-modified chitosan and silk-fibroin blend for target-specific delivery of antitumor agents. Carbohydrate Polymers. 2022;**277**: 118862

[91] Niu L, Shi M, Feng Y, Sun X, Wang Y, Cheng Z, et al. The interactions of quantum dot-labeled silk fibroin micro/nanoparticles with cells. Materials (Basel). 2020;**13**(15):17

[92] Tian D, Xu H, Xiao B, Zhou X, Liu X, Zhou Z, et al. Single-step formulation of levodopa-based nanotheranostics—Strategy for ultra-sensitive high longitudinal relaxivity MRI guided switchable therapeutics. Biomaterials Science. 2020;**8**(6):1615-1621

[93] Kucharczyk K, Rybka JD, Hilgendorff M, Krupinski M, Slachcinski M, Mackiewicz A, et al. Composite spheres made of bioengineered spider silk and iron oxide nanoparticles for theranostics applications. PLoS One. 2019;**14**(7):e0219790

[94] Wang J, Mao W, Lock LL, Tang J, Sui M, Sun W, et al. The role of micelle size in tumor accumulation, penetration, and treatment. ACS Nano. 2015;**9**(7):7195-7206

[95] Kinnear C, Moore TL, Rodriguez-Lorenzo L, Rothen-Rutishauser B, Petri-Fink A. Form follows function: Nanoparticle shape and its implications for nanomedicine. Chemical Reviews. 2017;**117**(17):11476-11521

[96] Zhang L, Feng Q, Wang J, Zhang S, Ding B, Wei Y, et al. Microfluidic synthesis of hybrid nanoparticles with controlled lipid layers: Understanding flexibility-regulated cell-nanoparticle interaction. ACS Nano. 2015;**9**(10):9912-9921

[97] Hui Y, Yi X, Hou F, Wibowo D, Zhang F, Zhao D, et al. Role of nanoparticle mechanical properties in cancer drug delivery. ACS Nano. 2019;**13**(7):7410-7424

[98] Beck-Broichsitter M, Nicolas J, Couvreur P. Design attributes of long-circulating polymeric drug delivery vehicles. European Journal of Pharmaceutics and Biopharmaceutics. 2015;**97**(Pt B):304-317

[99] Chen G, Roy I, Yang C, Prasad PN. Nanochemistry and nanomedicine for nanoparticle-based diagnostics and therapy. Chemical Reviews. 2016;**116**(5):2826-2885

[100] Liu J, Lan Y, Yu Z, Tan CS, Parker RM, Abell C, et al. Cucurbit[n]uril-based microcapsules self-assembled within microfluidic droplets: A versatile approach for supramolecular architectures and Materials. Accounts of Chemical Research. 2017;**50**(2):208-217

[101] Liu D, Cito S, Zhang Y, Wang CF, Sikanen TM, Santos HA. A versatile and robust microfluidic platform toward high throughput synthesis of homogeneous nanoparticles with tunable properties. Advanced Materials. 2015;**27**(14):2298-2304

[102] Sackmann EK, Fulton AL, Beebe DJ. The present and future role of microfluidics in biomedical research. Nature. 2014;**507**(7491):181-189

Chapter 4

Emerging Selenium Nanoparticles for CNS Intervention

Jonaid Ahmad Malik, Jeba AjgarAnsari, Sakeel Ahmed, Archana Rani, Shabana Yasmeen Ansari and Sirajudheen Anwar

Abstract

Central nervous system (CNS) diseases have seriously impacted human wellness for the past few decades, specifically in developing countries, due to the unavailability of successful treatment. Due to the blood-brain barrier (BBB) and blood-cerebrospinal fluid barrier transport of drug and treatment of CNS disorders has become difficult. Nanoscale materials like Selenium nanoparticles (SeNPs) offer a possible therapeutic strategy for treating brain diseases like Alzheimer's, Frontotemporal dementia, Amyotrophic lateral sclerosis, Epilepsy, Parkinson's disease, and Huntington's disease. After being functionalized with active targeting ligands, SeNPs are versatile and competent in conveying combinations of cargoes to certain targets. We shall pay close attention to the primarily targeted therapies for SeNPs in CNS diseases. The objective of this paper was to highlight new developments in the exploration of SeNP formation and their potential applications in the management of CNS diseases. Furthermore, we also discussed the mechanisms underlying management of CNS disease, several therapeutic potentials for SeNPs, and the results of their preclinical research using diverse animal models. These methods might lead to better clinical and diagnostic results.

Keywords: selenium, nanoparticles, CNS diseases, management of CNS disorders, Alzheimer's disease, Parkinson's disease, Huntington's disease, epilepsy

1. Introduction

The central nervous disordersare progressive degeneration of neurons in the central nervous system, which results in altered brain cellular function. The major symptoms begin from degeneration of neurons to loss of coordination and memory and ultimately result in complete loss of function in healthy individuals. The three crucial neurodegenerative disorders (NDs) have been recognized as Alzheimer's disease (AD), Parkinson's disease (PD), and Amyotrophic lateral sclerosis (ALS) [1], affecting millions of individuals worldwide. AD is one of the most prevalent NDs and has affected mostly 30% of the aged people [2]. The second most common neurodegenerative disorder is PD, which affects 10–15 individuals per 100,000 people yearly [3]. It is estimated that PD cases will increase worldwide and may cross 12 million cases in 2050 [4]. Similarly, cases of ALS are also increasing with 1–2.6 new cases per 100,000 persons. The average age of onset of this disease is 59–60 years, and the average time from diagnosis to death is 3–4 years [5]. Besides these disorders, Gliomas, and glioblastoma, intrinsic brain tumors arise from neuroglial cells [6].

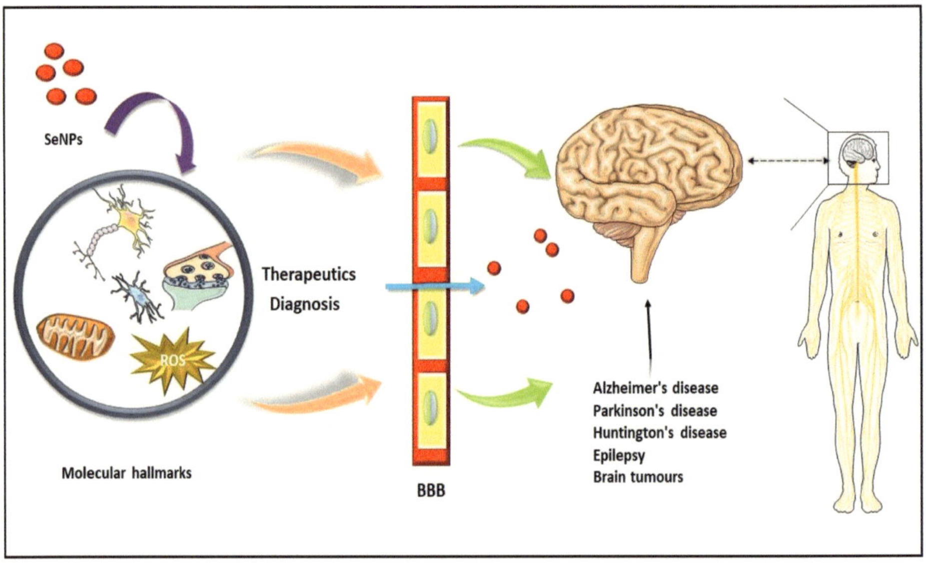

Other disorders include Huntington's disease (HD), depression, anxiety, autism spectrum disorders, seizures, etc. [7]. The establishment of innovative NDs treatments are urgently needed as the WHO has predicted that within the next 20 years NDs will surpass cancer and will become the second most common cause of death [8]. Natural products may offer great promises compared to classical therapies available to improve the symptoms but cannot prevent their progression. Therefore, researchers are continuously searching for new natural products that can potentially treat NDs without compromising the patient's health. However, all-natural products are not always safe; because natural products as drugs may have more adverse effects than their benefits as they are derived from various biological sources, their conversion into therapeutic formulations may face many hurdles such as safety issues, difficulty in identifying active ingredients because natural products contain various active phytochemicals such as flavonoids, alkaloids, etc. And it becomes difficult to identify which components of the herb have maximum therapeutic potential, low stability and high degradation, and difficulty crossing the blood-brain barrier [9]. However, till now, there are no effective treatments available that can alter the main symptoms of autism or which can improve the cognitive and deficit symptoms of schizophrenia; the majority of the people who have epilepsy, depression, brain injury, and posttraumatic stress disorder have acquired a few satisfaction from the current therapy available. The discovery of new effective medication for NDs has proven difficult compared to other diseases. Most pharmaceutical companies have shifted their interest from the field of neurology to other fields [10]. Chances of failure of clinical trial rate in the last stages are higher for a neurological and psychiatric disorders as compared to other diseases; due to the complex physiology of the brain, there are fewer animal models available that can effectively predict the safety and efficacy of the drug for the disorder that mainly affects the cerebral cortex [11]. Drug discovery is a costly and risky process, while drugs used to treat neurological or psychiatric disorders not only require a longer time to complete all stages of a clinical trial but also take much longer to complete the process of approval from a regulatory authority. The higher risk and high expenditure related to drug discovery and the development of neurological disorders are directly linked to scientific challenges. Several obstacles faced by pharmaceutical companies faces in discovering new drugs include: (a) Slow process of target identification and validation, (b) lack of appropriate animal models, (c) long duration of clinical trials paucity of knowledge regarding the disease's etiology [12]. One of the biggest hurdles in the establishment of the treatment of CNS disorder is because of the blood-brain barrier [13]. BBB restricts the permeation of therapeutic drugs to reach to the target site in the CNS therefore more than 98% of small molecular drugs remain ineffective in treating the CNS disorder [14].

Transportation of drugs through the BBB is a complex process requiring very sophisticated and nanosized particles. As nanoparticles have the potential to deliver drugs in various diseases, selenium nanoparticles (SeNPs) have been found to play a significant role in neurodegenerative diseases [15]. Various studies revealed that selenium nanoparticles have better bioavailability, improved antioxidant properties, and less toxicity than selenium-containing compounds. Thus selenium nanoparticles have free radical scavenging properties and improve behavior abnormalities and neurochemical alterations. Therefore, selenium nanoparticles have the potential to improve the impairment of memory and can be used as a potential therapy for NDs [16]. This book chapter provides a thorough overview of recent discoveries in the fields of investigation and use of nanoformulations for treating NCDs such as Parkinson's, Alzheimer's, ALS, and Huntington's, as well as the use of SeNPs for diagnosis and treatment.

2. Selenium nanoparticles in the management of CNS disorders

Due to the BBB and BCSFB, which prevents drug transport, treating ailments of the CNS is notoriously problematic. A promising clinical methodology for the intervention of some common NDs like, frontotemporal dementia, ALS, PD, and HD is provided by nanotechnology-based drug delivery methods, one of the new tactics to get around these obstacles and delivering medications to the CNS [17]. SeNPs could be a novel approach for treating such CNS disorders. Nanotechnology has emerged as an intriguing and promising new tool for treating NDs with considerable potential to solve issues with conventional methods. Nanostructured materials could traverse the BBB, target specific cells or signaling pathways, react to endogenous stimuli, transport genes, help axonal regeneration, and promote cell viability, among other specialized activities [18].

2.1 Role of SeNPs in treating CNS diseases

On the brain and neurons, selenium exhibits a direct antioxidant impact [19]. Low or moderate doses of selenium suppress cancer progression and have therapeutic benefits on NDs, including AD. Elevated levels of selenium increase the development of cancer cells and exhibit neurotoxicity. In *in vivo* and *in vitro*, selenium's antioxidant and anti-inflammatory activities have been established [20]. Selenium is a co-factor in the enzyme glutathione peroxidase, a scavenger. The catalytic properties of GSH-Px cause hydrogen peroxide (H_2O_2) to be transformed into water [19, 21–23]. A growing body of research shows that memory loss in AD patients is directly related to selenium deficiency in serum and hair samples. In animal studies and AD patients, selenium supplementation has reportedly been shown to reduce the likelihood of cognitive issues [20, 24–27]. The exploration of selenium and selenoproteins in neurological disorders, such as AD, has attracted much interest. Proteins known as selenoproteins include selenium as the amino acid selenocysteine. Since antioxidant mechanisms are crucial for delaying the emergence and spread of AD, these are primarily expressed in human brain tissue [28].

Additionally, certain recently developed selenoproteins and SeNPs with exceptional physiological characteristics exist. These particles may replace traditional therapeutic medications in managing AD because of their great efficiency and low toxic effect [20, 29]. SeNPs may be a promising therapeutic molecule for treating AD, according to the Nazrolu et al. report [20]. Using the whole-cell patch clamp method, Yuan et al. investigated the effects of SeNPs on sodium influxes and the excitation of DRG (dorsal root ganglion) neurons [30]. According to their study, SeNPs appeared to reduce sodium influx in a concentration and time-dependent way, raising the possibility that SeNPs may be neurotoxic [20, 31, 32]. Epigenetic, chronic stress, metabolic, and dietary factors all have a role in developing HD, an untreatable condition that causes a gradual loss of brain functionality. According to studies, selenium (Se) concentration in the brain are inadequate for HD illness, but restoring Se regulation there may lessen neuronal death and functioning. Most research showed that nano-Se reduced peroxidation, prevented huntingtin protein aggregation, and suppressed the production of histone deacetylase family members at the mRNA level. Nano-Se offers great promise as a treatment for HD. Therapy for HD disease will derive from nano-Se NPs' ability to heal neuronal processes and shield against degradation under stress [33]. Recent research has crucially demonstrated Se's beneficial impact on HD. For instance, sodium selenite may reduce mutant huntingtin clumping and rates of oxidized glutathione in HD mouse brains [34]. SeNPs were reported to offer neuroprotection by upregulating Nrf2 and HO-1, suppressing the inflammatory process, and apoptotic pathway, and avoiding the emergence of oxidative stress. Upon the onset of epileptic seizures, SeNPs can counteract alterations in the concentration and functionality of neuromodulators. SeNPs have strong antioxidant, anti-inflammatory, and neuromodulatory properties that make them a potential candidate for use as an anticonvulsant medication [35].

Selenium has been investigated as a screening tool for several neurological disorders, including epilepsy, AD, and PD. Se is available in high levels in the grey matter areas and the glandular sections of the brain. It participates in several neurotransmission and dopaminergic pathways [36, 37]. A few elements identified as potential influences for the involvement of Se in Alzheimer's pathogenicity include its antioxidant, neuroprotective effects, influence on the regulation of cytoskeletal elements assembly, affinity for several neurotoxic metals, and competence to mitigate Aβ accumulation and tau proteins hyperphosphorylation [34, 36, 38]. Se has been worn to prevent dopaminergic neurons by many selenoproteins, supporting its ability to fend off PD. Se levels were also linked to aggressive behaviour, anxiety, and mood swings. Se application in neurological diseases may be beneficial for individuals with profound Se insufficiency and/or mutants in genes related to Se transport or selenoproteins synthesis. At the same time, brain Se levels are usually low, and high Se levels might be detrimental (**Figure 1**) [39, 40].

SeNPs have shown great potential in managing various CNS disorders through particular mechanisms. The cellular signaling pathways that regulate the metabolic activity of neurons (TSC1/TSC2, p-mTOR, mTORC1), antioxidant (FoxO1, β-catenin/Wnt, Yap1), and inflammatory system (jak2/stat3, Adamts-1),

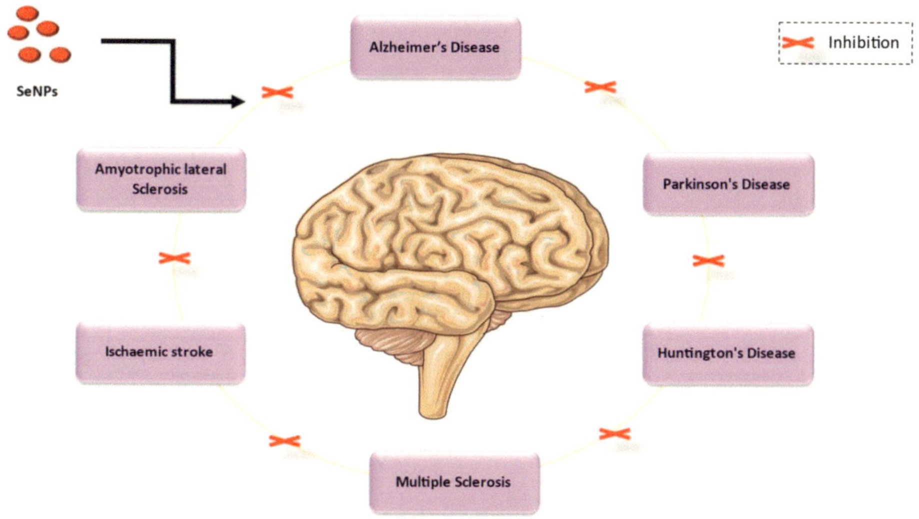

Figure 1.
Selenium nanoparticles in the management of several neurodegenerative diseases.

autophagy and induction of apoptosis (Mst1, ULK1, Bax, Caspase-3, and Bcl-2), and the preservation of hippocampal neurons (rictor/mTORC2) [41] these are the molecular mechanisms related with the neuroprotective activity of SeNPs. MST1 regulates neuronal cell death via Casp3 Signaling Pathway [42]. The JAK-STAT pathway prevents apoptosis in neurons. STAT3 is upregulated and stimulated selectively in regenerated neurons after axon damage. This pathway plays a direct role in forming glial scar tissue across lesions and neuronal restoration. After a CNS injury, STAT3 stimulation is required for the production of glial scars and the control of the spread of inflammation, both of which are essential for astrogliosis [43]. Numerous investigations have functionalized SeNPs with particular molecules, like sialic acid and epigallocatechin-3-gallate, to improve their penetrability toward the BBB. SeNPs are shown to minimize accumulation and stimulate their fragmentation to act as an antioxidant in the brain, either effectively or as part of GPx [20, 44] (**Figure 2**). Additionally, SeNPs were investigated in conjunction with substances that have demonstrated anti-disease Alzheimer's effects, such as resveratrol (Res) [45], curcumin (Cur) [46], chiral D-penicillamine (DPen) [47], and chlorogenic acid (CGA) [36, 48].

2.2 Types of SeNPs in the treatment of CNS disorders

SeNPs in conjugation with several molecules acts as an antioxidant and neuroprotective agent and has also been reported for various neurological diseases. Resveratrol's antioxidant and neuroprotective abilities, which counteract Aβ-aggregation and its oxidative consequences, have shown promise in the defense against AD. It has been demonstrated that Res-SeNPs selectively attach to Aβ through N-donors found in amino acids, forming a Se-N bond and enhancing Ressuppression on Cu^{2+}-induced Aβ aggregation. Res-SeNPs are non-toxic to neuroblastic cells (PC12 cells) and protect them against oxidative stress by reducing the apoptosis caused by Aβ, suggesting a potential synergism between SNPs and Res [45]. Curcumin also has demonstrated potential synergy with SeNPs in the therapy of AD. The antioxidant,

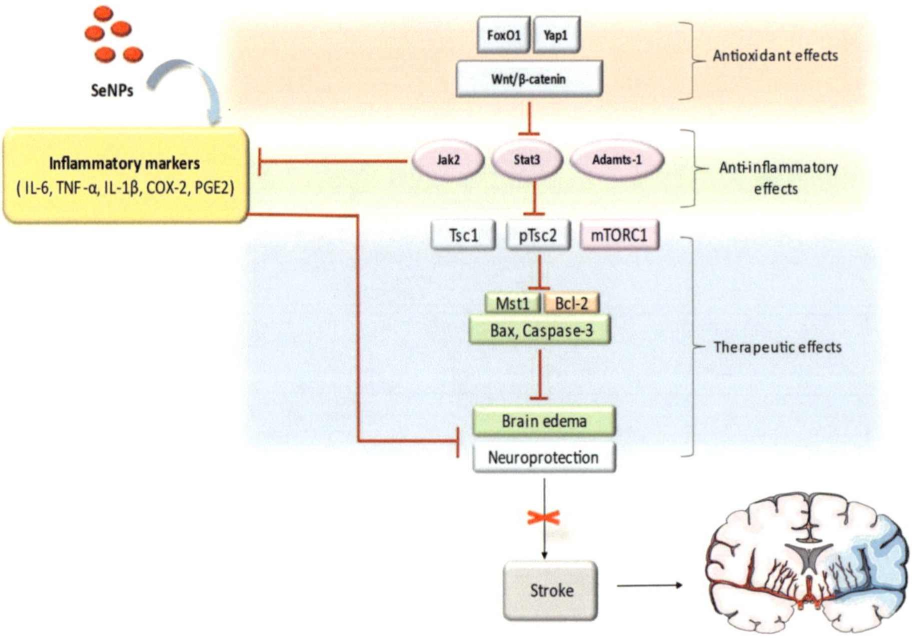

Figure 2.
Anti-inflammatory mechanism of SeNPs in management of central nervous system related diseases.

anti-Aβ inflammation and anti-Tau hyperphosphorylation capabilities of these NPs allowed them to pass through the BBB and bind to Aβ. It has been demonstrated that curcumin binds Aβ by hydrophobic contacts at the nonpolar regions of Aβ, enhancing their anti-inflammatory and antioxidant activities alongside Se [46]. As a result of their ability to bind to the N-donors of Aβ-proteins and form a Se-N bond, which prevents Aβ from aggregating them, CGA-SeNPs were shown to minimize Aβ-generated ROS in a dose-dependent manner, hindering their neurotoxicity and, consequently, lowering the rate of apoptosis. This was transcribed into a synergistic effect among CGA and SeNPs [49]. When given to transgenic mice 5XFAD with the mutation, the curcumin-loaded selenium-PLGA nanospheres developed by Huo et al. demonstrated a reduction in Aβ plaque production and inflammation [46]. Using an anti-Tfr receptor monoclonal antibody (OX26) as a functionalizing agent, PEG-SeNPs were also applied to treat stroke. OX26-PEG-SeNPs have been shown to play a function in preventing stroke in neuronal cells by minimizing the cellular edema brought on by aberrant Na^+ ion influx. When the middle cerebral artery was blocked in Wistar rats, SeNPs could reduces the infarction volumes, reduce the number of necrotic cells, increase the myelinated areas, and prevent the loss of axons in the hippocampal region [41]. SeNPs have also demonstrated promise in the battle against Huntington's disease. In HA759 mutant nematodes, SeNPs, in a dose-dependent manner, lowered neuronal death by reducing protein aggregates associated with HTT genetic variants and ROS, suppressing histone deacetylase mRNA, axonal degeneration, and improving reflexes [50]. Work on selenium-doped carbon quantum dots (Se-CQDs) has demonstrated their capacity to reduce ROS, and they have been successfully used to reduce secondary damage in TSCI. The findings showed that Se-CQDs had bioactivity and had a notable preventive role on astrocytes and PC12 cells toward H_2O_2-induced oxidative stress [51].

A frequent neurological condition called cerebral ischemia sets off a series of pathophysiological processes that include a drop in glucose and oxygen levels, an uncontrolled emission of glutamate, a fast rise in cellular calcium levels, and the production of free radicals. These occurrences consequently cause the endoplasmic reticulum and mitochondria's functions to be disrupted, which in turn causes the death of brain cells through apoptosis or necrosis [41, 52–54]. SeNPscan cross the BBB, builds up in the brain, and stops cell death from occurring. SeNPs are known to promote the production of BDNF and lower levels of Aβ and IL-6 in the hippocampus. Increased BDNF production reduces oxidative stress and prevents the degeneration of GABAergic neurons, particularly vulnerable to hypoxia and ischemia [55–57]. The mechanisms of the protective role of a modest dose of SeNPs on brain cells during OGD/R were examined by Turovsky et al. [58]. Additionally, the Bcl-2 family of proteins, the mechanisms of calcium homeostasis repair, suppression of mitochondrial and ER stress mechanisms, and eventually silencing of caspase-3 and inhibition of apoptosis are all part of SeNPs protective measure [58, 59].

Fei Gao et al. created selenium-chondroitin sulfate nanoparticles (CS@Se) as part of a multitargeted therapy for AD. Amyloid-ß (Aß) accumulation was successfully prevented by CS@Se, and SH-SY5Y cells were shielded against cytotoxicity brought on by $Aß_{1-42}$. In SH-SY5Y cells, okadaic acid-induced actin cytoskeleton disruption was dramatically reduced by CS@Se. The ROS and MDA levels were reduced by CS@Se, while the amounts of GSH-Px were elevated. This research shows that CS@Se might reduce tau protein hyperphosphorylation, ameliorate oxidative stress, prevent Aß from aggregating, and lessen cytoskeleton disruption. CS@Se is an effective multifunctional drug for the management of AD [60]. Xian Guo developed SeQDs, which have a multitarget therapeutic impact and can easily enter the BBB, to improve the therapeutic effect of pharmaceuticals through the BBB. SeQDs' unique fluorescence properties could be used to diagnose and manage AD. SeQDs are highly effective at scavenging free radicals and shielding cells from oxidative stress. By down-regulating PHF1 and CP13, the SeQDs can dramatically reduce tau protein phosphorylation and further neutralize free radicals, rebuild metabolic activity, preserve nerve cell solidity, and defend nerve cells from oxidative damage. These effects preclude Aß-mediated cytotoxicity and Aß aggregation, preventing the AD cascade reaction. Compared to conventional single-target medications, usingSeQDs in AD therapy has many benefits and offers a fresh approach to the co-managementof neurological illnesses [61]. The mechanism of SeNPs is depicted in **Figure 3**.

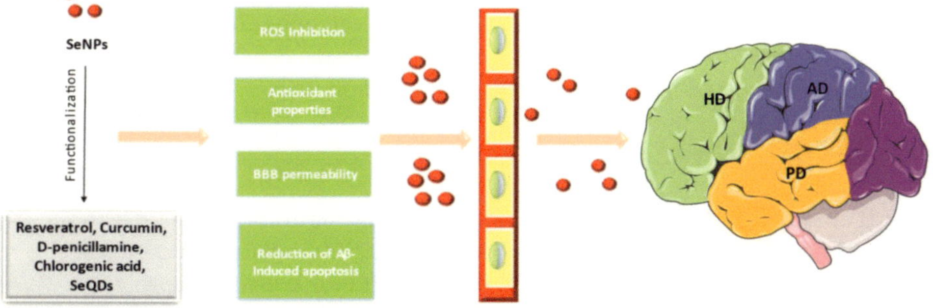

Figure 3.
SeNPs against various neurodegenerative disorders.

3. Therapeutic application of selenium nanoparticles in CNS disorder

3.1 Alzheimer's disease

AD is a progressive neurodegenerative disease; the pathological hallmark of this disease is the deposition of Aβ plaques. Various attempts have been made to develop therapies that potentially inhibit its deposition in the brain; in this regard, nanoparticles have shown promising results due to their distinctive physiochemical properties of small size and large surface area. Se is one of the most important mineral nutrients having a wide range of pharmacological actions. Studies have found that selenium has a neuroprotective effect. Due to the high stability of SeNPs, it has potential effects on the neurotoxicity of $A\beta_{42}$ in primary cultures of murine hippocampal neurons [62]. In human clinical trials, curcumin has been found as a highly efficacious compound without exerting any adverse effect, even when taken at a higher dose of 4 g/day. Curcumin forms intermolecular hydrogen bonds and binds effectively with Aβ plaques in AD. In AD, curcumin's drug delivery properties were modified by encapsulating SeNPs and changing the surface of the poly-lactide-co-glycolide (PLGA). In preclinical studies of memory impairment in mice models, it was found that the drug delivery system of curcumin-loaded SeNPs has the potential to decrease the aggregation of Aβ plaque in Alzheimer's disease mice model [46]. Moreover, selenium-chondroitin sulfate nanoparticles were also found to decrease the aggregation of amyloid-β and reduce the tau protein hyperphosphorylation by targeting the GSK-3β [63]. Another research revealed that chitosan-coated SeNPs (ChSeNPs) could enhance the effectiveness of stem cell-based therapy to attenuate the neurotoxicity in the streptozotocin-induced model in rats [64].

3.2 Parkinson's disease

PD is the second most prevalent CNS disorder after AD. Oxidative stress is considered one of the major factors responsible for this disease, causing neuronal death and apoptosis. Therefore, behavioral abnormalities in PD can be improved by decreasing the level of oxidative stress. As selenium possesses antioxidant properties, in preclinical studies, the neuroprotective effect of glycine-nano-selenium on oxidative stress was evaluated in PD rat model, and oxidative stress is induced by using 1-methyl-4-phenyl-1,2,3,6-tetrahydropyridine (MPTP) in the rat. It was found that oxidative stress in PD rat model was reduced by administering intragastric glycine nano-selenium, which ultimately reduced the neurobehavioural abnormalities in rats [65]. The administration of selenium in humans and animals has shown a rise in the level of glutathione and glutathione peroxidase, thus slowing the degeneration of neurons and preventing the depletion of dopamine levels. Hence it is regarded as an important micronutrients in Parkinson's disease as well [66].

3.3 Huntington's disease

Various studies have also been conducted for Huntington's disease by using selenium nanoparticles. HD is an inherited autosomal dominant disease caused by repeated trinucleotide sequence CAG that encodes for huntingtin protein [67]. Various laboratory findings suggest that oxidative stress is a major factor in Huntington's disease's pathogenesis. But due to poor knowledge of particular oxidative biomarkers, no antioxidants have effectively prevented neurodegeneration in Huntington's disease [68]. Studies have found Se to play a protective role in Huntington's disease, such as sodium selenite has the potential to lower mutant huntingtin aggregation and oxidized glutathione levels in the brain of HD mice model [34]. Recent studies have found a low level of Se in the brain of HD patients. Attenuating neuronal dysfunction can be achieved by maintaining the level of Se in the brain. In a preclinical study of HD models, Selenium nanoparticles have been found to prevent neuronal loss and improve behavioral dysfunction. Molecular testing has shown that selenium nanoparticles also prevent the damage produced by oxidative stress and attenuate the aggregation of huntingtin proteins. Thus Selenium nanoparticles are an effective therapy for Huntington's disease [50].

3.4 Amyotrophic lateral sclerosis

ALS is a chronic, deadly and irreparable NDs which involves the degradation of motor neurons in the motor cortex, brainstem, and spinal cord, which results in paralysis and death due to respiratory failure [69]. Because of the absence of efficient treatment, the majority of patients pass away within 3–5 years of assessment. Several cases of ALS are sporadic, while 15% of the cases are familial [70]. Genetic defects in SOD1 were found to be the first causative mutation involved in the pathology of ALS [71]. Apart from this, more than 50 genes were reported to be involved in ALS. The most common ones include mutations in chromosome 9 open reading frame 72 (*C9orf72*), TAR DNA-Binding (*TARDBP*), and fused in sarcoma (*FUS*) [72]. Recently in vitro studies have shown organo-selenium compound to potentially protect the neuronal damage and thus can be used as an alternative therapy in ALS [73].

3.5 Epilepsy

Selenium nanoparticles can also be used in other CNS disorders such as epilepsy. SeNP also has potential anticonvulsant activity due to its extensive antioxidant, anti-inflammatory, and neuromodulatory effect. Administration of SeNP decreases the duration of tonic, myoclonic and generalized seizures and can be used as an effective therapy in epilepsy [35]. SeNP can effectively cross the BBB and thus can be used to enhance the delivery of anti-cancer drugs in the brain, such as in the case of glioma in humans [74]. Drug delivery to cross the BBB is a complex process, and it requires a nanosized particle so that the drug can reach the brain. In this regard, SeNPs plays a vital role in the management of NDs (**Table 1**).

4. Mechanism of SeNPs in neurodegenerative disease

Selenium, a crucial trace element in both man and livestock, is essential in managing the biological stability of the brain and possesses neuroprotective properties. Various selenoproteins were also found to be involved in controlling NDs [80]. SeNPs significance in NDs has been widely reported in recent years (**Figure 4**), considering that neurons are highly vulnerable to damage from oxidative stress-related injury for variety of reasons, including excessive oxygen utilization (about 25% of the total body utilization). There is a substantial quantity of polyunsaturated fatty acids and low amount of antioxidant enzymes [81]. Natural antioxidants are frequently utilized to treat neurological illnesses since oxidative stress is one of the primary contributors to their etiology. Yet, they are ineffective [80], and as a result, using antioxidants

Disease	SeNPs	Animal model	Mechanism	Outcome	Reference
Alzheimer' disease	Cur/Se-PLGA nanospheres	The transgenic 5XFAD mice	Inhibited Aβ aggregation	Effective for AD treatment and to provide delivery at the site of target	[46]
	B6-SA-SeNPs	PC12 cells andbEnd.3 cells	B6-SA-SeNPs prevent the deposition of Aβ and effectively cross the BBB	B6-SA-SeNPs can be used to treat AD and have antioxidant and antiamyloid properties	[75]
	Selenium-chondroitin sulfate nanoparticles (CS@ Se)	SH-SY5Y cells	Prevent Aβ from clumping together, lessen cytoskeleton deterioration, combat oxidative stress, and lessen tau protein hyperphosphorylation	Potent multifunctional agent for AD	[63]
	Chitosan-coated Selenium nanoparticles (ChSeNPs)	Streptozotocin induced neurotoxicity in a rat model of AD	Decreases the neurotoxicity by increasing the antioxidant capacity	Decrease Aβ deposition and attenuate memory impairment	[64]
	SeNPs	Streptozotocin-induced neurotoxicity in the male rat of AD	SeNPs help in survival of neurons by regulating the system of oxidative stress, Cellular metabolite, and inflammatory reactions and maintaining the functional properties of the hippocampal neurons	SeNPs improve cognition by increasing the brain's antioxidant capacity, which inhibits Aβ aggregation pathways	[16]
	The combined therapy of SeNPs and stem cells	STZ-induced AD model	By lowering the accumulation of Aβ and raising the level of BDNF	SeNPs improve stem-cell-based therapy's ability to lessen cognitive deficits	[76]
Epilepsy	SeNPs	Pentylenetetrazole (PTZ)-mediated epileptic seizuresin mice	By inhibiting the apoptosis, oxidative stress and inflammatory response	Supplementation of SeNPs delays the onset and decreases the duration of tonic, myoclonic and generalized seizures	[35]
Parkinson's Disease	Glycine nano-selenium	MPTP induced neuronal abnormalities in a PD rat model	Decreased oxidative stress reduces neurocognitive disorders in rat brain	Improves behavior abnormalities and prevents the loss of dopaminergic neurons	[65]
Huntington's disease	Selenium nanoparticles	Transgenic HD models of *Caenorhabditis elegans*	SeNPs act by attenuating oxidative stress and inhibiting the aggregation of huntingtin proteins	Prevent neuronal damage and improves behavioral dysfunction	[50]

Disease	SeNPs	Animal model	Mechanism	Outcome	Reference
Amyotrophic Lateral Sclerosis	Selenium nanoparticle	—	Act by reducing oxidative stress	Inhibit amyloid-like aggregation of SOD1 in familial ALS	[77]
Spinal Cord Injury	TSIIA@SeNPs-APS	PC12 cells	Inhibited excessive ROS and reduced cell apoptosis	Potential therapeutic effects in the anti-oxidation therapy of SCI	[78]
	SeNPs@GM1/TMP	PC12 cells	Inhibiting excessive ROS and reducing cell apoptosis	Provide neuroprotective effect	[79]

Table 1.
Selenium nanoparticles showing promising results in neurodegenerative disease.

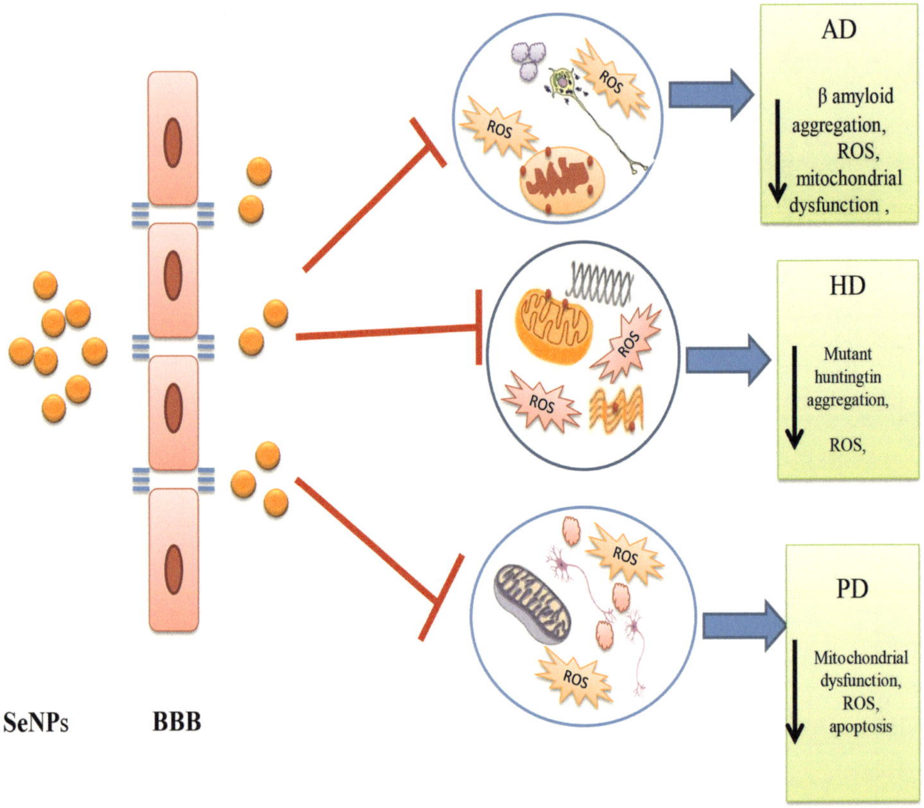

Figure 4.
Schematic representation of the mechanism of SeNPs in neurodegenerative disease.

in the form of nanoparticles is growing in popularity. Se's capacity to pass the BBB and suppress Aβ aggregation are two of its key effects in AD [82]. Se was found to have a beneficial impact on the activity of H_2O_2 absorption, the generation of intracellular ROS, and the aggregation of Aβ, which is found in the investigation of Se-containing clioquinol derivatives during the oxidation of Aβ caused by Cu^{2+} [83]. It is well recognized that the formation of improperly folded proteins in the brain and their aggregation is one of the primary causes of neurodegenerative disorders. Since Aβ may acquire several formats in AD, amyloid plaques are recognized due to improper protein folding and aggregation in the brain. According to studies, metal ions like Cu^{2+}, Zn^{2+}, and Fe^{2+} can bind to Aβ and co-localize with amyloid plaques in exceptionally high concentrations [84]. As a result, the use of metal chelators, like clioquinol (CQ), for AD treatment is of great interest. However, most chelators can also bind to other metal-containing proteins, which is undesirable and might disturb normal physiological functioning in the body [85].

SeNP has been shown to attach to Aβ, influence metal ions, and change their surfaces, such as ligands, charges, or reactivity [86]. As a result, it was demonstrated that l-Cys-modified SeNP (Cys-SeNP) could prevent $A\beta_{40}$ fibril formation caused by Zn^{2+} [87]. A multifunctional therapy for AD treatment has also been developed using chondroitin selenium sulfate (CS@Se) nanoparticles [63]. Chondroitin sulfate is a sulfated glycosaminoglycan (GAG) that binds to the protein core to generate the proteoglycan chondroitin sulfate (CSPG). The primary constituent of Perineuronal networks is

CSPG [88]. Aβ_{1-42} induced cytotoxicity in SH-SY5Y cells (human neuroblastoma) was prevented by CS@Se, which demonstrates its ability to prevent amyloid-β aggregation efficiently. Additionally, CS@Se lowered the hyperphosphorylation of tau (Ser396/Ser404), reduced the levels of ROS and MDA, and enhanced the levels of GSH-Px [63].

In HD, a hereditary neurodegenerative condition that results in the loss of brain cells and is linked to motor, cognitive, and behavioral impairments in adult patients, recent investigations have shown that Se has a protective function. Variable CAG trinucleotide repeats, found in the transcript that codes for the HTT, are a feature of this autosomal dominant disorder. The DNA repair system and mitochondrial malfunctions can be harmed by cumulative oxidative stress, which is thought to play a significant role in HD and other NDs. It has been demonstrated that sodium selenite can lessen mutant huntingtin aggregation and oxidize glutathione levels in HD mouse brains [34]. There is no viable treatment to stop HD from progressing or cure it. SeNP has been demonstrated to protect *C. elegans* from oxidative stress, ameliorate behavioral dysfunction, and prevent neuronal death at doses below 2 μM [50]. Nanoparticle therapy reduced the quantity of ROS, demonstrating their antioxidant properties, and stopped mutant HTT from aggregating in vivo.

After AD, PD is the second progressive neurological illness. The pathophysiology of PD is still unknown, although studies have revealed that oxidative stress, which causes neuronal death and apoptosis, is a key pathogenic component of PD [89]. An established neurotoxin called MPTP is used as a model for PD research. It has been demonstrated that MPTP can cause PD by boosting oxidative stress, which causes dopamine neurons to degenerate and causes neurobehavioral problems. By raising SOD and GSH-PX activity and lowering MDA levels, glycine-SeNP had an anti-oxidative effect on neurons. As a result, glycine-SeNP has the potential to treat Parkinson's disease [65]. Behavioral, molecular, and neurochemical alterations are the hallmarks of the persistent neurological condition known as epilepsy. Epilepsy is a neurological condition that affects between 0.5% and 1% of the world's population and is characterized by repeated, spontaneous seizures. Numerous conditions, such as cerebrovascular diseases, trauma, cancer, oxygen deprivation, infections, and genetic problems in brain development, can contribute to the development of seizures [90]. SeNP is a promising epilepsy treatment because of its excellent BBB-crossing capacity and few side effects. The malfunction of the mitochondria and endoplasmic reticulum causes oxidative stress, which increases the formation of free radicals and depletes neuronal antioxidant molecules. Oxidative stress is linked to both neuronal hyperexcitability and epileptogenesis [91]. The structure of selenoproteins and selenoenzymes include Se, which can inhibit ROS and, consequently, the onset of oxidative damage. Additionally, SeNP help to restore the levels of the neurotransmitters ACh, NE, DA, 5-HT, and GABA in brain tissue, which helps to restore neuronal connections and reduce apoptosis [35].

5. Diagnostic applications of SeNPs in CNS disorders

One of the main obstacles in treating and diagnosing CNS diseases is the inability of therapeutics to cross the BBB. A more comprehensive and accurate nanoparticle design is required to deliver therapeutic and diagnostic compounds to the CNS95 effectively. The evaluation and management of neurological conditions such as AD, PD, HD, head injuries, brain tumors, and epilepsy remain difficult tasks at this time. Numerous prospective medications have been studied to treat various neurological illnesses, but their efficacy is still constrained due to various difficulties [18].

For biomedical applications such as medication administration, bioimaging, and biosensing in CNS illnesses, a variety of inorganic NPs provide considerable efficiencies [92]. SeNPs offer a wide range of applications, including assessing and treating health-related problems that would otherwise be impossible to identify or address. Khalid et al. have examined the intrinsic fluorescence of SeNPs and their diagnostic potential. They discovered SeNPs' inherent fluorescence and its usefulness for nanoscale monitoring of cellular mechanisms. SeNPs' photoluminescence spectrum ranges from the visible to the near-infrared, making it useful for neuroblastoma cell tracking and their in vitro imaging. SeNPs have also been investigated as a peroxide biosensor. For accurate H_2O_2 sensing, Wang et al. produced semiconductor monoclinic SeNPs and they can be used to diagnose and assess the state of oxidative stress [93]. To produce an H_2O_2 sensor, plant-based rod-shaped SeNPs were created utilizing lemon fruit extract as a reducer and capping agent. Hydrogen peroxide sensing is a crucial component since it initiates a variety of cellular processes [94]. As selenium levels are said to be low in patients, SeNPs could be utilized to diagnose AD and HD.

Further research can be done on diagnostic techniques for detecting GPx or selenium levels. By focusing on many physiological pathways that control the metabolic status, inflammatory responses, oxidative defense system, and apoptosis, SeNPs functionalized with monoclonal antibodies (OX26) may be capable of defending against ischemic stroke [95]. The nano-based technique has significance for multiple sclerosis diagnosis and its involvement in treatments. The very sensitive DNA-carrier gold NPs-based coding technique can identify biomarkers in CSF or damaged brain tissue. This diagnostic test may be very helpful in diagnosing MS because radio imaging is the standard gold method for MS assessment (**Figure 5** and **Table 2**) [18].

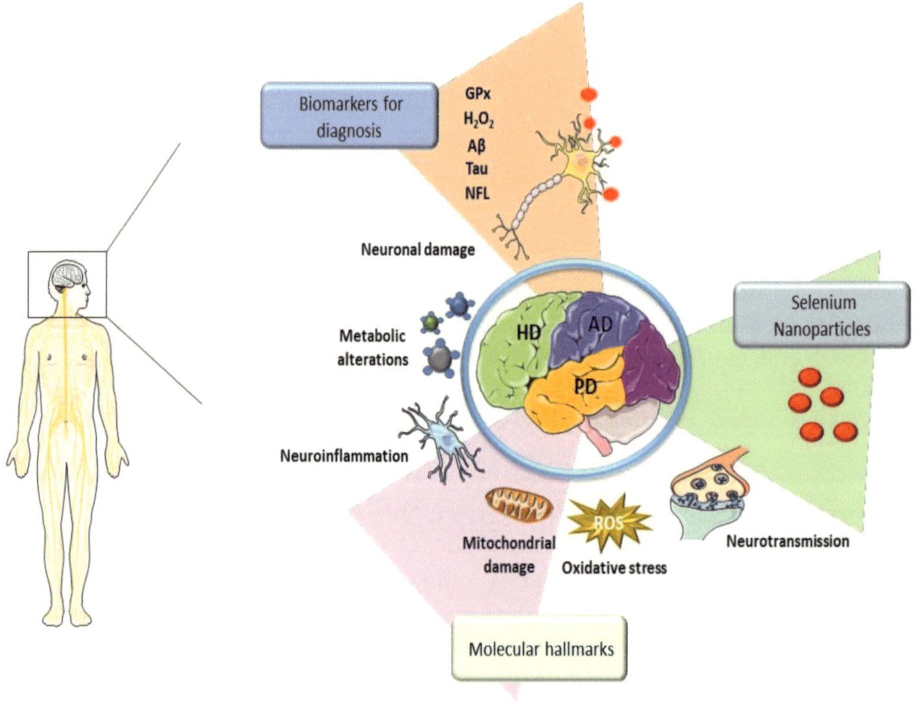

Figure 5.
Mechanism of selenium nanoparticles in the diagnosis of neurodegenerative disorders.

CNS disorder	SeNPs	Animal model	Outcome/inference	Reference
Brain diseases	Semiconductor monoclinic SeNPs	Mouse model	Useful for neuroblastoma cell tracking and in vitro imaging	[93]
	$Cu_{2-x}Se$ nanoparticles	Living mouse	Facilitated photothermal and single-photon emission computed tomography imaging tracking the restoration of the ruptured BBB by using ultrasound to infiltrate the brain	[96]
	Biogenicselenium nanorods	—	H_2O_2 sensor is developed, showing biomedical cellular peroxide sensing with low limits through the naked eye	[94]
Stroke	SeNPs functionalized with OX26	Murine model	For detecting GPx levels and being capable of defending against ischemic stroke	[95]
	OX26-PEG-Se NPs	Murine model	Using the crucially important for inflammation mTOR metabolic controller and associated signaling pathways such hippo, ERK5, Tsc1/Tsc2 complex, FoxO1, wnt/β- catenine signaling pathway, jak2/stat3 signaling pathways, and Adamts1	[41]
AD	State-of-the-art nanoparticles	Mouse model	Due to their efficacy as diagnostic tools, proteins like NFL, MMPs, p-tau217, and BACE are among the most potential biomarkers	[97]
	Selenium functionalized with Chondroitin sulfate	SH-SY5Y cells	Decreased the levels of ROSand MDA, increased the levelsof GSH-Px, and attenuatedthe hyperphosphorylation of tau by regulatingthe expression of GSK-3β	[63]

Table 2.
List of some selenium nanoparticles in the diagnosis of CNS diseases.

6. Future perspective

Nanotechnology-driven formulation techniques have enormous potential in the twenty-first century's drug research and discovery. Several nanotechnology-based preparations are available and can be easily accessible from the market, and this fact is not hidden from everyone. Se nanoparticles, an important trace element needed as a co-factor for various enzymes, have become an important tool in diagnostics and therapeutics for treating various illnesses, including neurodegenerative disorders. Indeed, the bulk of reported research has a significant SeNPs-based justification [98]. The development of nanotechnology has increased the number of possible therapeutic approaches to halt the course of AD. Oral/gastric barriers and the BBB are conventional neurotherapeutic obstacles that are effectively overcome by the proper design and production of NPs, improving the physicochemical characteristics of drugs in biological systems. However, the field of AD nano-therapeutics still has several limitations. There have been many in vitro experiments demonstrating the capability of SeNPs and its effectiveness, but there have been few in vivo trials. Therefore, future studies of these SeNPs may show systemic efficiency or toxicity in biological systems over the long run that can be contrasted to *in vitro* methods. Therefore, in the near future, potential, affordable AD treatments may result from evaluating the safety and efficacy of appropriate SeNPs in human clinical trials [99]. Recent research findings suggest that SeNPs can provide promising results in for the treating of HD through diets. In the future, an in-depth knowledge is required to know the mechanism of nano-Se in preventing the HD and their connection between physicochemical features and therapeutic potential will be beneficial for treating HD disease.

Furthermore, compared to other selenium species, the rational design of nanoSe may enhance dosage tolerance for HD therapy in the future [50]. It is abundantly evident that the current situation demands immediate and effective treatment for neuroprotection, neurorestorative, and neuroregeneration. Clinical translation in neurodegenerative disorders has become more challenging due to the lack of appropriate biomarkers, delayed diagnosis, incomplete understanding of molecular pathogeneses, lack of useful disease models, insufficient clinical protocol, and the generally asymptomatic nature of the disease. The deficiency of proper animal models that accurately reflect some crucial characteristics of ND in humans and a shortage of samples of patients are the two key limitations to therapeutic advancement. Therefore, the successful development of effective human disease-modifying medicines can be achieved through representative animal models.

Furthermore, inadequate knowledge of the molecular rationalization of aging and its biological impact and clinical consequences of neurodegenerative disorder contributes to delayed progress in translation. Success in therapeutic trials may result from shifting the focus from the primary pathogenic proteins to a plethora of disease-related proteins. In the future, using human CNS organoids to model neurological diseases is a realistic choice.3D brain organoids with the appropriate physicochemical signaling cues can be used to simulate patient-specific tissue patterns. Although organoids greatly improve the deep understanding of the development of brain and neurodegenerative illnesses, there are still several gaps in the field, including vascularization and non-neuronal cells. The targeting of particular brain cells in various NDs, such as in PD dopaminergic neurons are mainly targeted and this must be questioned when nanoformulation are prepared.

Additionally, it is important to consider adjusting pharmacokinetics and pharmacodynamics characteristics before administering NP [100]. Nanomedicine-based delivery systems raise concerns about their potential for toxicity, including the possibility of inflammation of neurons, excitotoxicity, mitochondrial and DNA damage, and some allergic reactions. Therefore, thoroughly researching the biocompatibility as well as biodegradability of nanodrugs is important [101]. The main function of the SeNPs in pharmacological defense against different types of inflammatory as well as oxidative stress-mediated situations is already discussed. However, nothing is known about how the SeNPs influence the pharmacokinetics and pharmacodynamics properties of selenoproteins. Most of the available research was not well structured and did not include comparisons to other Se sources. Future research should focus on understanding how selenoproteins contribute to the reported pharmacological effect and include relevant sources of Se [98].

7. Conclusion

Leading contributors to the world's disease burden are CNS illnesses, which encompass a wide range of brain diseases with both short- and long-term disabilities. Because of the shift in lifestyle and the swift, ongoing environmental degradation, CNS disorders like AD, PD, stroke, brain tumors, and neuroinflammation are distressingly damaging to humanity. With their complicated anatomy, specific microenvironment, and specificity to any foreign material such as drugs, BBB and BCSFB are the primary physiological barriers that pose a significant bottleneck for the effective therapy of CNS disorders and brain tumors. This presents the greatest hurdle to CNS drug discovery. SeNPs may help with the current issue of the lack of multifaceted drugs for various CNS disorders, which may contribute to various distinct biological mechanisms. SeNPs have led to cutting-edge nanoscale targeting approaches among the different treatment approaches. They are at the forefront of a new paradigm that could administer active agents with intriguing dynamics to treat these disorders. SeNPs offer superior medicinal qualities to selenium salts and lesser toxicity, despite their narrow therapeutic window. The current demand for effective nano-based treatments is concentrated on neuroprotection and restoration, which would greatly benefit from other nano-based strategies and advancements in the anatomy, pathology, and physiology of neuronal cells. Several nanoscale treatments (SeNPs) were found to treat neurological illnesses in AD, PD, and stroke models, including the suppression of Aβ oligomerization, reduction of ROS, and enhancement of functioning neural networks (**Figure 5**). SeNPs have allowed it to administer chemotherapy and antisense gene therapy in malignant brain tumors with pinpoint accuracy. This has led to a striking reduction in disease development in both *in vitro* and *in vivo* research.

SeNPs could completely alter how we address CNS-targeted therapeutics because of their competency to be nanoengineered so that the drug or carrier can encounter the BBB, diffuse inside the brain, and target specific cells or signaling systems for therapeutic delivery. This opens up new paths in the intervention of neurological diseases and has an extremely great prospect. It is very conceivable that SeNPs will alter how CNS disorders are treated. Shortly, the real objective of drastically increasing survival rates will be accomplished.

Abbreviations

AD	Alzheimer's disease
ALS	amyotrophic lateral sclerosis
APP	amyloid protein precursor
BBB	blood-brain barrier
BCSFB	blood-cerebrospinal fluid barrier
BDNF	brain-derived neurotropic factor
CNS	central nervous system
CSF	cerebrospinal fluid
DRG	dorsal root ganglion
FoxO1	Forkhead box protein O1
GSH-Px	glutathione peroxidase
GSK-3β	glycogen synthase kinase-3 beta
HO-1	heme oxygenase-1
HTT	Huntingtin protein
MDA	Malondialdehyde
MPTP	1-methyl-4-phenyl-1,2,3,6-tetrahydropyridine
mTORC1	mammalian target of rapamycin complex 1
NDs	neurodegenerative diseases
Nrf2	nuclear factor erythroid 2-related factor 2
OGD/R	oxygen glucose deprivation/re-oxygenation
PD	Parkinson's disease
PLGA	poly-lactide-co-glycolide
p-mTOR	phosphorylated mammalian target of rapamycin

Se-CQDs	selenium-doped carbon quantum dots
SeNPs	selenium nanoparticles
SOD	superoxide dismutase
TSC	tuberous sclerosis
ULK1	Unc-51 like autophagy activating kinase1
WHO	World Health Organization
Wnt/ß-catenin	Wingless/Integrated ß-catenin
Yap1	Yes-associated protein 1

Author details

Jonaid Ahmad Malik[1], Jeba AjgarAnsari[2], Sakeel Ahmed[3], Archana Rani[4], Shabana Yasmeen Ansari[5] and Sirajudheen Anwar[6*]

1 Department of Biomedical Engineering, Indian Institute of Technology Ropar, Rupnagar, Punjab, India

2 Department of Pharmaceutics, Government College of Pharmacy, Aurangabad, Dr. Babasaheb Ambedkar Marathwada University, Maharashtra, India

3 Department of Pharmacology and Toxicology, National Institute of Pharmaceutical Education and Research, Ahmedabad, India

4 Department of Pharmaceutical sciences, Guru Nanak Dev University, Amritsar, India

5 Pharmaceutical Unit, Department of Electronics, Chemistry and Industrial Engineering, University of Messina, Messina, Italy

6 Department of Pharmacology and Toxicology, College of Pharmacy, University of Hail, Hail, Saudi Arabia

*Address all correspondence to: si.anwar@uoh.edu.sa

References

[1] Jagaran K, Singh M. Nanomedicine for neurodegenerative disorders: Focus on Alzheimer's and Parkinson's diseases. International Journal of Molecular Sciences. 2021;**22**(16):9082. DOI: 10.3390/IJMS22169082

[2] Checkoway H, Lundin JI, Kelada SN. Neurodegenerative diseases. IARC Scientific Publications. 2011;**163**:407-419. DOI: 10.1071/rdv24n1ab251

[3] Tysnes OB, Storstein A. Epidemiology of Parkinson's disease. Journal of Neural Transmission. 2017;**124**(8):901-905. DOI: 10.1007/S00702-017-1686-Y

[4] Rocca WA. The burden of Parkinson's disease: A worldwide perspective. Lancet Neurology. 2018;**17**(11):928-929. DOI: 10.1016/S1474-4422(18)30355-7

[5] Talbott EO, Malek AM, Lacomis D. The epidemiology of amyotrophic lateral sclerosis. Handbook of Clinical Neurology. 2016;**138**:225-238. DOI: 10.1016/B978-0-12-802973-2.00013-6

[6] Hanif S, Muhammad P, Chesworth R, et al. Nanomedicine-based immunotherapy for central nervous system disorders. Acta Pharmacologica Sinica. 2020;**41**(7):936-953. DOI: 10.1038/s41401-020-0429-z

[7] Chapman CD, Frey WH, Craft S, et al. Intranasal treatment of central nervous system dysfunction in humans. Pharmaceutical Research. 2013;**30**(10):2475-2484. DOI: 10.1007/S11095-012-0915-1/FIGURES/1

[8] Durães F, Pinto M, Sousa E. Old drugs as new treatments for neurodegenerative diseases. Pharmaceuticals (Basel). 2018;**11**(2):44. DOI: 10.3390/PH11020044

[9] Di Paolo M, Papi L, Gori F, Turillazzi E. Natural products in neurodegenerative diseases: A great promise but an ethical challenge. International Journal of Molecular Sciences. 2019;**20**(20):5170. DOI: 10.3390/IJMS20205170

[10] Paul SM, Mytelka DS, Dunwiddie CT, et al. How to improve R&D productivity: The pharmaceutical industry's grand challenge. Nature Reviews. Drug Discovery. 2010;**9**(3):203-214. DOI: 10.1038/nrd3078

[11] Hyman SE. Revolution stalled. Science Translational Medicine. 2012;**4**:155. DOI: 10.1126/SCITRANSLMED.3003142

[12] Pankevich DE, Altevogt BM, Dunlop J, et al. Improving and accelerating drug development for nervous system disorders. Neuron. 2014;**84**(3):546-553. DOI: 10.1016/J.NEURON.2014.10.007

[13] Banks WA. From blood–brain barrier to blood–brain interface: new opportunities for CNS drug delivery. Nature Reviews Drug Discovery. 2016;**15**(4):275-292. DOI: 10.1038/nrd.2015.21

[14] Li G, Shao K, Umeshappa CS. Recent progress in blood-brain barrier transportation research. Brain Targeted Drug Delivery Systems: A Focus on Nanotechnology and Nanoparticulates. 2019:33-51. DOI: 10.1016/B978-0-12-814001-7.00003-2

[15] Rajeshkumar S, Ganesh L, Santhoshkumar J. Selenium nanoparticles as therapeutic agents in neurodegenerative diseases. Nanobiotechnology in Neurodegenerative Diseases. 2019:209-224. DOI: 10.1007/978-3-030-30930-5_8/COVER/

[16] Gholamigeravand B, Shahidi S, Amiri I, Samzadeh-kermani A, Abbasalipourkabir R, Soleimani AS. Administration of selenium nanoparticles reverses streptozotocin-induced neurotoxicity in the male rats. Metabolic Brain Disease. 2021;**36**(6):1259-1266. DOI: 10.1007/S11011-021-00713-8/FIGURES/7

[17] Nguyen TT, Dung Nguyen TT, Vo TK, et al. Nanotechnology-based drug delivery for central nervous system disorders. Biomedicine & Pharmacotherapy. 2021;**143**(July):112117. DOI: 10.1016/j.biopha.2021.112117

[18] Naqvi S, Panghal A, Flora SJS. Nanotechnology: A promising approach for delivery of neuroprotective drugs. Frontiers in Neuroscience. 2020;**14**(June):1-26. DOI: 10.3389/fnins.2020.00494

[19] Nazıroğlu M. Role of selenium on calcium signaling and oxidative stress-induced molecular pathways in epilepsy. Neurochemical Research. 2009;**34**(12):2181-2191. DOI: 10.1007/S11064-009-0015-8

[20] Nazıroğlu M, Muhamad S, Pecze L. Nanoparticles as potential clinical therapeutic agents in Alzheimer's disease: Focus on selenium nanoparticles. Expert Review of Clinical Pharmacology. 2017;**10**(7):773-782. DOI: 10.1080/17512433.2017.1324781

[21] Nogueira CW, Rocha JBT. Toxicology and pharmacology of selenium: emphasis on synthetic organoselenium compounds. Archives of Toxicology. 2011;**85**(11):1313-1359. DOI: 10.1007/S00204-011-0720-3

[22] Naziroğlu M. TRPM2 cation channels, oxidative stress and neurological diseases: Where are we now? Neurochemical Research. 2011;**36**(3):355-366. DOI: 10.1007/S11064-010-0347-4

[23] Nazıroglu M, Demirdas A. Psychiatric disorders and TRP channels: Focus on psychotropic drugs. Current Neuropharmacology. 2015;**13**(2):248-257. DOI: 10.2174/1570159X13666150304001606

[24] Vural H, Demirin H, Kara Y, Eren I, Delibas N. Alterations of plasma magnesium, copper, zinc, iron and selenium concentrations and some related erythrocyte antioxidant enzyme activities in patients with Alzheimer's disease. Journal of Trace Elements in Medicine and Biology. 2010;**24**(3):169-173. DOI: 10.1016/J.JTEMB.2010.02.002

[25] Rita Cardoso B, Silva Bandeira V, Jacob-Filho W, Franciscato Cozzolino SM. Selenium status in elderly: Relation to cognitive decline. Journal of Trace Elements in Medicine and Biology. 2014;**28**(4):422-426. DOI: 10.1016/J.JTEMB.2014.08.009

[26] Koç ER, İlhan A, Aytürk Z, et al. A comparison of hair and serum trace elements in patients with Alzheimer disease and healthy participants. Turkish Journal of Medical Sciences. 2015;**45**(5):1034-1039. DOI: 10.3906/SAG-1407-67

[27] Balaban H, Nazıroğlu M, Demirci K, Övey İS. The protective role of selenium on scopolamine-induced memory impairment, oxidative stress, and apoptosis in aged rats: The involvement of TRPM2 and TRPV1 channels. Molecular Neurobiology. 2017;**54**(4):2852-2868. DOI: 10.1007/S12035-016-9835-0

[28] Schweizer U, Bräuer AU, Köhrle J, Nitsch R, Savaskan NE. Selenium and brain function: A poorly recognized liaison. Brain Research. Brain Research

Reviews. 2004;**45**(3):164-178. DOI: 10.1016/J.BRAINRESREV.2004.03.004

[29] Fernandes AP, Gandin V. Selenium compounds as therapeutic agents in cancer. Biochimica et Biophysica Acta. 2015;**1850**(8):1642-1660. DOI: 10.1016/J.BBAGEN.2014.10.008

[30] Fan N, Sikand P, Donnelly DF, Ma C, LaMotte RH. Increased Na+ and K+ currents in small mouse dorsal root ganglion neurons after ganglion compression. Journal of Neurophysiology. 2011;**106**(1):211. DOI: 10.1152/JN.00065.2011

[31] Huijun Y, Jiarui L, Tonghan L. Effects of nano red elemental selenium on sodium currents in rat dorsal root ganglion neurons. Annual International Conference of the IEEE Engineering in Medicine and Biology—Proceedings. 2006:896-899. DOI: 10.1109/IEMBS.2006.260758

[32] Amini SM, Mahabadi VP. Selenium nanoparticles role in organ systems functionality and disorder. Nanomedicine Research Journal. 2018;**3**(3):117-124. DOI: 10.22034/NMRJ.2018.03.001

[33] Cong W, Bai R, Li Y, Wang L, Chen C. Biological and medical applications of materials and interfaces selenium nanoparticles as an efficient nanomedicine for the therapy of Huntington's disease selenium nanoparticles as an efficient nanomedicine for the therapy of Huntington's disease. 2019;**11**(38):34725-34735. DOI: 10.1021/acsami.9b12319

[34] Solovyev N, Drobyshev E, Bjørklund G, Dubrovskii Y, Lysiuk R, Rayman MP. Selenium, selenoprotein P, and Alzheimer's disease: Is there a link? Free Radical Biology & Medicine. 2018;**127**:124-133. DOI: 10.1016/J.FREERADBIOMED.2018.02.030

[35] Yuan X, Fu Z, Ji P, et al. Selenium nanoparticles pre-treatment reverse behavioral, oxidative damage, neuronal loss and neurochemical alterations in pentylenetetrazole-induced epileptic seizures in mice. International Journal of Nanomedicine. 2020;**15**:6339. DOI: 10.2147/IJN.S259134

[36] Vicente-Zurdo D, Romero-Sánchez I, Rosales-Conrado N, León-González ME, Madrid Y. Ability of selenium species to inhibit metal-induced Aβ aggregation involved in the development of Alzheimer's disease. Analytical and Bioanalytical Chemistry. 2020;**412**(24):6485-6497. DOI: 10.1007/S00216-020-02644-2

[37] Roman M, Jitaru P, Barbante C. Selenium biochemistry and its role for human health. Metallomics. 2014;**6**(1):25-54. DOI: 10.1039/C3MT00185G

[38] Solovyev N. Selenoprotein P and its potential role in Alzheimer's disease. Hormones (Athens, Greece). 2020;**19**(1):73-79. DOI: 10.1007/S42000-019-00112-W

[39] Ellwanger JH, Franke SIR, Bordin DL, Prá D, Henriques JAP. Biological functions of selenium and its potential influence on Parkinson's disease. Anais da Academia Brasileira de Ciências. 2016;**88**(3):1655-1674. DOI: 10.1590/0001-3765201620150595

[40] Zhang X, Liu RP, Cheng WH, Zhu JH. Prioritized brain selenium retention and selenoprotein expression: Nutritional insights into Parkinson's disease. Mechanisms of Ageing and Development. 2019;**180**:89-96. DOI: 10.1016/J.MAD.2019.04.004

[41] Amani H, Habibey R, Shokri F, et al. Selenium nanoparticles for targeted

stroke therapy through modulation of inflammatory and metabolic signaling. Scientific Reports. 2019;**9**(1):1-15. DOI: 10.1038/s41598-019-42633-9

[42] Khan M, Rutten BPF, Kim MO. MST1 regulates neuronal cell death via JNK/Casp3 signaling pathway in HFD mouse brain and HT22 cells. International Journal of Molecular Sciences. 2019;**20**(10):2504. DOI: 10.3390/IJMS20102504

[43] Nicolas CS, Amici M, Bortolotto ZA, et al. The role of JAK-STAT signaling within the CNS. JAK-STAT. 2013;**2**(1):e22925. DOI: 10.4161/JKST.22925

[44] Gupta J, Fatima MT, Islam Z, Khan RH, Uversky VN, Salahuddin P. Nanoparticle formulations in the diagnosis and therapy of Alzheimer's disease. International Journal of Biological Macromolecules. 2019;**130**:515-526. DOI: 10.1016/J.IJBIOMAC.2019.02.156

[45] Yang L, Wang W, Chen J, Wang N, Zheng G. A comparative study of resveratrol and resveratrol-functional selenium nanoparticles: Inhibiting amyloid β aggregation and reactive oxygen species formation properties. Journal of Biomedical Materials Research. Part A. 2018;**106**(12):3034-3041. DOI: 10.1002/JBM.A.36493

[46] Huo X, Zhang Y, Jin X, Li Y, Zhang L. A novel synthesis of selenium nanoparticles encapsulated PLGA nanospheres with curcumin molecules for the inhibition of amyloid β aggregation in Alzheimer's disease. Journal of Photochemistry and Photobiology. B. 2019;**190**:98-102. DOI: 10.1016/J.JPHOTOBIOL.2018.11.008

[47] Sun D, Zhang W, Yu Q, et al. Chiral penicillamine-modified selenium nanoparticles enantioselectively inhibit metal-induced amyloid β aggregation for treating Alzheimer's disease. Journal of Colloid and Interface Science. 2017;**505**:1001-1010. DOI: 10.1016/J.JCIS.2017.06.083

[48] Yang R, Zheng Y, Wang Q, Zhao L. Curcumin-loaded chitosan–bovine serum albumin nanoparticles potentially enhanced Aβ 42 phagocytosis and modulated macrophage polarization in Alzheimer's disease. Nanoscale Research Letters. 2018;**13**(1):1-9. DOI: 10.1186/S11671-018-2759-Z/FIGURES/6

[49] Yang L, Wang N, Zheng G. Enhanced effect of combining chlorogenic acid on selenium nanoparticles in inhibiting amyloid β aggregation and reactive oxygen species formation in vitro. Nanoscale Research Letters. 2018;**13**(1):265-76. DOI: 10.1186/S11671-018-2720-1

[50] Cong W, Bai R, Li YF, Wang L, Chen C. Selenium nanoparticles as an efficient nanomedicine for the therapy of Huntington's disease. ACS Applied Materials & Interfaces. 2019;**11**(38):34725-34735. DOI: 10.1021/ACSAMI.9B12319

[51] Luo W, Wang Y, Lin F, et al. Selenium-doped carbon quantum dots efficiently ameliorate secondary spinal cord injury via scavenging reactive oxygen species. International Journal of Nanomedicine. 2020;**15**:10113-10125. DOI: 10.2147/IJN.S282985

[52] Chamorro Á, Dirnagl U, Urra X, Planas AM. Neuroprotection in acute stroke: Targeting excitotoxicity, oxidative and nitrosative stress, and inflammation. Lancet Neurology. 2016;**15**(8):869-881. DOI: 10.1016/S1474-4422(16)00114-9

[53] Sekerdag E, Solaroglu I, Gursoy-Ozdemir Y. Cell death mechanisms

in stroke and novel molecular and cellular treatment options. Current Neuropharmacology. 2018;**16**(9): 1396-1415. DOI: 10.2174/ 1570159X16666180302115544

[54] Turovskaya MV, Gaidin SG, Vedunova MV, Babaev AA, Turovsky EA. BDNF overexpression enhances the preconditioning effect of brief episodes of hypoxia, promoting survival of GABAergic neurons. Neuroscience Bulletin. 2020;**36**(7):733-760. DOI: 10.1007/S12264-020-00480-Z

[55] Gaidin SG, Turovskaya MV, Gavrish MS, et al. The selective BDNF overexpression in neurons protects neuroglial networks against OGD and glutamate-induced excitotoxicity. The International Journal of Neuroscience. 2020;**130**(4):363-383. DOI: 10.1080/00207454.2019.1691205

[56] Turovskaya MV, Zinchenko VP, Babaev AA, Epifanova EA, Tarabykin VS, Turovsky EA. Mutation in the Sip1 transcription factor leads to a disturbance of the preconditioning of AMPA receptors by episodes of hypoxia in neurons of the cerebral cortex due to changes in their activity and subunit composition. The protective effects of interleukin-10. Archives of Biochemistry and Biophysics. 2018;**654**:126-135. DOI: 10.1016/J.ABB.2018.07.019

[57] Ahmed Ragab Abou Zaid O, Mohammed El-sonbaty S, Ezz El-arab Mohammed Barakat W. Ameliorative effect of selenium nanoparticles and ferulic acid on acrylamide-induced neurotoxicity in rats. Annals of Medical and Biomedical Sciences. 2017;**3**(2):35-45. Available from: https://www.researchgate.net/publication/335740590

[58] Turovsky EA, Mal'tseva VN, Sarimov RM, Simakin AV, Gudkov SV, Plotnikov EY. Features of the cytoprotective effect of selenium nanoparticles on primary cortical neurons and astrocytes during oxygen–glucose deprivation and reoxygenation. Scientific Reports. 2022;**12**(1):1-16. DOI: 10.1038/s41598-022-05674-1

[59] Yang J, Huang J, Shen C, et al. Resveratrol treatment in different time-attenuated neuronal apoptosis after oxygen and glucose deprivation/reoxygenation via enhancing the activation of Nrf-2 signaling pathway in vitro. Cell Transplantation. 2018;**27**(12):1789. DOI: 10.1177/0963689718780930

[60] Gao F, Zhao J, Liu P, et al. Preparation and in vitro evaluation of multi-target-directed selenium-chondroitin sulfate nanoparticles in protecting against the Alzheimer's disease. International Journal of Biological Macromolecules. 2019;**142**:265-276. DOI: 10.1016/j.ijbiomac.2019.09.098

[61] Guo X, Lie Q, Liu Y, Jia Z, Gong Y, Yuan X, et al. Multifunctional selenium quantum dots for the treatment of Alzheimer's disease by reducing Aβ-neurotoxicity and oxidative stress and alleviate neuroinflammation. ACS Applied Materials & Interfaces. 2021;**13**(26):30261-30273

[62] Yang L, Sun J, Xie W, Liu Y, Liu J. Dual-functional selenium nanoparticles bind to and inhibit amyloid β fiber formation in Alzheimer's disease. Journal of Materials Chemistry B. 2017;**5**(30):5954-5967. DOI: 10.1039/ C6TB02952C

[63] Gao F, Zhao J, Liu P, et al. Preparation and in vitro evaluation of multi-target-directed selenium-chondroitin sulfate nanoparticles in protecting against the Alzheimer's disease. International Journal of Biological Macromolecules.

2020;**142**:265-276. DOI: 10.1016/J.IJBIOMAC.2019.09.098

[64] Soleimani Asl S, Amiri I, Samzadeh-kermani A, Abbasalipourkabir R, Gholamigeravand B, Shahidi S. Chitosan-coated Selenium nanoparticles enhance the efficiency of stem cells in the neuroprotection of streptozotocin-induced neurotoxicity in male rats. The International Journal of Biochemistry & Cell Biology. 2021;**141**:106089. DOI: 10.1016/J.BIOCEL.2021.106089

[65] Yue D, Zeng C, Okyere SK, Chen Z, Hu Y. Glycine nano-selenium prevents brain oxidative stress and neurobehavioral abnormalities caused by MPTP in rats. Journal of Trace Elements in Medicine and Biology. 2021;**64**:126680 DOI: 10.1016/J.JTEMB.2020.126680

[66] Salaramoli S, Joshaghani H, Hashemy SI. Selenium effects on oxidative stress-induced calcium signaling pathways in Parkinson's disease. Indian Journal of Clinical Biochemistry. 2022;**37**(3):257-266. DOI: 10.1007/S12291-022-01031-1

[67] Bates GP, Dorsey R, Gusella JF, et al. Huntington disease. Nature Reviews Disease Primers. 2015;**1**(1):1-21. DOI: 10.1038/nrdp.2015.5

[68] Kumar A, Ratan RR. Oxidative stress and Huntington's disease: The good, the bad, and the ugly. Journal of Huntington's Disease. 2016;**5**(3):217-237. DOI: 10.3233/JHD-160205

[69] Al-Chalabi A, Van Den Berg LH, Veldink J. Gene discovery in amyotrophic lateral sclerosis: Implications for clinical management. Nature Reviews. Neurology. 2016;**13**(2):96-104. DOI: 10.1038/nrneurol.2016.182

[70] Paez-Colasante X, Figueroa-Romero C, Rumora AE, et al. Cytoplasmic TDP43 binds microRNAs: New disease targets in amyotrophic lateral sclerosis. Frontiers in Cellular Neuroscience. 2020;**14**:117. DOI: 10.3389/FNCEL.2020.00117/BIBTEX

[71] Deng HX, Hentati A, Tainer JA, et al. Amyotrophic lateral ssclerosis and structural defects in Cu,Zn superoxide dismutase. Science (80-). 1993;**261**(5124):1047-1051. DOI: 10.1126/SCIENCE.8351519

[72] Cunha-Oliveira T, Montezinho L, Mendes C, et al. Oxidative stress in amyotrophic lateral sclerosis: Pathophysiology and opportunities for pharmacological intervention. Oxidative Medicine and Cellular Longevity. 2020;**2020**. DOI: 10.1155/2020/5021694

[73] Amporndanai K, Rogers M, Watanabe S, Yamanaka K, O'Neill PM, Hasnain SS. Novel selenium-based compounds with therapeutic potential for SOD1-linked amyotrophic lateral sclerosis. eBioMedicine. 2020;**59**:102980. DOI: 10.1016/J.EBIOM.2020.102980

[74] Song Z, Liu T, Chen T. Overcoming blood–brain barrier by HER2-targeted nanosystem to suppress glioblastoma cell migration, invasion and tumor growth. Journal of Materials Chemistry B. 2018;**6**(4):568-579. DOI: 10.1039/C7TB02677C

[75] Yin T, Yang L, Liu Y, Zhou X, Sun J, Liu J. Sialic acid (SA)-modified selenium nanoparticles coated with a high blood–brain barrier permeability peptide-B6 peptide for potential use in Alzheimer's disease. Acta Biomaterialia. 2015;**25**:172-183. DOI: 10.1016/J.ACTBIO.2015.06.035

[76] Gholamigeravand B, Shahidi S, Afshar S, et al. Synergistic effects of adipose-derived mesenchymal stem cells and selenium nanoparticles on streptozotocin-induced memory

impairment in the rat. Life Sciences. 2021;**272**:119246. DOI: 10.1016/J.LFS.2021.119246

[77] Pichla M, Bartosz G, Sadowska-Bartosz I. The Antiaggregative and antiamyloidogenic properties of nanoparticles: A promising tool for the treatment and diagnostics of neurodegenerative diseases. Oxidative Medicine and Cellular Longevity. 2020;**2020**. DOI: 10.1155/2020/3534570

[78] Rao S, Lin Y, Lin R, et al. Traditional Chinese medicine active ingredients-based selenium nanoparticles regulate antioxidant selenoproteins for spinal cord injury treatment. Journal of Nanobiotechnology. 2022;**20**(1):1-15. DOI: 10.1186/S12951-022-01490-X

[79] Rao S, Lin Y, Du Y, et al. Designing multifunctionalized selenium nanoparticles to reverse oxidative stress-induced spinal cord injury by attenuating ROS overproduction and mitochondria dysfunction. Journal of Materials Chemistry B. 2019;**7**(16):2648-2656. DOI: 10.1039/C8TB02520G

[80] Khandel P, Yadaw RK, Soni DK, Kanwar L, Shahi SK. Biogenesis of metal nanoparticles and their pharmacological applications: present status and application prospects. Journal of Nanostructure in Chemistry. 2018;**8**(3):217-254. DOI: 10.1007/S40097-018-0267-4

[81] El-Zayat MM, Eraqi MM, Alrefai H, et al. The antimicrobial, antioxidant, and anticancer activity of greenly synthesized selenium and zinc composite nanoparticles using *Ephedra aphylla* extract. Biomolecules. 2021;**11**(3):470. DOI: 10.3390/BIOM11030470

[82] Magaldi S, Mata-Essayag S, Hartung De Capriles C, et al. Well diffusion for antifungal susceptibility testing. International Journal of Infectious Diseases. 2004;**8**(1):39-45. DOI: 10.1016/J.IJID.2003.03.002

[83] Sathiyamoorthy P, Lugasi-Evgi H, Van-Damme P, Abu-Rabia A, Gopas J, Golan-Goldhirsh A. Larvicidal activity in desert plants of the negev and bedouin market plant products. International Journal of Pharmacognosy. 2008;**35**(4):265-273. DOI: 10.1076/PHBI.35.4.265.13314

[84] Al-Hussaini R, Mahasneh AM. Microbial growth and quorum sensing antagonist activities of herbal plants extracts. Molecules. 2009;**14**(9):3425-3435. DOI: 10.3390/MOLECULES14093425

[85] Cuajungco MP, Fagét KY, Huang X, Tanzi RE, Bush AI. Metal chelation as a potential therapy for Alzheimer's disease. Annals of the New York Academy of Sciences. 2000;**920**(1):292-304. DOI: 10.1111/J.1749-6632.2000.TB06938.X

[86] Schomburg L. Dietary selenium and human health. Nutrients. 2016;**9**(1):22. DOI: 10.3390/NU9010022

[87] Zhuang C, Yao D, Li F, Zhang K, Feng Q, Gan Z. Study of micron-thick MgB2 films on niobium substrates. Superconductor Science and Technology. 2007;**20**(3):287. DOI: 10.1088/0953-2048/20/3/030

[88] Kwok JCF, Carulli D, Fawcett JW. In vitro modeling of perineuronal nets: Hyaluronan synthase and link protein are necessary for their formation and integrity. Journal of Neurochemistry. 2010;**114**(5):1447-1459. DOI: 10.1111/J.1471-4159.2010.06878.X

[89] Fedorova TN, Logvinenko AA, Poleshchuk VV, Illarioshkin SN. The

state of systemic oxidative stress during Parkinson's disease. Neurochemical Journal. 2017;**11**(4):340-345. DOI: 10.1134/S1819712417040031

[90] Liu S, Yu W, Lü Y. The causes of new-onset epilepsy and seizures in the elderly. Neuropsychiatric Disease and Treatment. 2016;**12**:1425. DOI: 10.2147/NDT.S107905

[91] Zhu X, Dong J, Han B, et al. Neuronal nitric oxide synthase contributes to PTZ kindling epilepsy-induced hippocampal endoplasmic reticulum stress and oxidative damage. Frontiers in Cellular Neuroscience. 2017;**11**:377. DOI: 10.3389/FNCEL.2017.00377/BIBTEX

[92] Feng L, Wang H, Xue X. Recent progress of nanomedicine in the treatment of central nervous system diseases. Advances in Therapy. 2020;**3**(5):1900159. DOI: 10.1002/ADTP.201900159

[93] Guan B, Yan R, Li R, Zhang X. Selenium as a pleiotropic agent for medical discovery and drug delivery. International Journal of Nanomedicine. 2018;**13**:7473-7490. DOI: 10.2147/IJN.S181343

[94] Sawant VJ, Sawant VJ. Biogenic capped selenium nano rods as naked eye and selective hydrogen peroxide spectrometric sensor. Sensing and Bio-Sensing Research. 2020;**27**(December):100314. DOI: 10.1016/j.sbsr.2019.100314

[95] Nayak V, Singh KR, Singh AK, Singh RP. Potentialities of selenium nanoparticles in biomedical science. New Journal of Chemistry. 2021;**45**(6):2849-2878. DOI: 10.1039/d0nj05884j

[96] Zhang H, Wang T, Qiu W, et al. Monitoring the opening and recovery of the blood-brain barrier with noninvasive molecular imaging by biodegradable ultrasmall Cu2- Se nanoparticles. Nano Letters. 2018;**18**(8):4985-4992. DOI: 10.1021/ACS.NANOLETT.8B01818/SUPPL_FILE/NL8B01818_SI_001.PDF

[97] Cano A, Turowski P, Ettcheto M, et al. Nanomedicine-based technologies and novel biomarkers for the diagnosis and treatment of Alzheimer's disease: From current to future challenges. Journal of Nanobiotechnology. 2021;**19**(1):1-30. DOI: 10.1186/S12951-021-00864-X

[98] Khurana A, Tekula S, Saifi MA, Venkatesh P, Godugu C. Therapeutic applications of selenium nanoparticles. Biomedicine & Pharmacotherapy. 2019;**111**(December):802-812. DOI: 10.1016/j.biopha.2018.12.146

[99] Karthivashan G, Ganesan P, Park SY, Kim JS, Choi DK. Therapeutic strategies and nano-drug delivery applications in management of ageing Alzheimer's disease. Drug Delivery. 2018;**25**(1):307-320. DOI: 10.1080/10717544.2018.1428243

[100] Mukherjee S, Madamsetty VS, Bhattacharya D, Roy Chowdhury S, Paul MK, Mukherjee A. Recent advancements of nanomedicine in neurodegenerative disorders theranostics. Advanced Functional Materials. 2020;**30**(35):2003054. DOI: 10.1002/ADFM.202003054

[101] Yavarpour-Bali H, Pirzadeh M, Ghasemi-Kasman M. Curcumin-loaded nanoparticles: A novel therapeutic strategy in treatment of central nervous system disorders. International Journal of Nanomedicine. 2019;**14**:4449. DOI: 10.2147/IJN.S208332

Chapter 5

Antibacterial Strategies: Photodynamic and Photothermal Treatments Based on Carbon-Based Materials

David Giancarlo García Vélez, Karina Janneri Lagos Álvarez and María Paulina Romero Obando

Abstract

The problem of bacterial resistance is based on the abuse of antibiotics such as trimethoprim, fluoroquinolones, chloramphenicol, and some carbapenems. For this reason, conventional treatments to treat diseases caused by bacteria have become ineffective. Therefore, developing new therapies with multifunctional materials to combat bacteria is mandatory. In this context, photodynamic treatment (PDT) and photothermal treatment (PTT) have been proposed to combat bacteria. These light-stimulated treatments are minimally invasive and have a low incidence of side effects. In addition, they are simple, fast, and profitable. The antibacterial effect of PDT, PTT, or synchronic PDT/PTT arises from the generation of reactive oxygen species (ROS) and heat caused by a photoactivated specific photosensitizer (PS) and photothermal agents (PTAs), respectively. The effectiveness of photoinduced treatment depends, among other parameters, on the nature and concentration of the PS/PTAs, light dose, and irradiation wavelength. PS/PTAs based on carbon-based materials (CBMs), such as graphene oxide, reduced graphene oxide, carbon dots, and carbon nanotubes as antibacterial agents, will be discussed in this chapter. These CBMs have emerged as excellent antibacterial alternatives due to their excellent physicochemical properties, biocompatibility, low toxicity in the dark, specificity, and excellent response to light. Moreover, several composites and hybrids employing polymers, metal oxides, and metals have been tested to enhance the antibacterial activity of the CBMs.

Keywords: photodynamic therapy, photothermal therapy, carbon-based materials, photosensitizers, photothermal agents

1. Introduction

Food and water for human consumption, medical equipment, lung walls, upper respiratory tract, and external wounds, that accommodate a small number of bacteria, can generate a potential health risk due to their high adaptability and bacterial proliferation [1]. Infections caused by bacterial pathogens have claimed many human and animal lives, mainly when the development of antibacterial treatments is deficient, for example, the plague pandemic coined as the "black death" in medieval Europe caused by the bacterium *Yersinia pestis*, cholera (*Vibrio cholera*), and tuberculosis (*Mycobacterium tuberculosis*), among others. Likewise, viral pathogens such as HIV and COVID-19 can suppress the immune system, leading to enhanced conditions for coinfection with bacterial pathogens [2, 3]. The shortage of drinking water and medical procedures exposed to bacterial pathogens in the air, or contaminated medical instruments, have become problems of great interest to the world because they are sources that produce bacterial infections that can lead to the death of people, mainly in developing countries [3, 4]. Bacterial infections significantly affect the health of people with cancer, diabetes, and HIV and transplant patients, a high-risk population. Likewise, they considerably affect the wound healing mechanism, reaching the amputation of affected regions or limbs [5, 6].

Antibiotics emerged in the previous era (1940–1980, the "Golden Age" of antibiotics) as an effective treatment for bacterial infectious diseases caused mainly by *Streptococcus pneumoniae* and *Staphylococcus aureus* [3, 6]. Undoubtedly, the general administration of antibiotics revolutionized the treatment of infections caused by pathogenic bacteria, saving countless lives, and they are still considered of great importance in modern bacterial therapies [1, 6]. Antibiotics are a subgroup of antimicrobial agents classified according to their effect, mechanisms of action, and spectrum. Antibiotics are designed to inhibit the growth and multiplication of susceptible bacterial cells selectively, interfering with the synthesis of the bacterial cell wall, protein synthesis, and nucleic acid synthesis, or affecting metabolic pathways [7, 8]. If antibiotic inhibits cell growth and multiplication, they are bacteriostatic, while when they cause internal mechanisms that lead to cell death, they are known as bactericides. However, they present specific mechanisms of action depending on the type of cell membrane of gram-positive or gram-negative bacteria. Likewise, they are broad spectrum when they can eliminate bacteria of both types [8]. Bacteria show the affinity of forming colonies in any solid or liquid substrate (catheters, prosthetics, human body parts, heart valves, and teeth), and proliferation in the presence of nutrients to release exopolysaccharides gives rise to biofilms [3, 6].

Biofilms are associated with the physiological states of bacteria and can be monostrains or multistrains, and there may be synergy or antagonism between the different strains [9, 10]. These biofilms are extremely difficult to eradicate because of the extracellular matrix (exopolysaccharides) that prevents the diffusion of antibiotics in the structure of the biofilm, as well as prevents the free entry and exit of nutrients and waste from bacteria. This situation leads to metabolic reduction (a subcritical condition that activates bacterial survival mechanisms). Therefore, most antibiotics become deficient because they were designed for exponential growth conditions [4, 6, 11].

Biofilms of multidrug-resistant (MR) bacteria are considered a source of infection that generates a high risk of affecting and causing death to people at any stage of life. For this reason, these biofilms are urgent public health problems in the world, charging 10 million human lives per year and costing 100 billion dollars by 2050 in the world economy. Thus, this strengthens the challenge to innovate current antibacterial treatments since the exchange processes of the genetic expression of bacterial pathogens are linked to the food chain, water sources, clinical care, and the environment in general, modifying the virulence of bacterial pathogens [3, 4, 6, 12], as shown in **Figure 1a**.

Searching for methods or treatments to control or eliminate resistant bacteria is not new, but there are limitations to their use in *in vivo* applications, such as selectivity

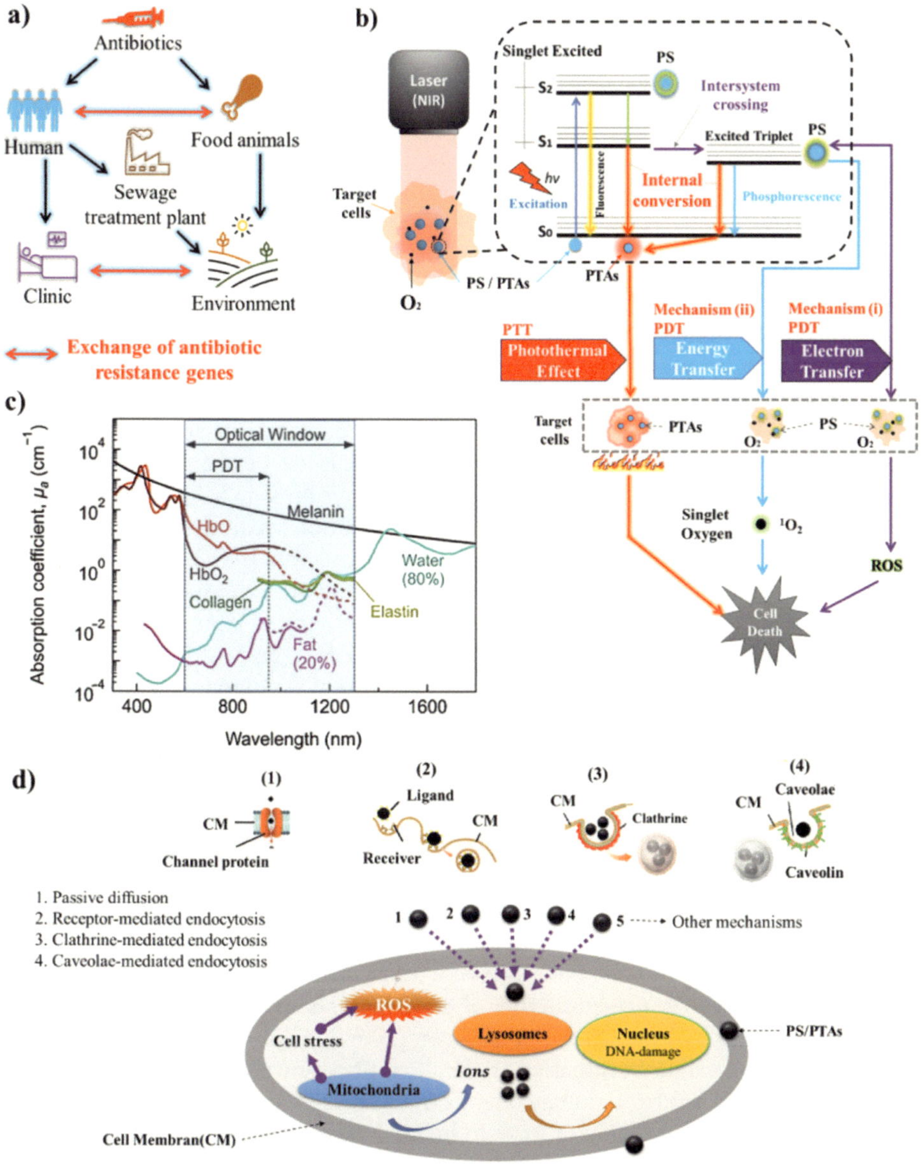

Figure 1.
(a) Exchange cycle of the genetic mutation of bacteria. (b) ROS and photothermal effect generation mechanisms for cell death in PDT and PTT. (c) Optical window of melamine, water, hemoglobin, and collagen, depending on the absorption coefficient [13]. Copyright 2022 MDPI. (d) Internalization mechanisms of the PS and PTAs in target cells.

and activation control [1]. Nanotechnology has presented successful solutions to this problem, such as metal nanoparticles (NPs) [9], metal oxide nanoparticles [10], carbon-based materials (CBM) [11], and nanocomposites [12], as antibacterial agents. In the CBM group, there are single-walled carbon nanotubes (SWCNTs), multiwalled carbon nanotubes (MWCNTs), graphene (G), graphene oxide (GO), reduced graphene oxide (r-GO), carbon dots (CDs), and fullerenes [4, 6]. These nanotechnology solutions have spread into applications such as water treatment [14–17], antimicrobial textiles [18], antimicrobial food packaging [19, 20], antibacterial coatings for medical instrumentation and equipment [21], bacterial distinguishment [22], regeneration of living tissues [12, 23], photocatalytic disinfection [1, 3, 12, 24], light-induced acidification [1, 25], photodynamic therapy (PDT) to antibacterial applications or antimicrobial PDT (APDT) [26–29], and photothermal treatment (PTT) or photothermal bacterial lysis (PTBL) [1, 30–33].

On the other hand, thermotherapy is a widely applied technique in medical treatments, mainly in the oncology area [34, 35]. It is based on using heat (conduction, convection, or conversion) to tissues (local, regional, or general) to induce damage to its cellular structure, causing death in target cells [3]. It also promotes an increase in blood flow that facilitates the supply of proteins, nutrients, and oxygen at the injury site. The rise of 1°C in the tissue temperature induces improvement between 10 and 15% of the local tissue metabolism [36]. Thermotherapy comprises two categories, that is, hyperthermia and thermal ablation, depending on the range of temperatures in the treatments. Hyperthermia encompasses a temperature range between 41 and 45°C, while thermal ablation encompasses temperatures above 46°C [37–39]. Thermotherapy supplies heat through different energy sources, for example, radio frequency, microwaves, high-intensity ultrasound, light (visible, near-infrared [NIR], and ultraviolet [UV]), and magnetic fields [40], and its name depends on the energy source.

PTT and PDT are antibacterial techniques that are derived from thermotherapy by using a light source (visible, NIR, and UV) to provide heat and reactive oxygen species (ROS) agents, and they differ mainly in the range of temperature and duration of treatment (PTT: > 46°C, 4–6 min; PDT: 41–45°C, 15–60 min) [13, 41], as well as by the mechanisms of action. PPT is based on the use of photothermal agents (PTAs) that produce heat in the presence of electromagnetic radiation, causing the rupture of cell membranes, protein denaturation, and irreversible cell destruction [5, 42]. If metal nanoparticles (NPs) or metal oxide NPs are used as PTAs, an effect known as "localized surface plasmonic resonance" (LSPR) [43] is produced, which allows the temperature of the nanoparticles to increase. When using CBM as PTAs, heating mechanisms are achieved through nonradiative relaxation pathways (internal conversion) (see **Figure 1b**). For this reason, CBM with high absorption and low fluorescence quantum yield will present higher photothermal conversion efficiency [44].

Antibacterial PDT employs three critical components for its application: photosensitizers (PSs), electromagnetic radiation (typically NIR region), and molecular oxygen (O_2). The PS absorbs light and donates electrons or energy interchange with surrounding O_2, promoting the formation of ROS, inducing irreversible damage to the cell membrane leading to the cell apoptosis or necrosis of the target cell [37]. Two different mechanisms achieve ROS generation as an agent of action in PDT. The first type of ROS is formed by the transfer of electrons between the PS and O_2 or substrate, generating oxygen radicals such as superoxide anion ($O^{\bullet-}$), hydroxyl radical ($HO^{\bullet}$), and hydroperoxyl radical ($HOO^{\bullet}$). This is done by transitioning PS molecules from a ground state (S_0) to a singlet excited state ($S_{1,2}$) and the excited triplet state, as shown in **Figure 1b**. In excited triplet states, these PS molecules exchange electrons with a target cell (O_2 mainly), producing free radicals that cause oxidative stress and cell death. The second type of mechanism to generate ROS consists of energy transfer between the PS in an excited triplet state and O_2, giving rise to singlet oxygen ($^1O_2$), which is more reactive and interacts more with proteins, lipids, and nucleic acids of target cells, in a perimeter of around 20 nm, producing cell apoptosis or necrosis [45, 46].

PPT and PDT are typically used under near-infrared (NIR, 700–950 nm) laser irradiation due to the optical window that this region presents (see **Figure 1c**), in which absorption by hemoglobin (HbO), melanin, and water is reduced, increasing in this way the availability of photons to interact with the PS or PTAs and greater penetration in tissues [13, 42, 47–49]. The treatment effectiveness improves under an incubation time when the PS and PTAs are internalized into the cells by different mechanisms [45] (see **Figure 1d**). The selectivity of this treatment can improve by tuning the PS and PTAs with agents related to the target cells to be treated [41, 50]. Several CBMs, such as CDs, SWCNTs, and MWCNTs, have been used in PDT and PTT applications as PS and PTAs due to their low toxicity, biocompatibility, tunable fluorescence properties, easy functionalization, and antimicrobial activity, which are ideal for *in vivo* applications [5, 7].

2. Graphene oxide and reduced-graphene oxide

Graphene, which is also called the "wonder material," constitutes a revolutionary discovery of the twenty-first century. This CBM has a two-dimensional planar structure like sheets of sp^2-hybridized carbon atoms packed into a hexagonal arrangement [48]. Graphene comprises isolated layers of graphite. Graphene possesses unique and fascinating properties such as large surface area, resistance, impermeability, hardness, lightweight, flexibility, and conductivity, which have encouraged its application in diverse and multidisciplinary fields [49]. Therefore, graphene has been employed in medicine, electronics, aerospace, energy, nanotechnology, and so on [50].

Graphene has been used in anticancer therapy, drug delivery, tissue engineering, and biomedical imaging. Nevertheless, pristine graphene is hydrophobic and relatively expensive to prepare. Thus, two alternatives of graphene derivatives have been proposed due to their better water affinity and the ability for mass production: graphene oxide (GO) and reduced-graphene oxide (r-GO) [51, 52]. GO is obtained by an oxidation procedure of graphite [53], generally carrying out the following routes of synthesis: Brodie method, Staudenmaier method, Tour method, Hofmann method, and Hummers method and its modification [54, 55]. On the other hand, r-GO is prepared by a reduction process of GO, commonly by chemical or thermal procedures [56]. The presence of carbon and oxygen functional groups, such as alkoxy, carboxylic, epoxy, hydroxyl, and carbonyl in the basal planes, and peripheries of these graphene derivatives, promotes a better hydrophilic character and solubility than graphene (see **Figure 2a**) and facilitates biointeractions with molecules like nucleic acids and proteins [58, 59]. Furthermore, these linked molecules determine the oxidative level of GO or r-GO [57, 60].

Regarding the antibacterial action of GO and r-GO, referring to the inhibition of growth and microorganism destruction, it is not only attributed to the photoinduced mechanisms like PDT by the generation of ROS or PTT by the generation of heat, but it is also a consequence of their two-dimensional structures. These CBMs physically kill the bacteria by direct contact with the sharp edge layers (thicknesses of 0.8–1.2 Å) of GO or r-GO and scrape the membrane, causing the rupture of the intracellular matrix and, consequently, the microbe's death [52]. The lateral size of GO and r-GO sheets influences the antibacterial effect. It covers the bacterial pathogen due to the electrostatic interaction between the functional groups of GO, r-GO (basal plane), and the bacterial membrane, thus inhibiting their nutrient absorption and proliferation mechanisms, leading to the death of the bacteria. **Figure 2b** shows AFM images of *S. Aureus* and *Escherichia coli* bacteria free of GO and with GO, showing that GO sheets superficially cover (folds formation) the bacteria; monolayer GO sheets

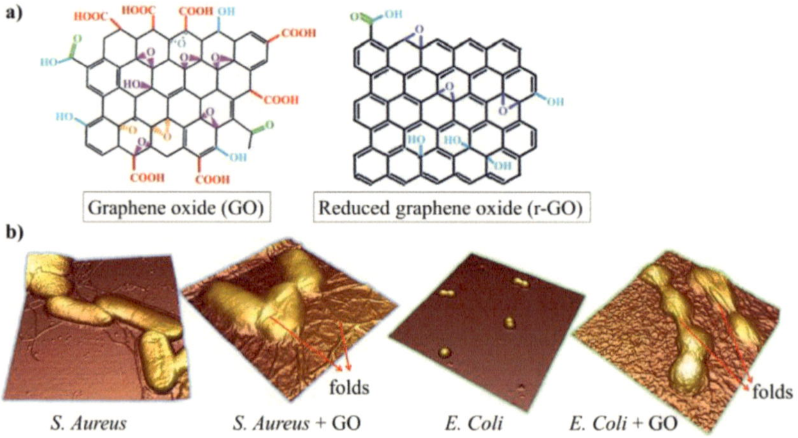

Figure 2.
(a) Structure of GO and r-GO. Adapted with permission [57]. Copyright 2019 Dove Medical. (b) Atomic force microscopy (AFM) images of S. aureus and E. coli with and without GO treatment [31]. Copyright 2020 Frontiers.

with an area > 0.4 μm^2 have higher antibacterial activity than GO sheets with an area < 0.2 μm^2 (nanographene oxide [NGO]) [61]. It is important to note that GO and r-GO have intrinsic antibacterial properties; additionally, they attack bacteria by two extra mechanisms: oxidative stress and cell entrapment [62].

2.1 Graphene oxide and reduced-graphene oxide in APDT

The antibacterial capacity of GO has been successfully tested against *S. aureus*, *E. coli*, *Klebsiella pneumoniae*, *Pseudomonas aeruginosa*, *Streptococcus mutans*, *Porphyromonas gingivalis*, *Fusobacterium nucleatum*, and *Pseudomonas putida*, among others [63]. It has been demonstrated that GO only produces singlet oxygen ($^1O_2$, energy transfer) under irradiation [64]; hence, it marginally contributes to the biocide capacity compared to the simultaneous generation of electron–hole pairs. In this context, GO with a lateral size of a few micrometers and a thickness of 1 nm reduced the percentage of bacterial survival to ~24.9% in *E. Coli* under irradiation-simulated sunlight exposure at 380 $mW{\cdot}cm^{-2}$, causing bacterial death mainly by the generation of $^1O_2$, to a lesser extent by ROS and intrinsic mechanisms of GO, but with a negligible photothermal effect [65]. The light exposure might reduce GO into r-GO (primarily by electron transfer), forming carbon-centered free radicals, which increase its antibacterial activity. Thus, r-GO performs better as a biocide than GO [63]. Within light-stimulated processes, GO and r-GO are considered ideal materials for the diagnosis and treatment by PDT and PTT because they can be absorbed in the first (650–950 nm) and second (1000–1350 nm) biological windows, where a sufficient tissue penetration of light is attained [31].

To improve the antibacterial capacity of these graphene derivatives, some researchers have proposed its usage along conventional PSs, such as indocyanine green, methylene blue, and toluidine blue [66]. Thus, several works have proved that these composites promoted higher ROS production and an enhanced antibacterial effect compared with the single components. For example, one work used a GO-based composite with indocyanine green to combat *Enterococcus faecalis* (an anaerobic gram-positive coccus bacterium). It demonstrated that GO upgraded the photodynamic action of indocyanine green, being 1.3 times more effective in the antibiofilm activity [67]. Likewise, nanoparticles of metals such as Ag have been employed to prepare composites showing excellent results in bacteria elimination [68, 69]. One work used GO along Ag nanoparticles (AgNPs) within *in vivo* subcutaneous tests and proved that after 20 min of irradiation with visible light (600 nm), an antibacterial efficacy of 96 and 99% for *E. coli* and *S. aureus* is obtained, respectively [70].

The NGO (~ 21.3 nm) functionalized with DNA-aptamer (short sequences of artificial DNA) is selective with *P. gingivalis*, reducing their viability in the order of 4.33 Log_{10} CFU under 980-nm irradiation (1 W for 1 min) and concentration of 1/2 MIC (minimum bacteriostatic concentration, 62.5 nM—obtained without irradiation), as seen in **Figure 3a**. The DNA-aptamer-NGO presents an intrinsic antibacterial activity and increases under irradiation. Their action mechanism reduces bacterial metabolic activity, as observed in **Figure 3b**, which leads to a higher rate of bacterial apoptosis as a function of concentration and irradiation (see **Figure 3c**) [71]. In APDT with GO, there is a dependence between the light dose and the loss of bacterial viability, for example, for the bacteria *E. coli*, its loss of viability is aggravated for 40–60 $J{\cdot}cm^{-2}$ at higher concentrations. However, a higher light dose is required for lower concentration, which also suggests the dependence on GO concentration (see **Figure 3d**). This behavior is similar in *S. aureus* with GO and NGO. That is, there is a threshold of light dose and concentrations of the PS, where the photons are available to excite GO and generate $^1O_2$ and the ROS is optimal for causing bacterial death and avoiding affecting healthy tissues.

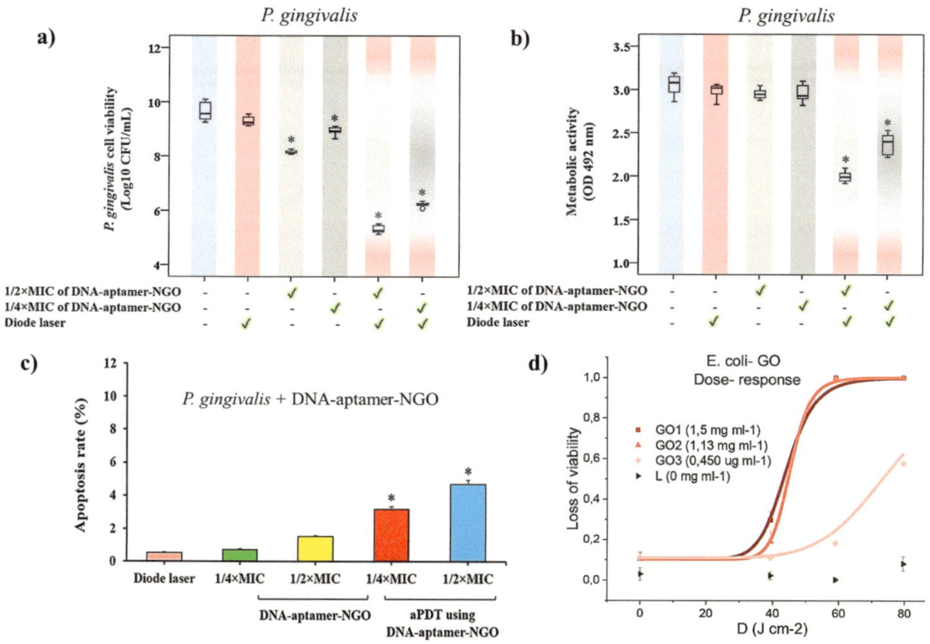

Figure 3.
(a) Bacterial viability of P. gingivalis with DNA-aptamer-NGO in APDT [71]. (b) Metabolic activity of P. gingivalis with DNA-aptamer-NGO in APDT [71]. (c) The apoptosis rate of P. gingivalis with DNA-aptamer-NGO in APDT [71]. Copyright 2022 Springer Nature. (d) Loss of viability in E. coli in APDT with GO as a function of light dose at 630 nm [31]. Copyright 2020 Frontiers.

2.2 Graphene oxide and reduced-graphene oxide in antibacterial PTT

The photothermal effect of GO and NGO in antibacterial PTT is a function of its concentration and size. In NGO, its smaller lateral size allows it to keep its temperature below 60°C (in aqueous solutions). In contrast, the temperature of GO having larger lateral size reaches above 60°C under the same conditions (630 nm, 65.5 $mW{\cdot}cm^{-2}$). Likewise, the heating efficiency is higher in GO (~1.45) than in NGO (~1.3) for a light dose of 60 $J{\cdot}cm^{-2}$ (see **Figure 4a**) [31]. This suggests that a larger surface has greater availability of photons for internal conversion. Therefore, GO and NGO are potential PTAs in antibacterial PTT. However, its selectivity can be improved by incorporating functionalizing agents that positively charge GO for better attraction to bacteria. The amino groups (NH_2) and polyethylene glycol (PEG) provide a positive charge to GO. Likewise, they can soften the sharp edges of GO, improving its cytotoxicity and biocompatibility but reducing its antimicrobial activity. Even so, nanocomposites, such as GO-PEG-NH_2, exhibit excellent antibacterial activity in PTT, as seen in **Figure 4b**. These GO nanocomposites inhibit susceptible bacteria such as *E. coli* and *S. aureus* at 50 $\mu m{\cdot}mL^{-1}$ under 808-nm irradiation and 1.5 $W{\cdot}cm^{-2}$ for 5 min. These nanocomposites partially damage the membrane in the two bacterial strains due to their intrinsic antimicrobial activity. When irradiated, the destruction of the bacterial membrane and a subsequent union of the sample bacteria are produced, as indicated in **Figure 4c** [72].

Furthermore, synergistic mechanisms have been achieved since GO and r-GO also present photothermal effects. One work used amino-functionalized GO and determined that it was easily targeted by electrostatic attraction into gram-negative and gram-positive bacteria surfaces. After the irradiation of 159 mW cm^{-2}, it was

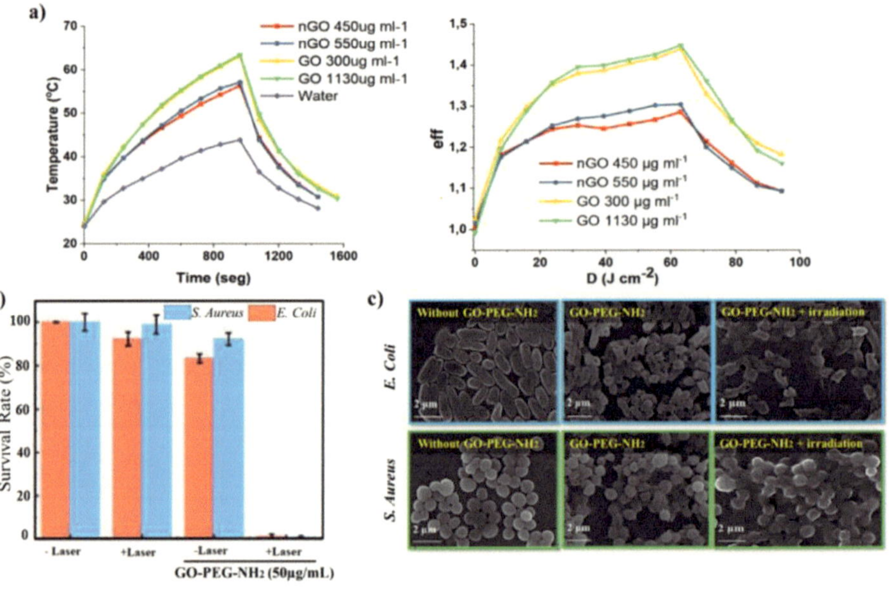

Figure 4.
(a) Thermal study of NGO and GO in aqueous solution for different irradiation times and concentrations [31]. Copyright 2020 Frontiers. (b) Bacterial survival rate of S. aureus and E. coli in antibacterial PTT using GO-PEG-NH_2 as PTAs [72]. (c) SEM images of the damage produced by GO-PEG-NH_2 in E. coli and S. aureus bacteria in antibacterial PTT [72]. Copyright 2020 MDPI.

proved that the temperature increased to 80°C using a concentration of 0.25 mg mL^{-1} [73]. Using GO and r-GO as PTAs and PS promises improved antimicrobial activity because bacteria are killed by oxidative stress (APDT) and photothermal effect (PTT), producing synergy between both therapies [74].

3. Carbon dots

Carbon dots (CDs) are considered nanospheres of diameter between 1 and 10 nm [54, 56] with carbonaceous nuclei that present sp^2 and sp^3 domains in crystalline and amorphous structures [52, 60, 75]. CDs are divided into the following four groups: carbon quantum dots (CQDs), graphene quantum dots (GQDs), carbon nanodots (CNDs), and carbon polymer dots (CPDs) (CQDs and GQDs exhibit quantum confinement) [60, 61, 65]. The electrons (lone pairs) of sp^2 domains absorb light (visible, NIR, and UV) [1, 57], passing from one energy level (π) to a higher one $\pi \rightarrow \pi^*$, surpassing the forbidden band. Likewise, the presence of functional groups in its structure (typically: -COOH, -OH, and $-NH_2$) [76–78] allow surface trap states (n) that reduce the bandgap, allowing electrons to absorb light, to reach a higher energy level $n \rightarrow \pi^*$. In this way, the CDs generate photoluminescence (PL) by different mechanisms: a photon emission due to the π-conjugated domains of the nucleus (CQDs and GQDs) and photon emission by their surface trap states and by the state of the molecule [76]. In recent years, CDs have gained significant attention for antibacterial applications [76] due to excellent photoluminescence properties, low toxicity, ease of surface functionalization, chemical stability, dispersion in aqueous media, and low cost [23, 61]. Studies of the antimicrobial activity with *S. aureus* and *E. coli* show that CDs have more permeability toward the bacterial cell membrane than traditional antibiotics [4, 62] due to their nanometric size, and it can be improved by reducing the size of CDs [79, 80]. As shown in **Figure 5a**, more significant numbers of small CDs cross the cell membrane than the larger ones.

A highlighting factor of antimicrobial activity is the surface charge of CDs. It must be positive to generate electrostatic attraction between CDs and teichoic and lipoteichoic acids in the gram-positive bacteria cell membrane, likewise with lipopolysaccharides (LPS) in the gram-negative bacteria cell membrane [81]. The surface charge of CDs is modified by their functionalization with suitable molecules, antibiotics, such as biguanide [80], levofloxacin [47], lysine, and folic acid [82], or antimicrobial nanoagents such as AgNPs [83], which increase the antibacterial activity. CDs (4.5–7 nm in size) passivated with amino, carbonyl, and hydroxyl functional groups can diffuse through the *S. aureus* and *E. coli* bacterial membranes without affecting them and continue until CDs disrupt the double helix of naked bacterial DNA, inhibiting bacterial proliferation [81] or activation of other bacteria-killing mechanism observed in **Figure 5b**. If CDs are functionalized with antibiotics, such as levofloxacin hydrochloride (which inhibits bacterial topoisomerase IV and DNA gyrase), the mechanisms to kill bacteria become more potent than the antibiotic action alone [84–86]. In such a way, the CDs induce ROS generation to damage the bacterial cell membrane partially, and the internalization of levofloxacin hydrochloride is easier, causing cytoplasmic leakage and early death of bacteria [47]. A superior feature of CDs is the low probability of causing bacterial resistance due to their excellent biodegradation (short time for resistant response, no efflux pump) [4, 65], and no known enzyme is capable of inhibiting the ROS as $HO^{\bullet}$ and $^1O_2$ [5].

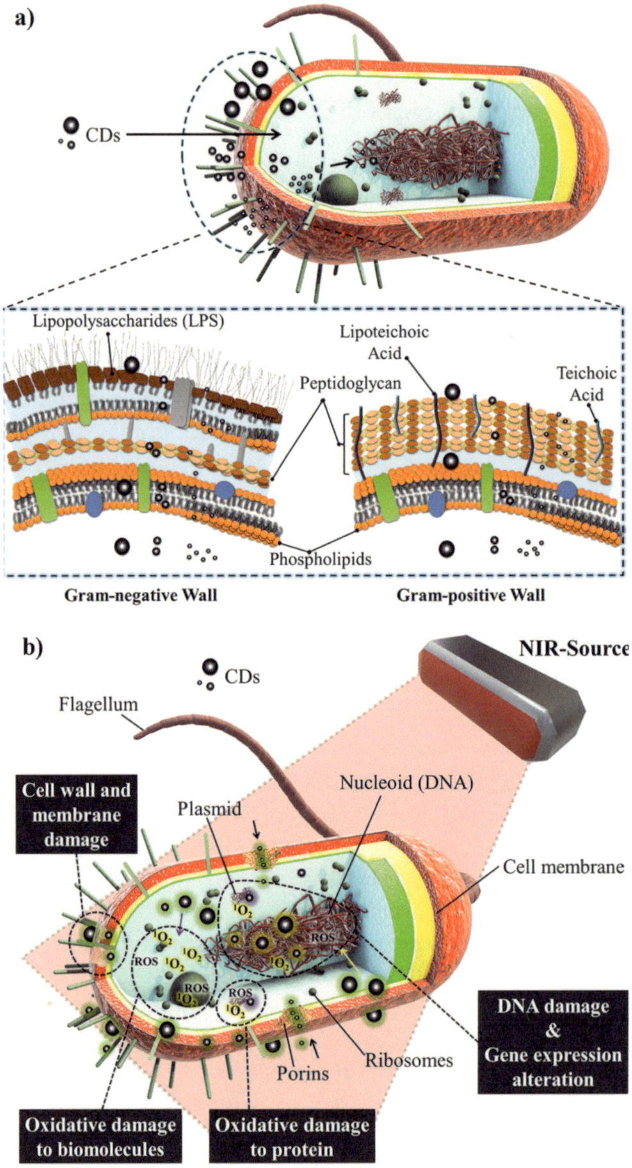

Figure 5.
(a) Diffusion of CDs of various sizes in the bacterial wall until reaching the DNA. (b) Mechanisms of bacterial death by generating ROS in the presence of CDs.

3.1 Carbon dots in APDT

The features of CDs follow the requirements of the new generation of PSs in APDT. CDs achieve inhibition of even MR bacteria such as multidrug-resistant *S. aureus* (MRSA) or multidrug-resistant *Acinetobacter baumannii* (MRAB) at relatively low concentrations (32–64 $\mu g \cdot mL^{-1}$) of CDs (~ 3 nm, obtained from 2,4-dihydroxybenzoic acid and 6-bromo-2-naphthol by the solvothermal method), irradiated with red light (590 nm) by a 30-$mW \cdot cm^{-2}$ source for 15 min [87, 88]. Likewise, susceptible bacteria, such as *E. coli*, are inhibited at 50 $\mu g \cdot mL^{-1}$ of Cl-GQDs (~3–5 nm, obtained from sucralose by the electrochemical method), irradiated with sunlight simulated in a 100-mW·cm^{-2} source for 2 hours [89]. However, the negligible cytotoxicity of these PSs makes them even more attractive for *in vivo* applications. The surface charge of CDs is a significant aspect of CDs-bacterial membrane coupling and must guarantee a positive charge. This surface charge depends on the treatment conditions, for example, pH and dispersion medium; *in vitro* or *in vivo* applications modify surface estate. Usually, the surface charge of CDs gives the Z potential [89]. The surface charge is related to the number of functional groups in CDs as PL emissive centers. With a negative or low charge, the antimicrobial activity decreases due to the low or null electrostatic attraction of the CDs-bacterial membrane and the few surface heteroatoms that promote the formation of ROS [5]. The irradiation times, environmental conditions, irradiation source, and estimation of the light dose administered are crucial in the correct development of APDT. However, in various studies, the incubation time is not reported [64, 77, 78, 90–93]. The incubation time refers to the process of internalization or endocytosis of CDs toward the membrane bacteria (see **Figure 1d**). In this sense, Liu

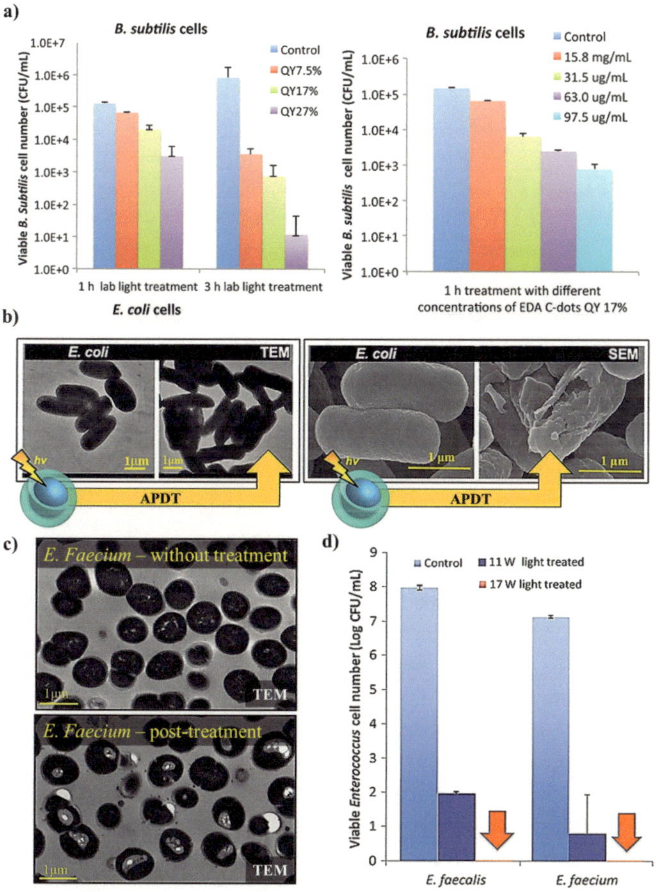

Figure 6.
(a) Cell viability by CFU number in Bacillus subtilis cells for samples with different concentrations and quantum yield (QY) of PL [94]. Copyright 2017 Royal Society of Chemistry. (b) SEM and TEM images of E. coli bacteria before and after APDT [5]. Copyright 2020 MDPI. (c) TEM image of E. faecium before and after APDT [95]. (d) Inhibition (orange arrows) of E. faecalis bacteria by increasing the irradiation dose in APDT [95]. Copyright 2020 Royal Society of Chemistry.

et al. [87] extended the antimicrobial activity of red carbon dots (R-CDs) against MRAB and MRSA by increasing the incubation time from 0 to 45 min, achieving bacterial survival rates of 97.7% at 0 min and 3.3% in MRAB and 7.5% in MRSA at 45 min, because CDs can effectively induce the formation of ROS once inside the bacteria.

An exciting aspect of CDs as PSs in APDT is the dependence between the antibacterial activity and its QY; **Figure 6a** shows this effect. The QY refers to the transformation of absorbed and emitted photons in the structure of the CDs. However, it is common to use the quinine sulfate standard that provides adequate information for the interchange of results [96]. **Figure 6a** indicates a relationship between the irradiation time (hours) and the bacterial viability (CFU mL^{-1}) in APDT, as well as a dependence between the bacterial activity (CFU mL^{-1}) and the concentration of CDs ($\mu g{\cdot}mL^{-1}$) in the treatment. Bacterial viability reduces as more CDs have free electron pairs that promote $\pi \rightarrow \pi^*$ or $n \rightarrow \pi^*$ transitions, which, in turn, induce more ROS, thus evidencing the ROS production mechanisms shown in **Figure 1a**, which is the characteristic of APDT. The doping and functionalization of CDs is an essential aspect of QY and significantly affects the antimicrobial activity in APDT and can induce new bacterial death mechanisms, as is the case of bromine-doped carbon nanodots (Br-CNDs). The Br-CNDs in a change of pH (basic-acid-basic) and darkness conditions induce reactive nitrogen species that generate dark toxicity [97].

SEM/TEM observations are a helpful tool to elucidate bacterial damage and are typically acquired before and after APDT. Together with staining assays and confocal laser scanning microscopy imaging techniques, it is possible to propose mechanisms of cell death [23]. **Figure 6b** shows the damage caused to the *E. coli* bacterial membrane before and after APDT. A change in morphology is evidenced by TEM and bacterial lysis by SEM. Similarly, TEM images in **Figure 6c** reveal damage caused to the cytosol of *Enterococcus faecium* bacteria without affecting the cell membrane after APDT [89]. The dose of light used in APDT is a parameter of significant consideration. **Figure 6d** shows that by increasing the power of the irradiation source (0, 11, and 17 W) in APDT, the antibacterial effect increases until viability is inhibited by the *E. faecium* bacteria [89].

3.2 Carbon dots in antibacterial PTT

In antibacterial PTT, the CDs cause initial damage to the bacterial membrane due to the absorption of photons and their internal conversion that increases their temperature, as indicated in **Figure 1b** and **5b**. CDs bind to the bacterial membrane mainly by electrostatic interactions and transfer heat to bacteria [98]. In this way, the bacterium becomes vulnerable to heat and allows the incoming of CDs that increase cell damage by inducing the ROS. Therefore, it is common for PTT to synergize with APDT to improve the bacteria-killing mechanism. The initial damage caused by the CDs is increased by incorporating an antibiotic or antibacterial agent, such as quaternary ammonium, which increases the damage to the bacterial membrane, allowing a more significant action than the action of the CDs alone [99]. The photothermal effect is effective in gram-positive and gram-negative bacteria due to the heat the CDs provide, affecting their different structures of peptidoglycan, phospholipids, and LPS. Therefore, CDs in antibacterial PTT present properties like those of broad-spectrum antibiotics. However, the antibacterial effect without irradiation may reduce effectiveness in gram-positive bacteria due to multiple layers of peptidoglycan. It is proved using nanohybrids of GQDs-AgNPs at a concentration of 2 $\mu g{\cdot}mL^{-1}$ and bacterial strains of *S. aureus* (gram-positive) and *E. coli* (gram-negative). Nevertheless, the total inhibition of bacterial strains is achieved by irradiation with red light (808 nm) from a 2-W·cm^{-2} source for 10 min and the same concentration of GQDs-AgNPs [83].

A relevant aspect of antibacterial PTT is the photothermal performance of CDs (conversion of photons into heat). In dispersion in a liquid medium, they act as heat-emitting sources, and depending on the medium to distribute this thermal energy efficiently, it is possible to reach high temperatures that cause damage to healthy tissue. The temperature reached in *in vivo* and *in vitro* applications is a function of photothermal performance, the concentration of CDs, irradiation time, the dispersion medium, and the dose of light supplied. **Figure 7a** shows the temperature dependence on the dispersion of CDs (doped with Fe), concentration, and irradiation time. The main parameter for temperature control is the concentration of CDs. However, in *in vivo* applications, temperature measurements are usually real time to avoid unwanted tissue damage. This procedure also depends on the depth of the treated infection [83]. CDs doped with Fe or Ag nanoparticles (AgNPs) can acquire a behavior like an enzyme peroxidase (POD) [83] interacting with H_2O_2, increasing its antimicrobial activity (99.85% inhibition *E. coli*) and promoting healing (see **Figure 7b** and **c**).

Antibacterial PTT with CDs allows bone infection treatment through hybrid nanomaterials such as chitosan (CS)-nanohydroxyapatite (nHA) scaffolds doped with CDs (CS-nHA-CDs). CS-nHA-CD scaffolds help as a base material for the new bone tissue with antibacterial features. These antibacterial scaffolds achieve an inhibition (*in vivo*) of up to 97 and 99% for *E. coli* and *S. aureus* bacteria, respectively, under red light irradiation (808 nm, 1 W·cm^{-2}) for 10 min. Healthy tissues reduce the dose of light reaching antibacterial scaffolds. Therefore, the temperature control and thermal performance of these materials are essential [23].

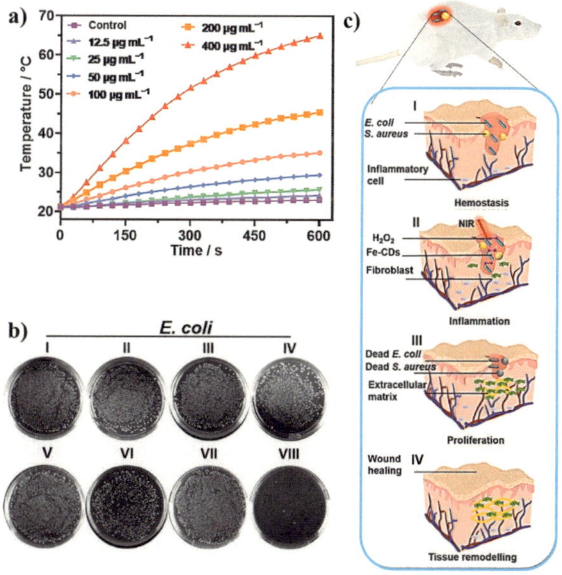

Figure 7.
(a) Thermal study of CDs doped with Fe for different concentrations as a function of irradiation time. (b) The antibacterial activity of Fe-CDs for a strain of E. coli, with an irradiation of 2 W·cm^{-2} at 808 nm, (I) control, (II) Fe-CDs, (III) H_2O_2, (IV) Fe-CDs + H_2O_2, (V) NIR, (VI) Fe-CDs + NIR, (VII) H_2O_2 + NIR, (VIII) Fe-CDs + H_2O_2 + NIR. (c) Scheme of antibacterial PTT in vivo, with wound healing effect. All the images were obtained from [30]. Copyright 2021 Elsevier.

4. Carbon nanotubes

CNTs comprise sheets of graphene rolled in the form of a tube, and transverse dimensions are in the nanometric range, but the length is over the nanometric scale. Therefore, CNTs are one-dimensional material. The number of graphene sheets forming CNTs allows the classification of this material into multiwalled carbon nanotubes (MWCNTs) with a diameter of ~10–100 nm and single-wall carbon nanotubes with a diameter of ~0.4–2 nm [100]. CNTs have crystalline sp^2 domains with graphene as a precursor. Their physicochemical structure makes them hydrophobic and cytotoxic. CDs are usually functionalized with noncovalent bonds to improve their biocompatibility and solubility in aqueous media [6]. The CNTs present a spontaneous interaction with bacteria and a strong absorbance in the NIR. For this reason, CNTs are suitable photothermal antibacterial agents in PTT [1]. CNTs are not eligible for APDT because they energetically reduce or inhibit singlet oxygen $^1O_2$ generation (SOG) and have QY below ~1 [101]. However, an appropriate functionalization or formation of a nanocomposite can modify their properties, allowing CNTs to perform as PSs. CNTs act like needles in the bacterial cell membrane, inducing damage according to the surface resistance of each bacterial strain. CNTs also act like a nanochannel once located in the bacterial membrane, the needle effect is more evident in SWCNTs due to reduced diameters, and the channel-like effect stands out in MWCNTs [100].

CNTs can be vertically directionally grown, producing a nanoforest of CNTs suitable to inhibit bacterial biofilms, causing immobilization due to their needle-like effect. The nanoforest of CNTs is a nanostructure like insects' wings (biomimetics) with excellent antimicrobial activity by their tower-like nanostructures. **Figure 8a** shows the antibacterial activity of the nanoforest of MWCNTs in a strain of *Klebsiella oxytoca*.

SWCNTs coupled with surfactants in antibacterial PTT show the inhibition of bacterial strains such as *E. soli* and *E. faecium* with more significant antimicrobial activity in *E. faecium*. The *E. faecium* bacterium is more susceptible to surfactants, allowing a better penetration of the CNTs in the cell membrane [103].

4.1 Carbon nanotubes in APDT and antibacterial PTT

CNTs' functionalizing agents, such as menthol-zinc phthalocyanine (ZnMintPc), zinc monocarboxyphenoxyphthalocyanine (ZnMCPPc), spermine, protoporphyrin IX, or nanocomposites of CNTs with a matrix such as GO and poly (N-vinyl caprolactam-co-poly (ethylene glycol diacrylate)) poly (VCL-co-PEGDA) polymer, significantly improve SOG in antibacterial PDT, generating oxidative damage or alterations in bacterial DNA [12, 27]. The photothermal effect (photons to heat) in MWCNTs is produced by internal conversion, just like GO, because they share the same carbon structure in a hexagonal honeycomb arrangement (see **Figure 8a**). However, magnetic nanoparticles (MNPs) that generate heat by the LSPR effect increase a slightly lower photothermal effect. Thus, MWCNTs (cyan curves) convert photons to heat more efficiently than MNPs (blue curves).

MWCNTs embedded in VCL/PEGDA (hydrogel) and ZnMintPc as PSs form a nanocomposite VCL/PEGDA-MWCNT-ZnMintPc with excellent antibacterial activity (see **Figure 8c**, C1). The complete inhibition of *E. coli* bacteria ascribes to the photothermal effects of MWCNTs (irradiated with a red light at 360 nm in a 65.5-$mW \cdot cm^{-2}$ source) and the generation of ROS by the PS (ZnMintPc). The action mechanism of this nanocomposite consists of cell membrane damage by direct contact (see **Figure 8d**) and oxidative damage. The gram-positive bacteria (*S. aureus*)

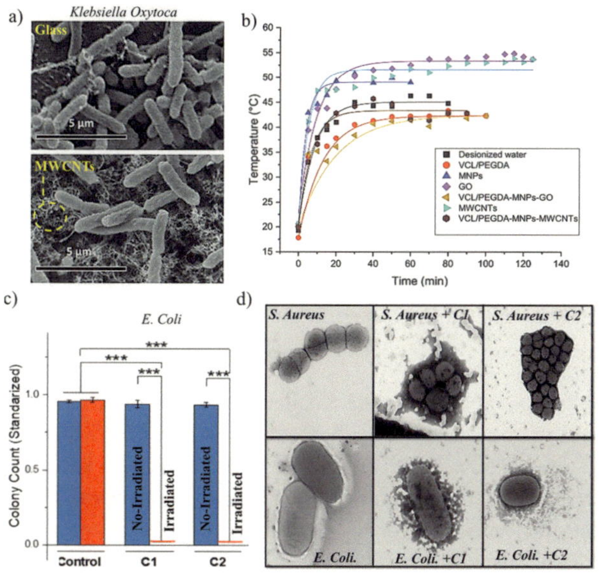

Figure 8.
(a) SEM observations of morphology and K. oxytoca bacteria in a glass substrate and vertical MWCNTs' forest [102]. (b) Evaluation of the photothermal effect of MWCNTs and other molecules of interest as a function of time [12]. (c) Extended plate count method, CFU, of a bacterial strain E. coli under NIR irradiation (630 nm, 65.5 mW·cm^{-2}) [12]. Copyright 2016 Royal Society of Chemistry. (d) STEM observations in bacterial strains S. aureus and E. coli before and after APDT-PTT with different nanocompounds (C1, C2) [12]. Copyright 2022 MDPI.

did not completely inhibit. However, the nanocomposite bacteria coupling mechanism is similar. This nanocomposite exhibits synergistic antimicrobial properties in APDT and antibacterial PTT, identical to a nanocomposite with GO and MNPs (C2 in **Figure 8c**) toward *E. coli* bacteria under the same conditions. The potential application of MWCNTs as PTAs and PS (within antibacterial PTT and APDT, respectively), has the advantage that they present an additional mechanism of coupling-bacterial death by its needle-like effect.

5. Perspectives

Carbon-based materials (CBMs), such as graphene oxide, reduced-graphene oxide, carbon dots, and carbon nanotubes, have promising possibilities as photosensitizers and photothermal agents within photodynamic and photothermal treatments to combat bacteria. These materials have been used as platforms and components to develop complex composites. Thus, to encourage their practical application in medicine, it is necessary to standardize large-scale production by maintaining high quality, reproducible, and uniform morphology and size of these CBMs.

Besides, exploring their killing or inhibition action against other microorganisms, like viruses and fungi, has become a topic of interest. Hence, it is important to continue the research on the toxicity of these materials in human health and the environment.

Antibacterial photodynamic and photothermal therapies have been extensively investigated in susceptible and multidrug-resistant (MR) monostrain bacteria and MR monostrain biofilms. However, MR dual-strain biofilms can proliferate synergistically under specific conditions studied in recent years. The wide variety of bacterial pathogens and their potential coupling in biofilms sustained new research to understand and combat this warning to health. In addition, the possibility of continuing to find new multistrain bacteria biofilms acting synergistically in hospital substrates and the obsolete antibiotics proves the seriousness of this bacterial risk. Therefore, it is essential to evaluate, propose, and develop proper culture conditions, as well as the new era of antibacterial agents, including CBMs, for their promising antibacterial activity in thermotherapy treatments.

6. Conclusion

This chapter summarizes the recent progress of carbon-based materials (CBMs) as a novel alternative to combat bacteria. The excessive use of antibiotics triggered bacterial resistance that caused severe diseases, even becoming a health risk. Therefore, developing new treatments has become mandatory to overcome this public problem. In this context, several authors have proposed CBMs as antibacterial agents, mainly focusing on their applications within light-assisted treatments as photodynamic and photothermal therapies since they are rapid, affordable, and minimally invasive and have less side effects. The main CBM employed to achieve this aim comprises graphene oxide, reduced-graphene oxide, carbon dots, and carbon nanotubes; nevertheless, the preparation of hybrids and composites has also been proposed to improve their antibacterial effect. Metal nanoparticles, biopolymers, metal oxide nanoparticles, and so on have been employed. We discussed some of the mechanisms whereby bacteria are inhibited or killed. Several works reported in the literature have achieved the complete elimination of bacteria. The most studied species are *E. coli* and *S. aureus*. Hence, this chapter evidences that CBM could be used as a benchmark antimicrobial agent.

Acknowledgements

The authors would like to thank Escuela Politécnica Nacional.

Conflict of interest

The authors declare no conflict of interest.

Author details

David Giancarlo García Vélez, Karina Janneri Lagos Álvarez
and María Paulina Romero Obando*
Materials Department, National Polytechnic School, Quito, Ecuador

*Address all correspondence to: maria.romerom@epn.edu.ec

References

[1] Feng Y, Liu L, Zhang J, Aslan H, Dong M. Photoactive antimicrobial nanomaterials. Journal of Materials Chemistry B. 2017;5(44):8631-8652

[2] Castañeda Gullot C, Ramos SG. Principales pandemias en la historia de la humanidad. Revista Cubana de Pediatria. 2020;**92**:1-24

[3] Kong X, Liu X, Zheng Y, Chu PK, Zhang Y, Wu S. Graphitic carbon nitride-based materials for photocatalytic antibacterial application. Materials Science and Engineering: R: Reports. 2021;**145**:100610

[4] Liang J, Li W, Chen J, Huang X, Liu Y, Zhang X, et al. Antibacterial activity and synergetic mechanism of carbon dots against gram-positive and -negative bacteria. ACS Applied Bio Materials. 2021;**4**(9):6937-6945

[5] Knoblauch R, Geddes CD. Carbon nanodots in photodynamic antimicrobial therapy: A review. Materials (Basel). 2020;**13**(18):2-35

[6] Paramanantham P, Anju VT, Dyavaiah M, Siddhardha B. Applications of carbon-based nanomaterials for antimicrobial photodynamic therapy. In: Prasad R, editor. Nanotechnology in the Life Sciences. Switzerland: Springer, Cham; 2019. pp. 237-259

[7] Ren Y, Liu H, Liu X, Zheng Y, Li Z, Li C, et al. Photoresponsive materials for antibacterial applications. Cell Reports Physical Science. 2020;**1**(11):100245

[8] Ortez JH, Rankin ID. Manual de pruebas de susceptibilidad antimicrobiana. In: Coyle MB, editor. Manual de Pruebas de Susceptibilidad Antimicrobiana. Seattle, Washington: American Society for Microbiology; 2005. p. 242

[9] Liu Y, Guo Z, Li F, Xiao Y, Zhang Y, Bu T, et al. Multifunctional magnetic copper ferrite nanoparticles as Fenton-like reaction and near-infrared photothermal agents for synergetic antibacterial therapy. ACS Applied Materials & Interfaces. 2019;**11**(35):31649-31660

[10] Al-Karagoly H, Rhyaf A, Naji H, Albukhaty S, Almalki FA, Alyamani AA, et al. Green synthesis, characterization, cytotoxicity, and antimicrobial activity of iron oxide nanoparticles using Nigella sativa seed extract. Green Processing and Synthesis. 2022;**11**(1):254-265

[11] Chen S, Saltikov C, Nichols F, Lu JE, Mercado R, Rojas-Andrade MD, et al. Antibacterial activity of Nitrogen-doped Carbon dots enhanced by atomic dispersion of copper. Langmuir. 2020;**36**(39):11629-11636

[12] Cuadrado CF, Díaz-Barrios A, Campaña KO, Romani EC, Quiroz F, Nardecchia S, et al. Broad-spectrum antimicrobial ZnMintPc encapsulated in magnetic-nanocomposites with graphene oxide/MWCNTs based on bimodal action of photodynamic and photothermal effects. Pharmaceutics. 2022;**14**(4):2-26

[13] McGhie BS, Aldrich-Wright JR. Photoactive and luminescent transition metal complexes as anticancer agents: A guiding light in the search for new and improved cancer treatments. Biomedicines - MDPI. 2022;**10**:2-4

[14] El-Gendy NS, Omran BA. Green synthesis of nanoparticles for water treatment. In: Fosso-Kankeu E, editor. Nano and Bio-Based Technologies for Wastewater Treatment: Prediction and Control Tools for the Dispersion of

Pollutants in the Environment. Vol. 1. 2019

[15] Saharan P, Chaudhary GR, Mehta SK, Umar A. Removal of water contaminants by iron oxide nanomaterials. Journal of Nanoscience and Nanotechnology. 2014;**14**(1):627-643

[16] Jani M, Arcos-Pareja JA, Ni M. Engineered zero-dimensional Fullerene/Carbon dots-polymer based nanocomposite membranes for wastewater treatment. Molecules. 2020;**25**(21):1-28

[17] Israr M, Iqbal J, Arshad A, Aisida SO, Ahmad I. A unique $ZnFe_2O_4$/graphene nanoplatelets nanocomposite for electrochemical energy storage and efficient visible light driven catalysis for the degradation of organic noxious in wastewater. Journal of Physics and Chemistry of Solids. 2020;**140**:109333

[18] Doganay D, Kanicioglu A, Coskun S, Akca G, Unalan HE. Silver-nanowire-modified fabrics for wide-spectrum antimicrobial applications. Journal of Materials Research. 2019;**34**(4):500-509

[19] Raul PK, Thakuria A, Das B, Devi RR, Tiwari G, Yellappa C, et al. Carbon nanostructures as antibacterials and active food-packaging materials: A review. ACS Omega. 2022;**7**:11555-11559

[20] Mousavi SM, Hashemi SA, Kalashgrani MY, Omidifar N, Bahrani S, Rao NV, et al. Bioactive graphene quantum dots based polymer composite for biomedical applications. Polymers (Basel). 2022;**14**(3):2-30

[21] Tonelli AM, Venturini J, Arcaro S, Henn JG, Moura DJ, da Viegas AC, et al. Novel core-shell nanocomposites based on TiO_2-covered magnetic Co_3O_4 for biomedical applications. Journal of Biomedical Materials Research Part B Applied Biomaterials. 2020;**108**(5):1879-1887

[22] Yang J, Zhang X, Ma YH, Gao G, Chen X, Jia HR, et al. Carbon dot-based platform for simultaneous bacterial distinguishment and antibacterial applications. ACS Applied Materials & Interfaces. 2016;**8**(47):32170-32181

[23] Lu Y, Li L, Li M, Lin Z, Wang L, Zhang Y, et al. Zero-dimensional carbon dots enhance bone regeneration, osteosarcoma ablation, and clinical bacterial eradication. Bioconjugate Chemistry. 2018;**29**(9):2982-2993

[24] Zhao X, Li J, Liu D, Yang M, Wang W, Zhu S, et al. Self-enhanced carbonized polymer dots for selective visualization of lysosomes and real-time apoptosis monitoring. iScience. 2020;**23**(4):100982

[25] Hamblin MR. Antimicrobial photodynamic inactivation: A bright new technique to kill resistant microbes. Current Opinion in Microbiology [Internet]. 2016;**33**:67-73 Available from: https://pubmed.ncbi.nlm.nih.gov/27421070

[26] Baltazar LM, Ray A, Santos DA, Cisalpino PS, Friedman AJ, Nosanchuk JD. Antimicrobial photodynamic therapy: An effective alternative approach to control fungal infections. Frontiers in Microbiology. 2015;**6**:1-11

[27] Qi M, Chi M, Sun X, Xie X, Weir MD, Oates TW, et al. Novel nanomaterial-based antibacterial photodynamic therapies to combat oral bacterial biofilms and infectious diseases. International Journal of Nanomedicine. 2019;**14**:6937-6956

[28] Romero MP, Alves F, Stringasci MD, Buzzá HH, Ciol H, Inada NM, et al. One-pot microwave-assisted synthesis of carbon dots and in vivo and in vitro antimicrobial photodynamic applications. Frontiers in Microbiology. 2021;**12**:1-13

[29] Chen Y, Gao Y, Chen Y, Liu L, Mo A, Peng Q. Nanomaterials-based photothermal therapy and its potentials in antibacterial treatment. Journal of Controlled Release [Internet]. 2020;**328**:251-262. Available from:. DOI: 10.1016/j.jconrel.2020.08.055

[30] Liu Y, Xu B, Lu M, Li S, Guo J, Chen F, et al. Ultrasmall Fe-doped carbon dots nanozymes for photoenhanced antibacterial therapy and wound healing. Bioactive Materials. [Internet]. 2022;**12**:246-256. Available from:. DOI: 10.1016/j.bioactmat.2021.10.023

[31] Romero MP, Marangoni VS, de Faria CG, Leite IS, de Silva CCC e, Maroneze CM, et al. Graphene oxide mediated broad-spectrum antibacterial based on bimodal action of photodynamic and photothermal effects. Frontiers in Microbiology. 2020;**10**:1-15

[32] Cui Q, Yuan H, Bao X, Ma G, Wu M, Xing C. synergistic photodynamic and photothermal antibacterial therapy based on a conjugated polymer nanoparticle-doped hydrogel. ACS Applied Bio Materials. 2020;**3**(7):4436-4443

[33] Yan H, Zhang B, Zhang Y, Su R, Li P, Su W. Fluorescent carbon dot-curcumin nanocomposites for remarkable antibacterial activity with synergistic photodynamic and photothermal abilities. ACS Applied Bio Materials. 2021;**4**(9):6703-6718

[34] Alan JF. Adaptative Phased Array Thermotherapy for Cancer. Vol. 2006. Damascus University Publications. United States of America: Artech House; 1999. pp. 1-6

[35] Kok HP, Cressman ENK, Ceelen W, Brace CL, Ivkov R, Grüll H, et al. Heating technology for malignant tumors: A review. International Journal of Hyperthermia. 2020;**37**(1):711-741

[36] Nadler SF, Weingand K, Kruse RJ. The physiologic basis and clinical applications of cryotherapy and thermotherapy for the pain practitioner. Pain Physician. 2004;**7**(3):395-399

[37] Gabriela NA, Grumezescu AM. Applied sciences photodynamic therapy — An up-to-date review. Applied Sciences. 2021;**11**:3626

[38] Hu T, Wang Z, Shen W, Liang R, Yan D, Wei M. Recent advances in innovative strategies for enhanced cancer photodynamic therapy. Theranostics. 2021;**11**(7):3278-3300

[39] Rhodes LE, De Rie M, Enström Y, Groves R, Morken T, Goulden V, et al. Photodynamic therapy using topical methyl aminolevulinate vs surgery for nodular basal cell carcinoma: results of a multicenter randomized prospective trial. Archives of Dermatology. 2004;**140**:17-23

[40] Estelrich J, Antònia BM. Iron oxide nanoparticles in photothermal therapy. Molecules. 2018;**23**(7):2-26

[41] Pedrosa P, Mendes R, Cabral R, Martins LMDRS, Baptista PV, Fernandes AR. Combination of chemotherapy and Au-nanoparticle photothermy in the visible light to tackle doxorubicin resistance in cancer cells. Scientific Reports. 2018;**8**(1):1-8

[42] Sun Q, Hou P, Wu S, Yu L, Dong L. The enhanced photocatalytic activity of Ag-Fe_2O_3-TiO_2 performed in Z-scheme route associated with localized surface plasmon resonance effect. Colloids and Surfaces A: Physicochemical and Engineering Aspects. 2021;**628**:127304

[43] Hu M, Chen J, Li ZY, Au L, Hartland GV, Li X, et al. Gold nanostructures: Engineering their plasmonic properties for biomedical

applications. Chemical Society Reviews. 2006;**35**(11):1084-1094

[44] Xin Q, Shah H, Nawaz A, Xie W, Akram MZ, Batool A, et al. Antibacterial carbon-based nanomaterials. Advanced Materials. 2019;**31**(45):1-15

[45] Patil RM, Thorat ND, Shete PB, Bedge PA. Comprehensive cytotoxicity studies of superparamagnetic iron oxide nanoparticles. Biochemistry and Biophysics Reports. 2018;**13**:63-72

[46] Lu X, Rycenga M, Skrabalak SE, Wiley B, Xia Y. Chemical synthesis of novel plasmonic nanoparticles. Annual Review of Physical Chemistry. 2009;**60**:167-192

[47] Wu LN, Yang YJ, Huang LX, Zhong Y, Chen Y, Gao YR, et al. Levofloxacin-based carbon dots to enhance antibacterial activities and combat antibiotic resistance. Carbon N Y. 2022;**186**:452-464

[48] Banerjee AN. Graphene and its derivatives as biomedical materials: Future prospects and challenges. Interface Focus. 2018;**8**(3):1-22

[49] Nurunnabi M, Parvez K, Nafiujjaman M, Revuri V, Khan HA, Feng X, et al. Bioapplication of graphene oxide derivatives: Drug/gene delivery, imaging, polymeric modification, toxicology, therapeutics and challenges. RSC Advances. 2015;**5**(52):42141-42161

[50] Dhinakaran V, Lavanya M, Vigneswari K, Ravichandran M, Vijayakumar MD. Review on exploration of graphene in diverse applications and its future horizon. Materials Today Proceedings [Internet]. 2020;**27**:1-5. DOI: 10.1016/j.matpr.2019.12.369

[51] Hu W, Peng C, Luo W, Lv M, Li X, Li D, et al. Graphene-based antibacterial paper. ACS Nano. 2010;**4**(7):4317-4323

[52] Kumar P, Huo P, Zhang R, Liu B. Antibacterial properties of graphene-based nanomaterials. Nanomaterials. [Internet]. 2019;**9**(5):737 Available from: https://www.mdpi.com/2079-4991/9/5/737

[53] Galván EM, Mateyca C, Ielpi L. Role of interspecies interactions in dual-species biofilms developed in vitro by uropathogens isolated from polymicrobial urinary catheter-associated bacteriuria. Biofouling. 2016;**32**(9):1067-1077

[54] Adetayo A, Runsewe D. Synthesis and fabrication of graphene and graphene oxide: A review. Open Journal of Composite Materials. 2019;**09**(02):207-229

[55] Yu W, Sisi L, Haiyan Y, Jie L. Progress in the functional modification of graphene/graphene oxide: A review. RSC Advances. 2020;**10**(26):15328-15345

[56] Gao W. Graphene Oxide: Reduction Recipes, Spectroscopy, and Applications [Internet]. Switzerland: Springer International Publishing; 2015. Available from: https://books.google.com.ec/books?id=-Z-4CQAAQBAJ

[57] Bai RG, Muthoosamy K, Manickam S, Hilal-Alnaqbi A. Graphene-based 3D scaffolds in tissue engineering: Fabrication, applications, and future scope in liver tissue engineering. International Journal of Nanomedicine. 2019;**14**:5753-5783

[58] Juarez GE, Galván EM. Role of nutrient limitation in the competition between uropathogenic strains of Klebsiella pneumoniae and *Escherichia coli* in mixed biofilms. Biofouling. 2018;**34**(3):287-298

[59] Verderosa AD, Totsika M, Fairfull-Smith KE. Bacterial biofilm eradication agents: A current review. Frontiers in Chemistry. 2019;**7**:1-17

[60] Harun SW. Handbook of graphene. In: Biomaterials [Internet]. Vol. 7. Malaya: Wiley; 2019. Available from: https://books.google.com.ec/books?id=ckKdDwAAQBAJ

[61] Liu S, Hu M, Zeng TH, Wu R, Jiang R, Wei J, et al. Lateral dimension-dependent antibacterial activity of graphene oxide sheets. Langmuir. 2012;**28**(33):12364-12372

[62] Pulingam T, Thong KL, Appaturi JN, Lai CW, Leo BF. Mechanistic actions and contributing factors affecting the antibacterial property and cytotoxicity of graphene oxide. Chemosphere. [Internet]. 2021;**281**:130739. Available from:. DOI: 10.1016/j.chemosphere.2021.130739

[63] Anand A, Unnikrishnan B, Wei SC, Chou CP, Zhang LZ, Huang CC. Graphene oxide and carbon dots as broad-spectrum antimicrobial agents-a minireview. Nanoscale Horizons. 2019;**4**(1):117-137

[64] Cui L, Ren X, Sun M, Liu H, Xia L. Carbon dots: Synthesis, properties and applications. Nanomaterials. 2021;**11**(12):1-38

[65] Chong Y, Ge C, Fang G, Wu R, Zhang H, Chai Z, et al. Light-enhanced antibacterial activity of graphene oxide, mainly via accelerated electron transfer. Environmental Science & Technology [Internet]. 2017;**51**(17):10154-10161. Available from:. DOI: 10.1021/acs.est.7b00663

[66] Kholikov K, Ilhom S, Sajjad M, Smith ME, Monroe JD, San O, et al. Improved singlet oxygen generation and antimicrobial activity of sulphur-doped graphene quantum dots coupled with methylene blue for photodynamic therapy applications. Photodiagnosis and Photodynamic Therapy. 2018;**24**:7-14

[67] Akbari T, Pourhajibagher M, Hosseini F, Chiniforush N, Gholibegloo E, Khoobi M, et al. The effect of indocyanine green loaded on a novel nano-graphene oxide for high performance of photodynamic therapy against Enterococcus faecalis. Photodiagnosis and Photodynamic Therapy. [Internet]. 2017;**20**:148-153. DOI: 10.1016/j.pdpdt.2017.08.017

[68] Ma J, Zhang J, Xiong Z, Yong Y, Zhao XS. Preparation, characterization and antibacterial properties of silver-modified graphene oxide. Journal of Materials Chemistry [Internet]. 2011;**21**(10):3350-3352. Available from:. DOI: 10.1039/C0JM02806A

[69] Xu WP, Zhang LC, Li JP, Lu Y, Li HH, Ma YN, et al. Facile synthesis of silver@graphene oxide nanocomposites and their enhanced antibacterial properties. Journal of Materials Chemistry [Internet]. 2011;**21**(12):4593-4597. Available from:. DOI: 10.1039/C0JM03376F

[70] Xie X, Mao C, Liu X, Zhang Y, Cui Z, Yang X, et al. Synergistic bacteria killing through photodynamic and physical actions of graphene oxide/Ag/collagen coating. ACS Applied Materials & Interfaces. 2017;**9**(31):26417-26428

[71] Pourhajibagher M, Moghadam SE, Alaeddini M. OPEN DNA - aptamer - nanographene oxide as a targeted bio - theragnostic system in antimicrobial photodynamic therapy against Porphyromonas gingivalis. Scientific Reports. 2022;**0123456789**:1-18

[72] Ma G, Qi J, Cui Q, Bao X, Gao D, Xing C. Graphene oxide composite for selective recognition, capturing, photothermal killing of bacteria over mammalian cells. Polymers (Basel). 2020;**12**(5):1-14

[73] Mei L, Lin C, Cao F, Yang D, Jia X, Hu S, et al. Amino-functionalized

graphene oxide for the capture and photothermal inhibition of bacteria. ACS Applied Nano Materials. 2019;**2**(5):2902-2908

[74] Magro M, Baratella D, Bonaiuto E, de Roger AJ, Vianello F. New perspectives on biomedical applications of iron oxide nanoparticles. Current Medicinal Chemistry. 2017;**25**(4):540-555

[75] Singh V, Joung D, Zhai L, Das S, Khondaker SI, Seal S. Graphene based materials: Past, present and future. Progress in Materials Science. [Internet]. 2011;**56**(8):1178-1271. Available from:. DOI: 10.1016/j.pmatsci.2011.03.003

[76] Chung YJ, Kim J, Park CB. Photonic carbon dots as an emerging nanoagent for biomedical and healthcare applications. ACS Nano. 2020;**14**(6):6470-6497

[77] Koutsogiannis P, Thomou E, Stamatis H, Gournis D, Rudolf P. Advances in fluorescent carbon dots for biomedical applications. Advances in Physics: X. 2020;**5**(1):2-37

[78] Das A, Kundelev EV, Vedernikova AA, Cherevkov SA, Danilov DV, Koroleva AV, et al. Revealing the nature of optical activity in carbon dots produced from different chiral precursor molecules. Light: Science & Applications. 2022;**11**(1):1-13

[79] Ran HH, Cheng X, Bao YW, Hua XW, Gao G, Zhang X, et al. Multifunctional quaternized carbon dots with enhanced biofilm penetration and eradication efficiencies. Journal of Materials Chemistry B. 2019;7(33):5104-5114

[80] Sun B, Wu F, Zhang Q, Chu X, Wang Z, Huang X, et al. Insight into the effect of particle size distribution differences on the antibacterial activity of carbon dots. Journal of Colloid and Interface Science. 2021;**584**:505-519

[81] Song Y, Lu F, Li H, Wang H, Zhang M, Liu Y, et al. Degradable carbon dots from cigarette smoking with broad-spectrum antimicrobial activities against drug-resistant bacteria. ACS Applied Bio Materials. 2018;**1**(6):1871-1879

[82] Yu M, Zhang G, Li P, Lu H, Tang W, Yang X, et al. Acid-activated ROS generator with folic acid targeting for bacterial biofilm elimination. Materials Science and Engineering: C. 2021;**127**:112225

[83] Chen S, Quan Y, Yu YL, Wang JH. Graphene quantum dot/silver nanoparticle hybrids with oxidase activities for antibacterial application. ACS Biomaterials Science & Engineering. 2017;**3**(3):313-321

[84] Jijie R, Barras A, Bouckaert J, Dumitrascu N, Szunerits S, Boukherroub R. Enhanced antibacterial activity of carbon dots functionalized with ampicillin combined with visible light triggered photodynamic effects. Colloids Surfaces B Biointerfaces. 2018;**170**:347-354

[85] Meziani MJ, Dong X, Zhu L, Jones LP, Lecroy GE, Yang F, et al. Visible-light-activated bactericidal functions of carbon "quantum" dots. ACS Applied Materials & Interfaces. 2016;**8**(17):10761-10766

[86] Kováčová M, Kleinová A, Vajďák J, Humpolíček P, Kubát P, Bodík M, et al. Photodynamic-active smart biocompatible material for an antibacterial surface coating. Journal of Photochemistry and Photobiology B: Biology. 2020;**211**:112012

[87] Liu W, Gu H, Ran B, Liu W, Sun W, Wang D, et al. Accelerated antibacterial red-carbon dots with photodynamic therapy against multidrug-resistant Acinetobacter baumannii. Science China Materials. 2022;**65**(3):845-854

[88] Xia C, Zhu S, Feng T, Yang M, Yang B. Evolution and synthesis of carbon dots: From carbon dots to carbonized polymer dots. Advancement of Science. 2019;**6**(23):2-23

[89] Wang L, Li Y, Wang Y, Kong W, Lu Q, Liu X, et al. Chlorine-doped graphene quantum dots with enhanced anti- and pro-oxidant properties. ACS Applied Materials & Interfaces. 2019;**11**(24):21822-21829

[90] Khan ZG, Patil PO. A comprehensive review on carbon dots and graphene quantum dots based fluorescent sensor for biothiols. Microchemical Journal. 2020;**157**:105011

[91] Li H, Kang Z, Liu Y, Lee ST. Carbon nanodots: Synthesis, properties and applications. Journal of Materials Chemistry. 2012;**22**(46):24230-24253

[92] Lim SY, Shen W, Gao Z. Carbon quantum dots and their applications. Chemical Society Reviews. 2015;**44**(1):362-381

[93] Sowers KL, Hou Z, Peterson JJ, Swartz B, Pal S, Prezhdo O, et al. Photophysical properties of CdSe/CdS core/shell quantum dots with tunable surface composition. Chemical Physics. 2016;**471**:24-31

[94] Al Awak MM, Wang P, Wang S, Tang Y, Sun YP, Yang L. Correlation of carbon dots' light-activated antimicrobial activities and fluorescence quantum yield. RSC Advances. 2017;**7**(48):30177-30184

[95] Abu Rabe DI, Mohammed OO, Dong X, Patel AK, Overton CM, Tang Y, et al. Carbon dots for highly effective photodynamic inactivation of multidrug-resistant bacteria. Materials Advances. 2020;**1**(3):321-325

[96] Dou Q, Fang X, Jiang S, Chee PL, Lee TC, Loh XJ. Multi-functional fluorescent carbon dots with antibacterial and gene delivery properties. RSC Advances. 2015;**5**(58):46817-46822

[97] Knoblauch R, Harvey A, Ra E, Greenberg KM, Lau J, Hawkins E, et al. Antimicrobial carbon nanodots: Photodynamic inactivation and dark antimicrobial effects on bacteria by brominated carbon nanodots. Nanoscale. 2021;**13**(1):85-99

[98] Qie X, Zan M, Gui P, Chen H, Wang J, Lin K, et al. Design, synthesis, and application of carbon dots with synergistic antibacterial activity. Frontiers in Bioengineering and Biotechnology. 2022;**10**:1-9

[99] Chu X, Zhang P, Wang Y, Sun B, Liu Y, Zhang Q, et al. Near-infrared carbon dot-based platform for bioimaging and photothermal/photodynamic/quaternary ammonium triple synergistic sterilization triggered by single NIR light source. Carbon N Y. 2021;**176**:126-138

[100] Mocan T, Matea CT, Pop T, Mosteanu O, Buzoianu AD, Suciu S, et al. Carbon nanotubes as anti-bacterial agents. In: Cellular and Molecular Life Sciences. Vol. 74. Switzerland: Birkhauser Verlag AG; 2017. pp. 3467-3479

[101] Albert K, Hsu HY. Carbon-based materials for photo-triggered theranostic applications. Molecules. 2016;**21**(11):1-29

[102] Malek I, Schaber CF, Heinlein T, Schneider JJ, Gorb SN, Schmitz RA. Vertically aligned multi walled carbon nanotubes prevent biofilm formation of medically relevant bacteria. Journal of Materials Chemistry B. 2016;**4**(31):5228-5235

[103] Dong L, Henderson A, Field C. Antimicrobial activity of single-walled carbon nanotubes suspended in different surfactants. Journal of Nanotechnology. 2012;**2012**:1-7

Section 3

Tissue Engineering

Chapter 6

Frontier Electrospun Fibers for Nanomedical Applications

Emilija Zdraveva and Budimir Mijovic

Abstract

Nanofibers fabrication nowadays has become unimaginable without mentioning or research involving the technique of electrospinning. Due to the vast possibilities that this technique offers in regard to nanofibers morphology, nanofibrous architecture, and application perspective, it has become the main interest of many scientists with various expertise profiles. Electrospun nanofibers are advantageous over conventional fibers due to their lightweight, high surface-to-volume ratio, adjustable fiber diameter/morphology, and well-controlled functionality. This chapter will highlight the possibilities of nanofibers' functionalization toward nanomedical applications including, drug delivery, wound healing systems, and tissue engineering scaffolds with a focus on bone and nerve tissue repair. The latest studies (from 2017 onwards) are discussed in terms of materials' composition, fabrication technologies, and significant performance of cultured cells *in vitro* and most importantly regenerated tissue after implantation *in vivo*.

Keywords: electrospinning, nanofibers, drug delivery, wound healing, scaffolds

1. Introduction

The principle of the electrospinning process is based on the physical phenomenon that describes the cone stretching of a water droplet when charged amber is held above it. Similarly, an electrostatic force stretches a polymer solution or melt, to form the popular Taylor cone while generating nano-scaled fibers with the aid of a pump and a high voltage power supply. As produced nanofibers have advantageous properties over conventional ones including, small diameter (few nm to 1 μm), high surface-to-volume ratio, light weight structures with interconnected pores, high porosity, and ease of controlled functionality. Their uniqueness makes them perfect candidates for the application in nanomedicine. This review will focus on the most reported nanomedical application areas such as tissue engineering, thus the development of scaffolds for both hard (bones) and soft tissues (cartilage, skin, tendons and ligaments, blood vessels, muscles, and nerves), drug delivery, and wound healing. Special attention will be given to the latest frontiers in functional nanofibers, facing challenges and future aspects.

Apart from mimicking the extracellular matrix (ECM) in native tissues, electrospun nanofibrous materials can meet the requirements in scaffold development that is, biodegradability and biocompatibility, adjustable porosity and mechanical integrity, as well as sufficient delivery of oxygen and nutrients through permeability. Functionalization of these structures by means of medical/biological species incorporation or surface modification provides them to act as delivery systems for fast, sustained, or tunable drug release. Further, in wound healing the electrospun dressing due to its porous structure allows both breathability and waterproofness. Finally, the promotion of rapid wound healing is achieved by specific cell type culture while supporting cell adhesion, migration, and proliferation.

2. Latest progress in nanofibrous drug delivery systems

A drug delivery system is an engineered formulation or device that can carry a therapeutic agent and deliver it at a specific time, with a controlled release rate and dosage, to a target site in the body. The advantages of a drug delivery system over conventional drug administration routes are improved efficiency or sustained drug concentration over time, reliability or reproducible release rates, low cytotoxicity, no side-effects or repetitive dosing, therapy optimization, elimination of the drug stability issue, and conserved agent activity [1, 2]. These systems are designed to treat cancer, and certain ailments and for tissue regeneration [3]. With emerge of nanotechnology, modern drug delivery systems like electrospun nanofibers have attracted much attention within the scope of smart therapies. Electrospun nanofibers offer suitable polymer utilization, controllable process conditions, and ease of drug release design in a protracted, stimulus-activated, and biphasic manner. In order to fabricate this specific on-demand drug delivery nanofibrous material many variables need to be optimized in terms of composition, process requirements, and possible post-processing treatment [4].

2.1 Drug incorporation technologies

Both medical and biological components (single or multiple) are incorporated into nanofibers by blend electrospinning [5], emulsion electrospinning [6], bi-component (co-axial, side by side) [7], tri-axial electrospinning [8] or post-treatment via surface immobilization [3]. Blend electrospinning combines the same solvent-dissolving components, while heterogeneous systems are spun through emulsion electrospinning. Core-shell fibers can be fabricated by both co-axial or emulsion electrospinning where usually the core contains the drug while the shell is a polymer. The first is made by co-axial nozzles while the latter is spontaneous due to the different polymeric solutions' inherent surface energies [9]. A special nozzle is designed for side-by-side electrospinning where the resulting fiber contains the polymer matrix and the drug separately. This approach is used when the drug is not soluble in the polymer solvent [10]. Target drug molecules are loaded in the nanofibers through simple physical adsorption, nanoparticle encapsulation, or surface coating via layer-by-layer deposition. Bioactive molecules can also be immobilized on the fibers' surface through chemical modification by covalent bonding (i.e., carboxyl, amine, and hydroxyl groups) [11].

Blend electrospun chitosan/polyvinyl alcohol (CS/PVA) nanofibers were loaded with ofloxacin to be used as an ocular drug delivery system for a sustained release of up to 96 h. The fabrication process involved multilayered electrospinning to result in a sandwich structure, thus the CS/PVA was covered with a hydrophobic polymer, Eudragit RL100, on top and bottom and further cross-linked by glutaraldehyde [12].

An anticancer agent doxorubicin was successfully incorporated into polyvinyl alcohol/polycaprolactone core-shell nanofibers by coaxial electrospinning. These pH-responsive drug-release fibers were efficient against cervical cancer cell attachment and proliferation [13]. Similarly, core-shell drug-loaded polyvinyl pyrrolidone (PVP)/polylactic acid (PLA) nanofibers were prepared by electrospinning an emulsion through a single nozzle set-up. The drugs loaded were procyanidins, a natural antioxidant, and *Apocynum Venetum* cellulose nanofibers. The core was the water phase and was prepared by adding the fourth-mentioned hydrophilic drugs into the PVP aqueous solution. These were added into the oil phase (the sheath) containing the PLA/emulsifier solution. The antioxidant efficiency of the drug delivery system was 88.62%, as confirmed by the 2,2-diphenyl-1-picrylhydrazyl assay [14]. A side-by-side setup was used to electrospin the so-called Janus beads-on-a-string fibers comprising polyvinylpyrrolidone K90 on one side and ethyl cellulose on the other side loaded with two model drugs, methylene blue and ketoprofen, respectively. In this complex double drug-loaded system, the bead sides were the insoluble ethyl cellulose and the ketoprofen drug. These nanofibers performed with double drugs controlled-release profiles [15]. Multiple functional molecules with different therapeutic activities can be produced by multiaxial electrospinning. For example, triaxial fibers consisted of a polyvinylpyrrolidone core, a polycaprolactone (PCL) intermediate layer, and an outermost PCL layer as well (the two were dissolved in different solvents). The model molecules used were keyacid blue dye, loaded in the core and keyacid uranine loaded in the sheath layer. In this type of drug delivery vehicles, the intermediate layer serves as the barrier as to prevent leaching from the core [16]. Generally, polymers mostly used in drug delivery systems include natural proteins or polysaccharides like gelatin, collagen, albumin or chitosan, dextran, alginic acid, respectively. Synthetic polymers used include both biodegradable and non-biodegradable ones, that is, polyesters, poly(ortho esters), poly(alkyl cyanoacrylates), acrylic polymers (poly(methyl methacrylate), and poly(hydroxyl ethyl methacrylate)), respectively. Cellulose derivatives including, ethyl cellulose and hydroxypropyl methylcellulose are used as well [17]. A range of natural and synthetic drugs have been reported to be incorporated into electrospun nanofibers such as: antibiotics, antioxidants, antibodies, anticancer drugs, proteins, etc., as well as compounds like growth factors, DNA, and RNA [17, 18].

2.2 Drug release pace mechanisms

Drugs can be released from the electrospun nanofibers immediately, through the so-called burst effect. When immediate action is required the drug is released shortly. Further, the release can be sustained or slow through a certain period of time. This means low drug concentration delivered within several hours, days, or years. Finally, the release can be triggered by external stimuli, which will result in the first two release rates. External stimuli may involve changes in pH, temperature, light, solvent, ionic strength, etc. [19]. In electrospun nanofibers drug release rate can be controlled by polymer type (i.e., hydrophilic, hydrophobic, biodegradable), drug properties (solubility, stability, loading locations), fabrication technology, fibers morphology, and micro-structure [19, 20]. The mechanisms involved in the drug release profiles include diffusion and swelling followed by diffusion, degradation, or erosion (surface or bulk erosion) [19, 21]. An additive is released from a polymer matrix through diffusion, relaxation, or degradation. When diffusion occurs the added component diffuses down due to concentration difference, while in the relaxation state, it moves out from the polymer as a result of the chain relaxation. The diffusion depends on the drug's molecular weight, drug concentration, and solubility, also the diffusion coefficient and diffusion distance. A medium can cause the polymer to dissolve/degrade thus will release the added component. In surface erosion, the polymer surface erodes due to chain scission, while in bulk erosion the whole matrix erodes. In electrospun fibers, drugs are released from the fibers into the pores due to relaxation/degradation mechanisms or both, as well as from the pores via the diffusion mechanism [19, 22, 23]. As forth-mentioned, the fabrication technology will certainly affect the drug release profile thus the delivery system will be designed in compliance with the therapeutic requirements. The immediate burst release effect is achieved with the blend of electrospun fibers, while co-axial fibers will prolong the drug release as usually, the drug is within the fiber's core. The burst release of the drug is usually in the initial stage and if necessary this can be delayed by the side-by-side electrospinning technique. Emulsion electrospun fibers will also reduce the initial burst release effect. To further slow down the drug release rate several polymer layers/barriers can be added in the co-axial fibers resulting from the multi-axial approach. These fibers can provide a combination of several drugs and both immediate and sustained release and thus will be used for short or long-term therapies.

Unlike covalent bonding in the fiber's surface chemical modification approach, the physical adsorption results in fast drug release due to weak hydrogen bonds, and electrostatic or hydrophobic interactions [24]. Special vehicles encapsulating the drug can be added into electrospun fibers thus promoting drug-sustained release. Such vehicles include: liposomes, micelles, nanoparticles, nanotubes, nanospheres, microspheres, etc., which provide intracellular drug delivery [25–27]. These hybrid formulations can eliminate the burst release problem, improve drug loading efficiency, and kidney excretion and minimize drug fluctuations [28].

A study described in detail a three-stage long-term drug release profile of electrospun poly(D,L-lactide-co-glycolide) (PLGA) combined with one of the following polymers, poly(ethylene glycol) (PEG), poly(ethylene glycol)–b-poly(D,L -lactide) (PEG-b-PDLLA), polyglycolide (PGA), poly(dioxanone) (PDO), or poly(trimethylene carbonate) (PTMC), as well as ciprofloxacin hydrochloride (CiH) as an antimicrobial agent. Stage I is characterized by a burst release effect due to fiber swelling and the second polymer hydrophilicity. The second stage results in drug release through diffusion into the gel-continuous structure and in this stage the drug is released slowly. Stage III involves remaining drug release due to the polymer's hydrolytic degradation [29]. Another release mechanism was described for core-sheath polyvinylpyrrolidone-curcumin/poly (3-hydroxybutyric acid-co-3-hydroxyvaleric acid) (PVP-Cur/PHBV) electrospun fibers with the aid to delay the release of poorly water-soluble drugs. In a water-based medium this composite system increased the drug diffusion path between the core and the dissolution medium. More precisely, the water penetrates the PHBV layer first, then reaches the PVP core and returns back through the same diffusion path with a small amount of the core molecules. When the shell is fully penetrated by the water a large amount of the PVP layer along with the drug molecules passes to the release medium [30]. These are examples of drug delivery systems designed for the release of drugs in a perfectly controlled manner.

2.3 Most significant therapeutic performances

This section gives a brief overview of recently reported studies that concern mostly discussed therapies in drug delivery applications of electrospun nanofibers.

As forth-mentioned three groups of therapy areas were recognized, including cancer treatment, tissue repair, and some minor ailments.

As reported by the WHO, cancer is the second leading death worldwide and is the cause of one in six deaths, in 2018. The most common cancer in men and women are lung, prostate, colorectal, stomach and liver, and breast, colorectal, lung, cervical, and thyroid, respectively [31]. Recently, drug delivery systems are found promising in local chemotherapy and thus can provide high therapeutic performance, but also can be used in diagnostics. **Table 1** lists some examples of recent research on electrospun fibers for the treatment of hepatoma (liver), colon, breast, prostatic, and pancreatic cancer. Besides blend, emulsion, and co-axial electrospinning the drug delivery, nanofibers were fabricated by consecutive layering. Orthotopic hepatoma [32] and subcutaneous hepatoma [33] were successfully treated by emulsion and blend nanofibers composed of methoxy poly(ethylene glycol)-*block*-poly(lactide-*co*-glycolide) (mPEG-b-PLGA)/Dextran/10-Hydroxycamptothecin/Tea polyphenols and Poly(ethylene oxide) (PEO)/Poly(L-lactide) (PLA)/Doxorubicin hydrochloride, respectively. The sequential release and synergistic effect of the two drugs in the first study resulted in tumor inhibition as well as metastasis prevention [32]. In the second study, the authors highlighted the biphasic release profile of the PEO/PLA/Doxorubicin hydrochloride nanofibers which satisfied the demand for suppression of the initial excessive drug release, as to avoid its blood toxic level, as well as the demand to reach constant high drug level over a long period of time [33]. Tri-layered sandwich NFs composed of poly methyl methacrylate/polycaprolactone (PMMA/PCL), PCL/PMMA/6-Mercaptopurine, and PCL/PMMA layers showed high selective index for breast cancer and reduced cancer cells viability by 10% [34]. In another study, human prostatic cancer PC3 cells exhibited 38% alive cells, while breast cancer cells were death by 40–50%, when core-shell PCL or PCL/PVA NFs were used in combination with 5-fluorouracil or/and paclitaxel [35]. Pancreatic cancer treatment was also studied in terms of NFs compositions (PCL with 5-fluorouracil or methotrexate) and processing condition influences on the drugs release profiles [36].

Some of the diseases treated by drug-loaded electrospun Nfs reported were malaria, allergies, corneal abrasion, prosthesis, and gastroenteritis infections (**Table 1**). In the treatment of malaria, electrospun PCL/collagen/Artemisinin (ART) NFs have overcome the limitations of a neat ART, that is, short half-life, poor solubility, limited bioavailability, re-crystallization, and performed with a sustained release *in vivo* [37]. Soft tissue and bone infections caused by methicillin-resistant *Staphylococcus aureus* were studied by the local administration of Linezolid blended within electrospun PLGA/PCL. *In vivo* results in infected (due to tibia fracture) rats showed healing acceleration (cell growth and proliferation) due to drug sustained release over a longer time, minimized side effects, and reduced drug dose by 37-fold [38]. Bletilla striata polysaccharide (BSP) porous wafer was coated by electrospun PCL and loaded with Levofloxacin hydrochloride for the healing of acute infectious gastroenteritis. The system showed minor cytotoxicity to human gastric epithelial cells and high effective clearance of *Helicobacter pylori* compared to the pure drug [39]. The main problem in the administration of eye drugs is their short lifespan, due to immediate clearance as the eye cannot accommodate additional liquids. Authors developed a new solid *in situ* gelling system composed of electrospun gellan gum/pullulan NFs that forms gel immediately after contact with the ocular tissue. When compared to commercial eye drops in *in vivo* porcine corneas study the results showed higher fluorescein (dye model) signal intensity (up to 40%), thus prolonged residence time due to fiber higher viscosity, as well as more homogenous surface distribution due to lens curvature simulation [40].

Polymers	Drug	Electrospinning	Therapy	References
mPEG-b-PLGA/ Dextran	10- Hydroxycamptothecin / Tea polyphenols	Emulsion core-shell NFs	Orthotopic hepatic cancer	[32]
PEO/PLA	Doxorubicin hydrochloride	Blend NFs	Subcutaneous hepatic cancer	[33]
PMMA/PCL	6-Mercaptopurine	3-layered NFs	Colon, liver and breast cancer	[34]
PCL/PVA	5-Fluorouracil/ Paclitaxel	Core-shell NFs	Breast and prostatic cancer	[35]
PCL	5-Fluorouracil / Methotrexate	Core-shell NFs	Pancreatic cancer	[36]
PCL/Col	Artemisinin	Blend NFs	Malaria treatment	[37]
PLGA/PCL	Linezolid	Blend NFs	Prosthesis related infections	[38]
PCL/BSP	Levofloxacin hydrochloride	Porous wafer coated with electrospun NFs	Acute infectious gastroenteritis	[39]
Gellan gum/ pullulan	Dye model	Blend NFs	Ocular diseases	[40]
PVP/PLGA	Moxifloxacin/ Pirfenidone	Core-shell NFs	Corneal abrasion	[41]
(SF/PCL)/ PVA	Bone morphogenetic protein 2/ connective tissue growth factor	Core-shell/LBL NFs	Bone regeneration	[42]
PDLLA/ PLGA	Nerve growth factor/ Glial cell line-derived neurotrophic factor	Dual emulsion electrospun NFs	Nerve regeneration	[43]
PCL/gelatin	Vascular endothelial growth factor	NFs-encapsulated gelatin micro or sub-micro particles	Vascular tissue regeneration	[44]
PCL	Cilostazol	Blend NFs – tubular structure	Cardiovascular tissue regeneration	[45]
Chitosan/ poly-cyclodextrin	Simvastatin	Stent coated with NFs	Restenosis prevention	[46]

Table 1.
Recent studies in drug delivery NFs for cancer treatment, tissue repair, and some minor ailments.

A similar purpose was demonstrated by core-shell PVP/PLGA NFs encapsulating two types of drugs pirfenidone and moxifloxacin in the inner and outer layers, respectively. The system proved sustained drug release over a period of hours due to the effective entrapment of the same in the two compartments [41]. Besides medications generally in the field of tissue repair drug delivery systems incorporate genetic materials as well as bioactive molecules. Here studies reported on the fabrication of electrospun NFs with the combination of co-axial and layer-by-layer techniques, NFs with incorporated micro and sub/micro vehicles as well as NFs fabricated into tubular structures. In bone tissue regeneration, co-axially electrospun NFs were electrospun with the incorporation of bone morphogenetic protein 2 (BMP 2) into silk fibroin (SF)/PCL as the core, while pure PVA was the shell. PVA was further coated with surface immobilization of connective tissue growth factor (CTGF) with LBL assembly. Time controlled dual bioactive compound delivery resulted in enhanced bone formation with a pro-angiogenic effect (promotion of vessel formation) on bone healing due to sustained release of BMP2 and transient release of CTGF, respectively, when implanted subcutaneously in the abdominal midline of a nude mice [42]. For peripheral nerve regeneration poly(D, L-lactic acid) (PDLLA)/PLGA NFs incorporating nerve growth factor (NGF) and glial cell line-derived neurotrophic factor (GDNF), separately, were prepared by dual emulsion electrospinning. In the *in vitro* studies with rat pheochromocytoma cell line culture, both released incorporated components induced high neurite outgrowth and neural differentiation, **Figure 1**. The bioactivity of the growth factors was well preserved as indicated by the minor decrease (about 10%) of the degree of neural differentiation [43]. Each year 17.9 million deaths result due to cardiovascular diseases such as coronary heart disease, cerebrovascular disease, rheumatic heart disease, etc. [31]. Stent structures were prepared by tubular electrospinning of blend PCL/Cilostazol [45] and by coating self-expandable nitinol stent with blend electrospun Chitosan/β-cyclodextrin/Simvastatin [46]. In the first study, according to reported diffusion and polymer relaxation, the cilostazol release mechanism of the delivery system will facilitate the reendothelialization process [45]. In the second study, the cell viability tests showed that endothelial cells were less affected by the simvastatin than smooth muscle cells, thus confirming the system's selective activity toward the two types of cells present in the vessel wall [46]. When gelatin micro- or sub-micro particles carrying vascular endothelial growth factor, were incorporated into electrospun PCL, the drug delivery system showed to induce mesenchymal stem cells differentiation to endothelial cells and maintain angiogenesis for long periods of time as the nodes and tubes remained around 54.7% and 50.3%, respectively, from the original number after 24 h [44].

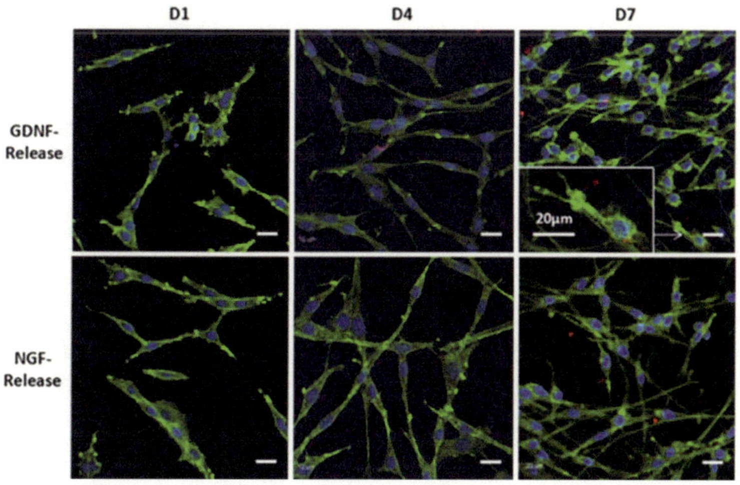

Figure 1.
The release of GDNF (upper images), the release of NGF (lower images) from PDLLA/PLGA scaffold [43].

3. Advances in nanofibrous wound dressings

This section will partially overlap with the previous one due to medicated electrospun nanofibers used in the wound healing process, thus the focus will be on their performance with no detailed discussion on materials or fabrication technologies.

A wound is cellular/tissue structure (i.e., skin) damage caused by traumatic injuries from external mechanical forces, surgeries, burns, chemical agents, or chronic disease ulcers, which results in normal tissue function loss [47, 48]. The process of wound healing involves complex interactions between cells and "mediators" and undergoes through the stages of hemostasis (days 1–3), and inflammation (days 4–6), chemotaxis and activation, proliferation (epithelization, angiogenesis, and provisional matrix formation, days 4–14) and maturation and remodeling (days 8–1 year) [49]. There are acute and chronic wounds, but also some chronic wounds are non-healable and result from acute ones [50]. Wound dressings are used in the process of wound healing and are categorized as passive, interactive, advanced, and bioactive wound dressings. An ideal wound dressing should provide: moisturized, clean and warm environment, hydration, protection to the peri-wound area, elimination of excess exudates, is permeable to gas exchange and impermeable to microorganisms, its nontoxic, nontraumatic, wound shape conforming, eliminates discomfort, easy to handle, and cheap [51, 52]. Electrospun nanofibrous dressings can be categorized as advanced or bioactive wound dressings due to the usual drugs/bioactive components incorporation and the fact that these modern materials offer advanced features. Primarily they can prevent wound desiccation by liquid exudation, provide controlled evaporation, excellent oxygen permeability, and promote fluid drainage capacity, as well as microorganism invasion inhibition [53]. Secondarily, encapsulated active components will enhance the healing process.

Recently reported studies are discussed in terms of *in vitro* and *in vivo* materials performances as well as to-date clinical trials.

Table 2 summarizes some of the most significant studies reported recently concerning wound dressings categorized as antibacterial, diabetic wounds, skin burns, and general or non-categorized dressings. Chronic wounds in diabetic patients are due to uncontrolled blood sugar, thus delaying the healing process and developing foot ulcers due to intense inflammation [54, 55]. The study of electrospun cellulose acetate (CA)/gelatin (Gel)/berberine (Beri) concerns the fourth-mentioned issue. It reveals that the developed dressing has both microbial and antibacterial effects as the number of the bacterial colony after 7 days of incubation was less than two and the minimum inhibition concentration for the *S. aureus* and *Pseudomonas aeruginosa* was 44 ± 5.7, respectively. The MTT assay also confirms the highest cell proliferation, while the histopathology and histomorphometry on the animal wounds showed the highest re-epithelialization after 16 days, healthier-looking skin with a thin epidermis, normal rete ridges, and skin thickness [54]. The incorporation of the alkannin/shikonin mixture in electrospun CA/PCL [56] and beta-glucan in hydroxypropyl methylcellulose/PEO [57] confirmed great cytocompatibility to fibroblast and keratinocyte cells, respectively. In the first study, the Hs27 fibroblast cells were elongated, well distributed, and in the formation of clusters [56]. The latter proved that the excipients in the *in situ* gel-forming of the βG-nanofibers had a positive effect on wound healing, while the mats improved the repair physically as well [57]. A multi-layered, chitosan/PVA + chitosan/PVA + nanobioglass (nBG), sequentially electrospun dressing provided a multifunctional activity thus accelerating the repair of chronic diabetic wounds. The authors reported that the nBG reduced the effect

Composition	Wound therapy	Cell performance	Animal study	References
CA/Gelatin/ Berberine	Diabetic foot ulcer (DFU) treatment	Enhance proliferation of L929 murine fibroblastic cells	Diabetic Wistar rats, highest collagen density, 88.8 ± 6.7%, highest angiogenesis, 19.8 ± 3.8	[54]
CA/PCL/Alkannin/ Shikonin	Skin chronic wounds	Elicit Hs27 fibroblast cells migration of 77.1%, high wound closure of 117.9%		[55]
Chitosan/ PVA + Chitosan/ PVA + nBG	Skin chronic (diabetic) wounds	The bioactive ions promoted human dermal fibroblastic cells proliferation, and expression of bFGF and VEGF	Wound closure acceleration in Sprague-Dawley diabetic model rats, through upregulation of VEGF, TGF-β, and downregulation of the inflammatory cytokines	[56]
PLGA/ Lipopolysaccharide (LPS)/IFN-γ activated mice RAW264.7 or human THP-1 cell membrane /load of BMMSCs or hBMMSCs	Skin chronic (diabetic) wounds	Augment BMMSCs or hBMMSCs proliferation, promotes resistance to oxidative stress, gene expressions, and keratinocyte (JB6 or HaCaT) migration	In situ immunostimulation capacity (rapid re-epithelialization, collagen remodeling, antioxidant stress, better angiogenesis)	[57]
Hydroxypropyl methylcellulose/ PEO/beta-glucan	Hard-to-heal diabetic wounds	Cytocompatible to (HaCaT) keratinocyte cells	Faster wound closure in diabetic db/db mice of 76.8–82.3%, day 4	[58]
PVA/ Chitosan/ Starch	Antibacterial skin wound treatment	L929 mouse fibroblast cells viability of 68–98% after 48 h, PVA/chitosan/starch (90/10/10) wound healing effectiveness of 100%		[59]
PCL/Surfactin +Gelatin/ Curcumin	Antibacterial skin wound treatment	Cytocompatible with L929 fibroblast cells, round-like cell morphology	Wistar rats' skin wounds model completely healed at day 14	[60]
PVA/Lysine/ Ibuprofen + coated Lavender oil	Antibacterial skin wound treatment	Excellent biocompatibility with human dermal fibroblasts, remaining: viable cells with a number increase along time; with normal phenotype and biological activity		[61]

Composition	Wound therapy	Cell performance	Animal study	References
L1 - PCL L2 - PVA/ Collagen/ Momordica charantia	Antibacterial skin wound treatment	Highest bitter melon extract content significantly increased L929 murine fibroblastic cells proliferation	Highest wound closure in the Wistar rats with the value of 94.01 ± 8.12%	[62]
PLA/PVA/ Sodium alginate	Anti-inflammatory and antibacterial skin wounds treatment	After 7 days the L929 mouse fibroblast cell number was higher	At day 16, the wound in Sprague Dawley male rats completely healed, with denser and ordered collagen fibers, and significant enhancement of angiogenesis	[63]
CA/Gelatin/ Nanohydroxyapatite	Skin wounds	Highest proliferation of L929 murine fibroblastic cell line at 25 mg nHA	Highest wound closure in the Wistar rats with the value of 93.5 ± 1.6%	[64]
PCL/Gelatin/ Naringin	Skin wounds	Highest proliferation of L929 murine fibroblastic cell line at 6% of Naringin	Highest wound closure in the Wistar rats with the value of 89.82 ± 3.30% and 99.39 ± 0.58% 7 and 14 days post-wounding	[65]
Chitosan/ Bromelin	Skin burns	Low cytotoxicity of the non-crosslinked membrane on human dermal fibroblasts	Greater re-epithelialization, debridement, and reduction of necrosis in second-degree burn rat's skin	[66]
POCA/PPF	Mild skin burns		Marked reduction by about 99% of the epidermis thickness and by about 85% in derma cell density in male UVB-burned C57BL/6 J mice	[67]
PVA/Birch bark dry extract (TE)	Skin burns		Porcine ex-vivo greatest wound area reduction and complete re-epithelialization at the lowest TE	[68]

Table 2.
In vitro/in vivo studies in wound healing.

of the inflammatory cytokines, the chitosan was infection protective, while the PVA maintained the microenvironment moisture [58]. An extensive study was conducted on cell membrane modified electrospun PLGA with an *in situ* immunostimulatory capacity for exaggerated bone marrow-derived mesenchymal stem cells (BMMSCs) biofunctions.

PLGA was modified by LPS/IFN-γ activated mice RAW264.7 or human THP-1 macrophages cytomembrane for an *in vitro* study, while the further load of BMMSCs or hBMMSCs ameliorate healing in diabetic wound mice, **Figure 2**. This novel living dressing supports the cells under oxidative stress and favors immunoregulation, collagen remodeling, and neovascularization in the wounded areas [59]. Concerning antibacterial properties in wound dressings two studies were conducted *in vitro* for the evaluation of mouse and human fibroblast viability and proliferation. The first study used electrospun PVA/chitosan/starch with an antibacterial effect against Gram-positive and Gram-negative bacteria and an absolute wound healing effectiveness in a scratch assay [60].

In the second study, the ibuprofen, which was directly loaded into the electrospun PVA/lysine, and the coated lavender oil showed high radical scavenging values of 17.68 ± 3.99% and 38.54 ± 5.58%, respectively. This activity is important in order to avoid the oxidative stress that promotes the inflammatory process in the wound [61].

Therapeutic and regenerative effects were evaluated in a hybrid PCL/Sur + Gel/Cur electrospun dressing. The material resulted in the rapid development of the dermis and collagen arrangement, thus a complete wound repair at day 14 [62]. Similar successful results were obtained in the studies that concern the production of two-layered electrospun PCL and PVA/Collagen/Momordica charantia [63], CA/gelatin/ nanohydroxyapatite [64], and PCL/gelatin/naringin [65]. The studies have shown that the concentration of the active components relates t to the fibroblast cells proliferation rate. Also all of the dressings performed with a wound closure efficiency higher than 90%.

In the study of PLA/PVA/sodium alginate (SA) membranes, the wound area revealed denser and ordered collagen fibers, with a high number of thick blood vessels and reduced levels of the inflammatory cytokine factor [66].

In the treatment of skin burn injuries, electrospun chitosan was loaded with the pineapple enzyme, bromelain. Lower bromelain concentration has shown both lower cytotoxicity against human fibroblasts and accelerated wound healing in the rat model [67]. Electrospun poly(octyl cyanoacrylate) (POCA)/polypropylene fumarate (PPF) acted as a strong anti-inflammatory dressing with an 80% cytokines reduction, comparable skin thickness, and dermal cells density with the native ones [68]. Another interesting bioactive component used was birch bark dry extract, which accelerated wound repair of an *ex vivo* porcine skin, when compared to the same composition of oleogel [69].

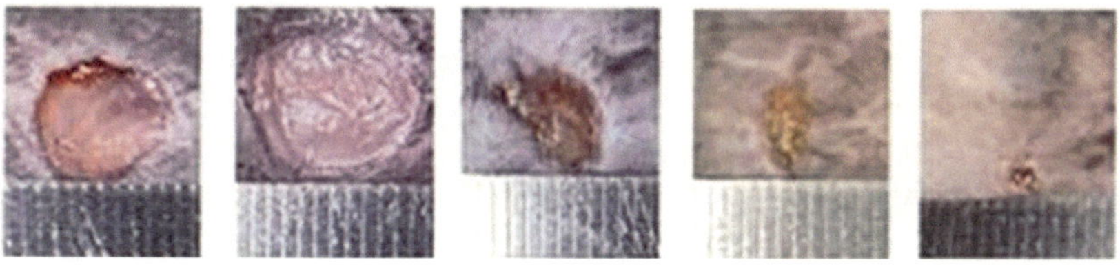

Figure 2.
Wound closure in diabetic mice from day 0–15 [59].

A recent clinical trial in the wound healing application field reported on the usage of electrospun PCL/CS-Zein-Curcumin dressing in the treatment of emergency burn patients. The efficiency of the same was compared with traditional gauze wound dressing and a silver ion alginate dressing. Selected patients were evaluated in terms of wound healing and scar repair, bacteria rates, patients' pain assessment, and overall nursing satisfaction. The dressing has shown antibacterial superiority with a decreased infection of 78.2%, as well as a total healing rate of 97%. Patients have had a great relief of pain in all burn treatment stages with a lessened experience of anxiety and depression [70]. Phase I clinical trial of electrospun fibrin dressing (commercial name SurgiClot) has demonstrated its efficacy in cancellous bone bleeding. The same was superior to standard US army dressing and is suggested to be used in different surgical procedures such as orthopedic, spine, cardiovascular, or head/neck surgery [71]. Phoenix wound matrix is a 3D electrospun material out of various synthetic polymers for the treatment of acute, chronic, and burn wounds. This commercial product was reported to be successful in different clinical cases, including patients with pressure injuries, DFU, chronic lower extremity, complex chronic or acute, surgical, trauma, or burn wounds [72]. Clinical case studies are also available for the market product SpinCare concerning donor site wounds [73] and partial thickness (or mixed) burns [74, 75]. In the case of split-skin graft surgery, dyslipidemia and diabetes mellitus type 2 patient was treated with the spincare layer, which served as a temporary skin and wound epithelization support [73].

4. Electrospun nanofibers in tissue repair

Probably the most investigated application field of functional electrospun materials would be the field of electrospun scaffolds for tissue engineering. A scaffold act as a temporary structural support for cell attachment, growth, and proliferation and most importantly for subsequent tissue regeneration. In order to credible mimic, the native extracellular matrix (ECM), the scaffold should possess functions that consider its architecture, mechanical property, cyto- and tissue compatibility, and bioactivity. Briefly, porous structure, certain void volume as well as mechanical integrity that will match the host tissue would facilitate tissue repair. The scaffold material should be biocompatible with both cells and tissues and when necessary should match the tissue degradability rate. Incorporated bioactive components should provide cell cues for their direction and differentiation [76]. The following section will discuss the latest studies on electrospun fibrous scaffolds that consider the application in hard tissues or bone and soft non-connective or nervous tissue regeneration.

4.1 Bone tissue engineering

The bones are natural composite materials with two types of bone tissues, that is, the compact and the spongy tissue. The structure of the compact tissue is regularly organized in lamellae and differs in thickness or form depending on the bone type it constitutes. About 80% of the total human bone mass belongs to this tissue. Collagen fibers have an irregular arrangement in the lamellae of the spongy bone tissue. This tissue has higher metabolic activity than the compact one [77]. There are also five types of bone cells, including osteoprogenitor cells, osteoblasts, bone lining cells, osteocytes, and osteoclasts. Osteoblasts, osteocytes, and bone lining cells originate from the osteoprogenitor or the mesenchymal stem cells, while the osteoclasts originate from the hemopoietic stem cells [78]. In terms of chemical composition the bone organic ECM consists of collagen type I and non-collagenous proteins, while the bone inorganic ECM is represented by minerals with a hydroxyapatite structure, that is, calcium, phosphate, and carbonate. There are different types of bones pathological conditions categorized as: fracture injuries, orthopedic surgeries, ischemic, hereditary, and metabolic diseases as well as bone cancer [77]. Bone disorders such as arthritis, bone cancer, osteosarcoma, and osteoarthritis are addressed as one of the greatest causes of humans deaths [79].

This section will discuss the latest studies concerning electrospun scaffolds for bone repair with a focus on the type of cells used for the *in vitro* experiments and the development of 2D or 3D architectures.

4.1.1 In vitro studies of osteoblasts performance

Electrospun scaffolds were prepared from PCL and pomegranate peel extract as a natural food waste with high phenolic content. Such scaffold revealed enhanced antioxidant activity by 96% and the cultured osteoblasts showed complete embedment in the scaffolds, with an active osteogenic matrix secretion for bone ingrowth [80]. Nano-hydroxyapatite (nHA) was deposited onto electrospun poly-3-hydroxybutyrate-co-3-hydroxyvalerate/polyaspartic acid resulting in a porous, hydrophilic, and good-strength scaffold. The scaffold was promising in the new bone formation due to enhanced human fetal osteoblast adhesion and osteointegration as well as tissue mineralization with Ca and P deposition [81]. Silver-doped nHA incorporated into electrospun PCL showed initial high burst release for bacteria clearance, while further silver content reduction resulted in a booster effect of the rat mesenchymal stem cells' (MSCs)metabolic activity on the 21st day of culture. This result is in accordance with the start of the healing process [82]. Bone marrow mesenchymal stem cell behavior was observed depending on electrospun PLLA nanofibrous scaffold surface topography, that is, random or aligned structure. The cells exhibited increased migration speed due to the contact guidance with the orientation of the aligned nanofibers. This alignment also improved osteogenic differentiation and significantly promoted osteogenic gene expressions [83]. PLGA scaffold with a bioactive interface was prepared by electrospinning and post-layer-by-layer deposition of hematite nanoparticles (αFeNPs). The performance of rat adipose-derived stem cells was enhanced in terms of cell spreading, cytoskeletal organization, osteogenic differentiation, and bone matrix mineral synthesis [84].

4.1.2 Bioactive scaffolds incorporating growth factors, genes and ECM vesicles

Electrospun PCL was loaded with nHA and chitosan-based nanoparticles carrying NELL-1 protein. The system had a double barrier which resulted in a sustained release of the growth factor. The highest osteogenic activity of MC3T3-E1 cells was observed on the 21st day of cell culture. The scaffold promoted cells' adhesion, proliferation, osteogenic differentiation, and maturation [85]. Three-component simultaneously electrospun layers scaffold incorporated PLGA and PEG or both with recombinant human vein endothelial growth factor (rhVEGF), recombinant human BMP-2 (rhBMP-2) factor, and Ca nanoparticles. The scaffold resembled an ECM-like hybrid microstructure with a high initial rhVEGF release and high bioactivity to promote vascularity in the first and further bone formation stages. The scaffolds induced human umbilical vein endothelial cells proliferation as well as murine pluripotent mesenchymal cells osteogenic differentiation with high alkaline phosphatase expression, calcium deposition, and gene expression [86]. ECM vesicles are proteolipid-packaged particles with bioactive cell contents that have a crucial role in the bone mineralization process, but also mediate the cell-cell signaling in the mechanical load bone formation process [87, 88]. Collagen-coated electrospun PCL scaffold was functionalized by osteocyte-derived mechanically activated-extracellular vesicles to enhance bone tissue regeneration. As fabricated scaffolds resulted in a significant increase in the alkaline phosphatase (ALP) activity and a continuous increase in matrix mineralization, thus promoting osteoblast differentiation [89]. Preosteoblast and endothelial cells exhibited improved osteogenic differentiation and angiogenic activity when cultured on mesenchymal stem cells-derived ECM vesicles loaded SF/PCL scaffolds. The same also promoted the regeneration of a calvarial rat bone defect [90]. Plasmid DNA (pDNA) polyplexes were loaded into electrospun gelatin/polyethylene glycol scaffolds to obtain a functional gene expression system. Further loading of the scaffolds with BMP-2 protein resulted in osteogenic ALP activity with an absorbance rate above 8 or 4 mAbs/min in human myoblast C2C12 and mouse osteoblast MC3T3-E1 cells, respectively [91].

4.1.3 3D electrospun structures for in vivo implants

Electrospun 3D cell/mesh complexes were fabricated via layer-by-layer deposition of PLLA/gelatin in a random and nestlike structure. Each layer (four in total) was prior cultured with bone mesenchymal stromal cells and afterward implanted into rat cranial defects. The *in vitro* study showed the promotion of the cells' osteogenic differentiation, while the *in vivo* study showed new calcified bone formation with greater bone healing in case of the nestlike cell/mesh complexes [92]. Similarly, the cell/scaffold complex was fabricated by culturing human fetal osteoblasts onto PCL/nHA, which was electrospun onto stainless steel mesh via layer-by-layer assembly. This 3D system showed cell migration between the adjacent layers. The complex was further functionalized by alendronate to promote osteogenic differentiation [93]. Successful osteoblast adhesion and proliferation, as well as gene expression for their differentiation and mineralization, was accomplished by electrospun poly(butylene-adipate-co-terephthalate)] /nHAp. The scaffolds resulted in the highest bone volume formation in rat tibia defect model after six weeks of implantation [94]. Electrospun poly(D, L-lactic acid)-poly(ethylene glycol)-poly(D, L-lactic acid) (PELA) immobilized with BMP-2 was used for the treatment of secondary hip osteoarthritis. Cultured human bone marrow-derived mesenchymal stem cells (hMSCs) were well attached with minor cytotoxicity, **Figure 3a**. The study revealed a positive effect on the osteogenesis, the complete acetabular defects repair, **Figure 3b**, but also an incomplete bone formation which is influenced by the insufficient BMP-2 dose [95]. Several techniques were combined along with electrospinning to fabricate nHA/PLA/gelatin 3D scaffold. Firstly, short nanofibers were formed from the nHA/PLA/gelatin mat by homogenating, while the 3D nanofibrous scaffold was fabricated by freeze drying and thermo-crosslinking of the nanofibers. The final structure was obtained by immobilization of BMP-2 with a prior polydopamine coating and subsequent freeze drying. The as-prepared architectures enhanced BMSCs osteogenic differentiation and most importantly, the volume and rate of growth of the newly formed bone in the rat cranial bone defect model was significant, from the fourth to the eighth week of implantation [96]. Tibia bone defects in rabbit models were treated by a tubular 3D structure consisting of a shell and a core from electrospun collagen/PCL/HA

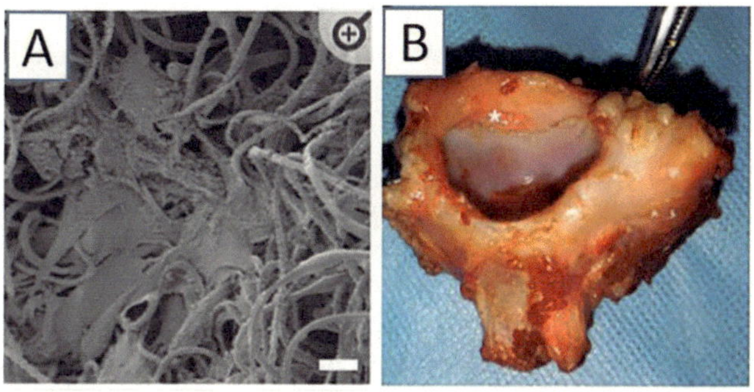

Figure 3.
a) hMSCs cultured on PELA/BMP-2 scaffolds and b) PELA/BMP-2 implants at 24 weeks [95].

and freeze-dried collagen/icariin-loaded chitosan microspheres, respectively. These scaffolds had excellent osteoinductivity and osteoconductivity due to the simultaneous effect of the HA and the icariin [97]. The scaffolds' pore structure provided bone remodeling, while the matching degradation rate with the host defect [98] provided bone matrix mineralization [99].

4.2 Nervous tissue engineering

Trauma, neurological disorder, or tumor excisions are the known causes of nerve injuries. Peripheral nerve injuries (or neuropathy) are a major problem for millions of people yearly, resulting in a painful condition due to motor function and sensory perception mitigation. Nerve tissue engineering as the current alternative to autologous nerve transplantation, is the solution to provide a biomimetic material for both mechanical and chemical support of the native tissue repair [100, 101].

4.2.1 Peripheral nerve and spinal cord injury

3D hollow cylinder structure was electrospun combining uniaxially aligned PCL and PCL/PLGA nanofibers as the outer and the inner layer, respectively. The scaffolds were implanted into the sciatic nerve gaps of a rat model with a peripheral nerve injury. The study revealed enhanced nerve fiber regeneration and myelin repair, as well as increased cell count due to the open structure which provided migration/penetration into the scaffolds [102]. Aligned glycosaminoglycans (from aorta porcine tissue) functionalized electrospun PCL scaffolds were used as well in peripheral nervous system repair. The scaffolds have been shown to effectively influence Schwann cells (SCs) adhesion, proliferation, and differentiation [103]. Peripheral nerve injury was also suggested to be repaired by electrospun piezoelectric polyvinylidene fluoride (PVDF) or polyvinylidene fluoride trifluoroethylene (P(VDF-TrFE)) scaffolds. These polymers are known for their piezoelectricity which stimulates cells ingrowth induced by electrical activity upon mechanical force application. Having this in mind, the scaffolds showed biocompatibility with cultured SCs and supported sensory neurite outgrowth [104]. When layered double hydroxides (LDH) nanoclay particles were incorporated into electrospun PCL/gelatin, increased viability, and proliferation of the cultured human neuroblastoma SH-SY5Y cells were observed at higher LDHs, but there was no increased differentiation [105]. Photocrosslinked gelatin methacryloyl was used for the fabrication of an aligned hydrogel electrospun microfiber bundle to be used in the repair of spinal cord injury (SCI). The viability of the cultured bone marrow-derived mesenchymal stem cells (similar to neural stem cells) was reported to be 95%, with cell infiltration depth of 197.5 ± 18.1 μm into the scaffold on the 3rd day of culture, while SCI rats were observed to have significant functional recovery of the lower limbs. Generally the *in vivo* study resulted in the increase of the neural stem cells, neurons, synaptic connections, and vascular endothelial cells, as well as an inhibition of glial scar formation [106]. Electrospun (fiber aligned) aminated PLA, loaded with nerve growth factor, was grafted with liposomes pDNA vehicles to be used as well in SCI repair. The scaffold in the rat *in vivo* study performed with a great reduction of the inflammation and scar formation, while simultaneously promoting angiogenesis, neurogenic bioability and resulting in significant neuromotor function recovery of the hindlimb [107]. Nerve growth guidance and peripheral nerve regeneration were supported by electrospun PCL/collagen/nano bioglass. The *in vitro* study reported the promotion of human endometrial stem cells (hEnSCs) adhesion and proliferation [108]. Scaffolds were also fabricated by combining several techniques including electrospinning of PLA, spin coating, and ammonia-induced solid-state polymerization to obtain eumelanin-coated PLA for the treatment of neurodegenerative disorder. The scaffolds supported human-derived cell line SH-SY5Y, from neuroblastoma to survive and differentiate into a neuronal phenotype [109]. Electrospun PCL-amnion nanofibrous membranes were also proposed to be used after neurolysis and as reported it resulted in the reduction of the peripheral nerve adhesion, lessen intraneural macrophage invasion, high gastrocnemius muscle weight, and muscle bundle area as well as improved sense and movement of the model rats limbs [110]. Another condition proposed to be treated with electrospun biomimetic polybenzyl glutamate, was optic neuropathy which causes irreversible blindness due to retinal ganglion cell degeneration [111, 112]. Prepared scaffolds with cultured induced pluripotent stem cells promoted neurite outgrowth or neuronal differentiation, neuronal maturation, and retinal differentiation and maturation [112].

4.2.2 Nerve guidance conduits

Nerve guidance conduits (NGC) are a physical barrier and guiding tool for regenerative axons across gap lesions. In peripheral nerve injuries, NGC is frequently used as simple hollow structures, but also in a more advanced manner functionalized (or filled) with molecular (growth factors) and gene or cell-based (i.e., Schwann cells, bone marrow stromal cells, human umbilical-cord stem cells, neural stem cells, etc.) therapies [113]. Earlier multichannel conduits [114] and recently micro-channeled tubular structures, **Figure 4a** [115], are also proposed, as well as optimal aligned architectures with conductive features to stimulate neural regeneration [117]. PLGA electrospun yarns were encapsulated by electrospun PLGA tubular wall to form NGC which were further coated by laminin through covalent binding. As prepared nerve guidance conduits showed a synergistic effect of the 3D yarns structure and the laminin layer that resulted in significant Schwann cells (SCs) proliferation and migration, respectively [118]. Similarly, NGC was prepared by wrapping a sponge-like electrospun/freeze-dried PLCL/SF structure via electrospinning the same composition outer layer, **Figure 4b**. *In vitro* study showed that the SCs grew on the NGC surface but also infiltrated the sponge structure for 1000 in-depth on the 7th day of culture. The *in vivo* study in rat sciatic nerve defect model revealed good nerve regeneration, especially when compared to a hollow NGC structure [116].

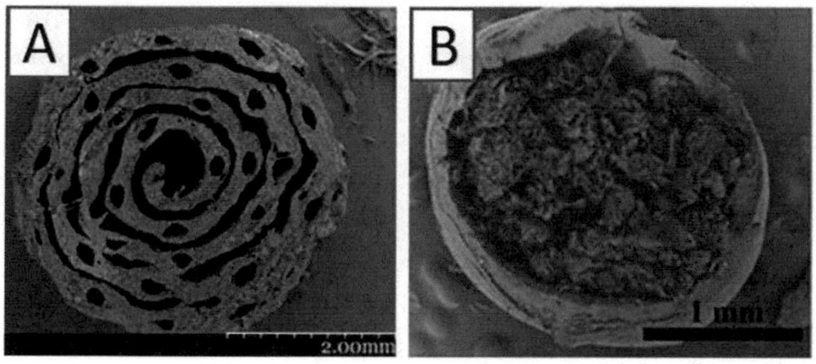

Figure 4.
a) NGC with microchannels [115] and b) NGC with nanofibrous sponges [116]. Reprinted (adapted) with permission from [116]. Copyright 2017 American Chemical Society.

Electrospun PVDF-TrFE and matrigel coated conduit, with random or aligned fibrous inner walls, cultured with SCs were used for the rat spinal cord repair *in vivo*. The conduits supported cell viability and axon regeneration with a greater uniform distribution in the aligned structures [119]. Mussel adhesive proteins fused with biofunctional peptides from the extracellular matrix were incorporated into PLGA to form electrospun-aligned nanofiber conduit. The scaffold supported a stable environment for both neural and Schwann cells differentiation. Dense and large axon- and endoneurium-like structures similar to one of the autografts were also observed for the fabricated scaffold which further provided volume and functional accelerated nerve regeneration [120]. Nerve guides tubular structures with microchannels and parallel fibers walls were electrospun from PCL or PLLA to act as cell delivery vehicle of stromal vascular fraction for a 10 mm nerve gap bridging. The PLLA nerve guides supported axonal regeneration but were not as good as an autologous nerve graft [115]. Dual electrospun PLGA and poly (d,l-lactic acid) (PDLLA) incorporating glial cell line-derived growth factor and nerve growth factor (NGF), respectively, were produced with an aligned fibrous topography. Rat pheochromocytoma cells (PC12) migrated on the aligned scaffolds in a well-organized manner with greater neurite sprouting and elongation. Fiber topography and biochemical cues resulted in enhanced neurite outgrowth and neural differentiation [121]. Similarly, aligned electrospun PCL/chitosan scaffolds with an immobilized effective nerve regeneration agent NGF showed elongated PC12 cell morphology with increased neuronal viability as well as enhanced differentiation and neurite outgrowth [122]. Despite the traditional application of the NGF for PC12 differentiation, authors proposed an electroconductive electrospun SF/reduced graphene oxide (rGO) scaffold that provides superior levels of cell differentiation into neural phenotypes due to electrical stimulation ability. Cell viability was >95%, while attached cells appeared with thin filopodia and neuritis. Increased differentiation of the cells was noticed in case of applied electrostimulation [123]. Good cell attachment and proliferation were obtained in a similar electroconductive scaffold PCL/gelatin incorporating graphene as well as tetracycline hydroxide drug with antibacterial properties [124].

5. Challenges and future aspects

Recent studies of electrospun nanofibrous materials for nanomedical applications consider synthetic or natural biomaterials, as well as inorganic or composite-based multifunctional systems that most often incorporate bioactive compounds in order to offer mechanical support, cells viability, and life activities support, compatibility with ECM interactions and new tissue formation. Although there are advances in terms of materials diversity, fabrication procedure, therapeutic/healing performance, challenges are still present and need to be addressed. When new materials or bioactive compounds are encapsulated some studies mention possible cell toxicity, thus further research needs to be carried out on the issue of an optimal concentration set that will not affect cell viability. *In vitro* studies apply animal-based or commercial cell lines, some of which are cancer-derived, thus human-based cells would be a better choice in order to simulate the natural surroundings. Also, since human cells are more sensitive and slower to accommodate in a new environment, the same can reflect the original conditions more closely. There is still a lack of clinical trials, as most of the studies are conducted on rat animal models. Standardization of the procedures as well as materials ethical approval is imperative in order to transfer the product to a commercial market. Within this respect, the fabrication processes need industrialization, although many researchers have demonstrated large-scale electrospinning machines (i.e., needleless electrospinning) continuous production of a nanomedical material still remains a challenge.

6. Conclusion

The research in the field of electrospun nanofibrous materials for nanomedical applications has nt decreased, it is actually still growing and the last few years have shown more challenges and frontiers. This chapter discusses selected studies, from 2017 onwards, that are dealing with the development and performance of electrospun nanofibrous materials for drug delivery, wound healing, and tissue engineering with a focus on bone and nerve tissue repair. In drug delivery systems, materials' composition as well as fabrication technologies affect drugs' pace mechanisms, which finally affects therapeutic performance. On demand, drug delivery materials are mainly dealing with cancer treatment, tissue repair, and some minor ailments. Wound therapeutic electrospun materials are focusing on chronic diabetic wounds and skin burn wounds while generally comprising natural or synthetic remedies with antibacterial and healing functions. Diverse tissue engineering scaffolds are designed to carry growth factors, genetic materials, or target selected vesicles to boost systems' regenerative performance. Scaffolds are fabricated as 2D or 3D architectures via simple or modified electrospinning or in a combination with other techniques (i.e., freeze drying, L-B-L, coating, wrapping, etc.). Versatile cells are cultured on the scaffolds, after or during the fabrication procedure with a general successful result in terms of cell viability, growth, migration, and specific differentiation. *In vivo* studies are still based on animal models, although there are some clinical trials as well. Peripheral and spinal cord injuries are reported to have better outcomes in case of aligned structures with biochemical as well as electro-conductive cues for neurite outgrowth and neural differentiation. Some of the challenges and future aspects are also addressed.

Acknowledgements

The OAPF was funded by the University of Zagreb Faculty of Textile Technology under the short-term financial support for research for 2022, TP23/22, PI: Emilija Zdraveva.

DOI: http://dx.doi.org/10.5772/intechopen.109389

Conflict of interest

The authors declare no conflict of interest.

Author details

Emilija Zdraveva* and Budimir Mijovic
University of Zagreb Faculty of Textile Technology, Zagreb, Croatia

*Address all correspondence to: emilija.zdraveva@ttf.unizg.hr

References

[1] Martínez-Pérez CA. Electrospinning: A promising technique for drug delivery systems. Reviews on Advanced Materials Science. 2020;**59**(1):441-454

[2] Bajpai AK, Shukla SK, Bhanu S, Kankane S. Responsive polymers in controlled drug delivery. Progress in Polymer Science. 2008;**33**(11):1088-1118

[3] Balaji A, Vellayappan M, John A, Subramanian A, Jaganathan S, Supriyanto E, et al. An insight on electrospun-nanofibers-inspired modern drug delivery system in the treatment of deadly cancers. RSC Advances. 2015;**5**(71):57984-58004

[4] Singh B, Kim K, Park M-H. On-demand drug delivery systems using nanofibers. Nanomaterials. 2021;**11**(12):3411

[5] Li H, Liu K, Sang Q, Williams GR, Wu J, Wang H, et al. A thermosensitive drug delivery system prepared by blend electrospinning. Colloids and Surfaces B: Biointerfaces. 2017;**159**:277-283

[6] Shibata T, Yoshimura N, Kobayashi A, Ito T, Hara K, Tahara K. Emulsion-electrospun polyvinyl alcohol nanofibers as a solid dispersion system to improve solubility and control the release of probucol, a poorly water-soluble drug. Journal of Drug Delivery Science and Technology. 2022;**67**:102953

[7] Wang J, Windbergs M. Controlled dual drug release by coaxial electrospun fibers–impact of the core fluid on drug encapsulation and release. International Journal of Pharmaceutics. 2019;**556**:363-371

[8] Ghosal K, Augustine R, Zaszczynska A, Barman M, Jain A, Hasan A, et al. Novel drug delivery systems based on triaxial electrospinning based nanofibers. Reactive and Functional Polymers. 2021;**163**:104895

[9] Wei K, Li Y, Lei X, Yang H, Teramoto A, Yao J, et al. Emulsion electrospinning of a collagen-like protein/PLGA fibrous scaffold: Empirical modeling and preliminary release assessment of encapsulated protein. Macromolecular Bioscience. 2011;**11**(11):1526-1536

[10] Gupta P, Wilkes GL. Some investigations on the fiber formation by utilizing a side-by-side bicomponent electrospinning approach. Polymer. 2003;**44**(20):6353-6359

[11] Yoo HS, Kim TG, Park TG. Surface-functionalized electrospun nanofibers for tissue engineering and drug delivery. Advanced Drug Delivery Reviews. 2009;**61**(12):1033-1042

[12] Mirzaeei S, Taghe S, Asare-Addo K, Nokhodchi A. Polyvinyl alcohol/chitosan single-layered and polyvinyl alcohol/chitosan/eudragit rl100 multi-layered electrospun nanofibers as an ocular matrix for the controlled release of ofloxacin: An in vitro and in vivo evaluation. AAPS PharmSciTech. 2021;**22**(5):1-13

[13] Yan E, Jiang J, Yang X, Fan L, Wang Y, An Q, et al. pH-sensitive core-shell electrospun nanofibers based on polyvinyl alcohol/polycaprolactone as a potential drug delivery system for the chemotherapy against cervical cancer. Journal of Drug Delivery Science and Technology. 2020;**55**:101455

[14] Wang L, Wang C, Wang L, Zhang Q, Wang Y, Xia X. Emulsion electrospun

polylactic acid/Apocynum venetum nanocellulose nanofiber membranes with controlled sea buckthorn extract release as a drug delivery system. Textile Research Journal. 2021;**91**(9-10):1046-1055

[15] Li D, Wang M, Song W-L, Yu D-G, Bligh SWA. Electrospun Janus beads-on-a-string structures for different types of controlled release profiles of double drugs. Biomolecules. 2021;**11**(5):635

[16] Han D, Steckl AJ. Triaxial electrospun nanofiber membranes for controlled dual release of functional molecules. ACS Applied Materials & Interfaces. 2013;**5**(16):8241-8245

[17] Coelho JF, Ferreira PC, Alves P, Cordeiro R, Fonseca AC, Góis JR, et al. Drug delivery systems: Advanced technologies potentially applicable in personalized treatments. EPMA Journal. 2010;**1**(1):164-209

[18] Nagarajan S, Pochat-Bohatier C, Balme S, Miele P, Kalkura SN, Bechelany M. Electrospun fibers in regenerative tissue engineering and drug delivery. Pure and Applied Chemistry. 2017;**89**(12):1799-1808

[19] Rathinamoorthy R. Nanofiber for drug delivery system–principle and application. Pakistan Textile Journal. 2012;**61**:45-48

[20] Khalf A, Madihally SV. Recent advances in multiaxial electrospinning for drug delivery. European Journal of Pharmaceutics and Biopharmaceutics. 2017;**112**:1-17

[21] Tiwari SK, Tzezana R, Zussman E, Venkatraman SS. Optimizing partition-controlled drug release from electrospun core–shell fibers. International Journal of Pharmaceutics. 2010;**392**(1-2):209-217

[22] Munj HR, Lannutti JJ, Tomasko DL. Understanding drug release from PCL/gelatin electrospun blends. Journal of Biomaterials Applications. 2017;**31**(6):933-949

[23] Tamani F, Bassand C, Hamoudi M, Danede F, Willart J-F, Siepmann F, et al. Mechanistic explanation of the (up to) 3 release phases of PLGA microparticles: Diprophylline dispersions. International Journal of Pharmaceutics. 2019;**572**:118819

[24] Ye K, Kuang H, You Z, Morsi Y, Mo X. Electrospun nanofibers for tissue engineering with drug loading and release. Pharmaceutics. 2019;**11**(4):182

[25] Mickova A, Buzgo M, Benada O, Rampichova M, Fisar Z, Filova E, et al. Core/shell nanofibers with embedded liposomes as a drug delivery system. Biomacromolecules. 2012;**13**(4):952-962

[26] Hu J, Zeng F, Wei J, Chen Y, Chen Y. Novel controlled drug delivery system for multiple drugs based on electrospun nanofibers containing nanomicelles. Journal of Biomaterials Science, Polymer Edition. 2014;**25**(3):257-268

[27] Haroosh H, Chaudhary D, Dong Y, Hawkins B. Electrospun PLA: PCL/halloysite nanotube nanocomposites fibers for drug delivery. In: Processing and Fabrication of Advanced Materials XIX. 2011. pp. 847-858

[28] Xie C, Li X, Luo X, Yang Y, Cui W, Zou J, et al. Release modulation and cytotoxicity of hydroxycamptothecin-loaded electrospun fibers with 2-hydroxypropyl-β-cyclodextrin inoculations. International Journal of Pharmaceutics. 2010;**391**(1-2):55-64

[29] Wu J, Zhang Z, Zhou W, Liang X, Zhou G, Han CC, et al. Mechanism of a long-term controlled drug release system

based on simple blended electrospun fibers. Journal of Controlled Release. 2020;**320**:337-346

[30] Liu Y, Chen X, Liu Y, Gao Y, Liu P. Electrospun coaxial fibers to optimize the release of poorly water-soluble drug. Polymers. 2022;**14**(3):469

[31] WHO. 2022. Available from: https://www.who.int/health-topics/cancer#tab=tab_1.

[32] Li J, Xu W, Li D, Liu T, Zhang YS, Ding J, et al. Locally deployable nanofiber patch for sequential drug delivery in treatment of primary and advanced orthotopic hepatomas. ACS Nano. 2018;**12**(7):6685-6699

[33] Kuang G, Zhang Z, Liu S, Zhou D, Lu X, Jing X, et al. Biphasic drug release from electrospun polyblend nanofibers for optimized local cancer treatment. Biomaterials Science. 2018;**6**(2):324-331

[34] Salim SA, Kamoun EA, Evans S, El-Moslamy SH, El-Fakharany EM, Elmazar MM, et al. Mercaptopurine-loaded sandwiched tri-layered composed of electrospun polycaprolactone/poly (methyl methacrylate) nanofibrous scaffolds as anticancer carrier with antimicrobial and antibiotic features: Sandwich configuration nanofibers, release study and in vitro bioevaluation tests. International Journal of Nanomedicine. 2021;**16**:6937

[35] Iqbal S, Rashid MH, Arbab AS, Khan M. Encapsulation of anticancer drugs (5-fluorouracil and paclitaxel) into polycaprolactone (PCL) nanofibers and in vitro testing for sustained and targeted therapy. Journal of biomedical nanotechnology. 2017;**13**(4):355-366

[36] Mitxelena-Iribarren O, Riera-Pons M, Pereira S, Calero-Castro FJ, Castillo Tuñón JM, Padillo-Ruiz J, et al. Drug-loaded PCL electrospun nanofibers as anti-pancreatic cancer drug delivery systems. Polymer Bulletin. 2022;**79**(8):1-16

[37] Huo P, Han X, Zhang W, Zhang J, Kumar P, Liu B. Electrospun nanofibers of Polycaprolactone/collagen as a sustained-release drug delivery system for artemisinin. Pharmaceutics. 2021;**13**(8):1228

[38] Boncu TE, Guclu AU, Catma MF, Savaser A, Gokce A, Ozdemir N. In vitro and in vivo evaluation of linezolid loaded electrospun PLGA and PLGA/PCL fiber mats for prophylaxis and treatment of MRSA induced prosthetic infections. International Journal of Pharmaceutics. 2020;**573**:118758

[39] Li Z, Zeng R, Yang L, Ren X, Maffucci KG, Qu Y. Development and characterization of PCL electrospun membrane-coated Bletilla striata polysaccharide-based gastroretentive drug delivery system. AAPS PharmSciTech. 2020;**21**(2):1-14

[40] Göttel B, e Silva JMS, de Oliveira CS, Syrowatka F, Fiorentzis M, Viestenz A, et al. Electrospun nanofibers–a promising solid in-situ gelling alternative for ocular drug delivery. European Journal of Pharmaceutics and Biopharmaceutics. 2020;**146**:125-132

[41] Tawfik EA, Craig DQ, Barker SA. Dual drug-loaded coaxial nanofibers for the treatment of corneal abrasion. International Journal of Pharmaceutics. 2020;**581**:119296

[42] Cheng G, Yin C, Tu H, Jiang S, Wang Q, Zhou X, et al. Controlled co-delivery of growth factors through layer-by-layer assembly of core–shell nanofibers for improving bone regeneration. ACS Nano. 2019;**13**(6):6372-6382

[43] Liu C, Wang C, Zhao Q, Li X, Xu F, Yao X, et al. Incorporation and

release of dual growth factors for nerve tissue engineering using nanofibrous bicomponent scaffolds. Biomedical Materials. 2018;**13**(4):044107

[44] Jiang Y-C, Wang X-F, Xu Y-Y, Qiao Y-H, Guo X, Wang D-F, et al. Polycaprolactone nanofibers containing vascular endothelial growth factor-encapsulated gelatin particles enhance mesenchymal stem cell differentiation and angiogenesis of endothelial cells. Biomacromolecules. 2018;**19**(9):3747-3753

[45] Rychter M, Baranowska-Korczyc A, Milanowski B, Jarek M, Maciejewska BM, Coy EL, et al. Cilostazol-loaded poly (ε-Caprolactone) electrospun drug delivery system for cardiovascular applications. Pharmaceutical Research. 2018;**35**(2):1-20

[46] Kersani D, Mougin J, Lopez M, Degoutin S, Tabary N, Cazaux F, et al. Stent coating by electrospinning with chitosan/poly-cyclodextrin based nanofibers loaded with simvastatin for restenosis prevention. European Journal of Pharmaceutics and Biopharmaceutics. 2020;**150**:156-167

[47] Fatehi P, Abbasi M. Medicinal plants used in wound dressings made of electrospun nanofibers. Journal of Tissue Engineering and Regenerative Medicine. 2020;**14**(11):1527-1548

[48] Strodtbeck F. Physiology of wound healing. Newborn and Infant Nursing Reviews. 2001;**1**(1):43-52

[49] George Broughton I, Janis JE, Attinger CE. The basic science of wound healing. Plastic and Reconstructive Surgery. 2006;**117**(7S):12S-34S

[50] Landén NX, Li D, Ståhle M. Transition from inflammation to proliferation: A critical step during wound healing. Cellular and Molecular Life Sciences. 2016;**73**(20):3861-3885

[51] Sood A, Granick MS, Tomaselli NL. Wound dressings and comparative effectiveness data. Advances in wound care. 2014;**3**(8):511-529

[52] Gizaw M, Thompson J, Faglie A, Lee S-Y, Neuenschwander P, Chou S-F. Electrospun fibers as a dressing material for drug and biological agent delivery in wound healing applications. Bioengineering. 2018;**5**(1):9

[53] He J-H, Liu Y, Mo L-F, Wan Y-Q, Xu L. Electrospun Nanofibres and their Applications. UK: iSmithers Shawbury, Lightning Source Inc.; 2008

[54] Samadian H, Zamiri S, Ehterami A, Farzamfar S, Vaez A, Khastar H, et al. Electrospun cellulose acetate/gelatin nanofibrous wound dressing containing berberine for diabetic foot ulcer healing: in vitro and in vivo studies. Scientific Reports. 2020;**10**(1):1-12

[55] Wynn M, Freeman S. The efficacy of negative pressure wound therapy for diabetic foot ulcers: A systematised review. Journal of Tissue Viability. 2019;**28**(3):152-160

[56] Arampatzis AS, Kontogiannopoulos KN, Theodoridis K, Aggelidou E, Rat A, Willems A, et al. Electrospun wound dressings containing bioactive natural products: Physico-chemical characterization and biological assessment. Biomaterials Research. 2021;**25**(1):1-21

[57] Grip J, Engstad RE, Skjæveland I, Škalko-Basnet N, Isaksson J, Basnet P, et al. Beta-glucan-loaded nanofiber dressing improves wound healing in diabetic mice. European Journal of Pharmaceutical Sciences. 2018;**121**:269-280

[58] Chen Q, Wu J, Liu Y, Li Y, Zhang C, Qi W, et al. Electrospun chitosan/PVA/bioglass Nanofibrous membrane with spatially designed structure for accelerating chronic wound healing. Materials Science and Engineering: C. 2019;**105**:110083

[59] Gao S, Chen T, Wang Z, Ji P, Xu L, Cui W, et al. Immuno-activated mescenchymal stem cell living electrospun nanofibers for promoting diabetic wound repair. 2022;**20**. Article Number: 294

[60] Adeli H, Khorasani MT, Parvazinia M. Wound dressing based on electrospun PVA/chitosan/starch nanofibrous mats: Fabrication, antibacterial and cytocompatibility evaluation and in vitro healing assay. International Journal of Biological Macromolecules. 2019;**122**:238-254

[61] Sequeira RS, Miguel SP, Cabral CS, Moreira AF, Ferreira P, Correia IJ. Development of a poly (vinyl alcohol)/lysine electrospun membrane-based drug delivery system for improved skin regeneration. International Journal of Pharmaceutics. 2019;**570**:118640

[62] Hadizadeh M, Naeimi M, Rafienia M, Karkhaneh A. A bifunctional electrospun nanocomposite wound dressing containing surfactin and curcumin: in vitro and in vivo studies. Materials Science and Engineering: C. 2021;**129**:112362

[63] Salami MS, Bahrami G, Arkan E, Izadi Z, Miraghaee S, Samadian H. Co-electrospun nanofibrous mats loaded with bitter gourd (Momordica charantia) extract as the wound dressing materials: in vitro and in vivo study. BMC Complementary Medicine and Therapies. 2021;**21**(1):1-12

[64] Samadian H, Salehi M, Farzamfar S, Vaez A, Ehterami A, Sahrapeyma H, et al. In vitro and in vivo evaluation of electrospun cellulose acetate/gelatin/hydroxyapatite nanocomposite mats for wound dressing applications. Artificial cells, Nanomedicine, and Biotechnology. 2018;**46**(sup1):964-974

[65] Salehi M, Vaez A, Naseri-Nosar M, Farzamfar S, Ai A, Ai J, et al. Naringin-loaded poly (ε-caprolactone)/gelatin electrospun mat as a potential wound dressing: in vitro and in vivo evaluation. Fibers and Polymers. 2018;**19**(1):125-134

[66] Bi H, Feng T, Li B, Han Y. In vitro and in vivo comparison study of electrospun PLA and PLA/PVA/SA fiber membranes for wound healing. Polymers. 2020;**12**(4):839

[67] Bayat S, Amiri N, Pishavar E, Kalalinia F, Movaffagh J, Hashemi M. Bromelain-loaded chitosan nanofibers prepared by electrospinning method for burn wound healing in animal models. Life sciences. 2019;**229**:57-66

[68] Romano I, Summa M, Heredia-Guerrero JA, Spanò R, Ceseracciu L, Pignatelli C, et al. Fumarate-loaded electrospun nanofibers with anti-inflammatory activity for fast recovery of mild skin burns. Biomedical Materials. 2016;**11**(4):041001

[69] Mwiiri FK, Brandner JM, Daniels R. Electrospun bioactive wound dressing containing colloidal dispersions of birch bark dry extract. Pharmaceutics. 2020;**12**(8):770

[70] Liu Q, Yang L, Peng Q. Study on scar repair and wound nursing of chitosan-based composite electrospun nanofibers in first aid of burn. Materials Express. 2021;**11**(8):1420-1427

[71] Balain B, Craig N, Gnanalingham K, Madan S, Sharma H, Wynne-Jones G, et al. Safety and efficacy of a novel

fibrin dressing on bleeding cancellous bone. Journal of Clinical Experimental Orthopaedics. 2018;**4**:50

[72] Nanomedic. RenovoDerm [cited 2022 20.09]. Available from: https://www.renovoderm.tech/Phoenix-Wound-Matrix/Clinical-Outcomes/Clinical-Round-Table/

[73] Nanomedic. SpinCare - Donor Site Wounds [cited 2022 20.09]. Available from: https://nanomedic.com/research/donor-site-wounds/

[74] Nanomedic. SpinCare - Partial Thickness Wounds [cited 2022 20.09]. Available from: https://nanomedic.com/research/partial-thickness-burns/

[75] Nanomedic. SpinCare - Partial Thickness Mixed Burn [cited 2022 20.09]. Available from: https://nanomedic.com/research/using-the-spincare-system-for-partial-thickness-mixed-burn-mrk0121/

[76] Chan B, Leong K. Scaffolding in tissue engineering: General approaches and tissue-specific considerations. European spine journal. 2008;**17**(4):467-479

[77] Bilgiç E, Boyacıoğlu Ö, Gizer M, Korkusuz P, Korkusuz F. Architecture of bone tissue and its adaptation to pathological conditions. In: Comparative Kinesiology of the Human Body. UK: Elsevier; 2020. pp. 71-90

[78] Downey PA, Siegel MI. Bone biology and the clinical implications for osteoporosis. Physical therapy. 2006;**86**(1):77-91

[79] Anjum S, Rahman F, Pandey P, Arya DK, Alam M, Rajinikanth PS, et al. Electrospun biomimetic Nanofibrous scaffolds: A promising Prospect for bone tissue engineering and regenerative medicine. International Journal of Molecular Sciences. 2022;**23**(16):9206

[80] Sadek KM, Mamdouh W, Habib SI, El Deftar M, Habib ANA. In vitro biological evaluation of a fabricated Polycaprolactone/pomegranate electrospun scaffold for bone regeneration. ACS Omega. 2021;**6**(50):34447-34459

[81] Zhang S, Jiang G, Prabhakaran MP, Qin X, Ramakrishna S. Evaluation of electrospun biomimetic substrate surface-decorated with nanohydroxyapatite precipitation for osteoblasts behavior. Materials Science and Engineering: C. 2017;**79**:687-696

[82] Paterson TE, Shi R, Tian J, Harrison CJ, De Sousa MM, Hatton PV, et al. Electrospun scaffolds containing silver-doped hydroxyapatite with antimicrobial properties for applications in orthopedic and dental bone surgery. Journal of functional biomaterials. 2020;**11**(3):58

[83] Xie J, Shen H, Yuan G, Lin K, Su J. The effects of alignment and diameter of electrospun fibers on the cellular behaviors and osteogenesis of BMSCs. Materials Science and Engineering: C. 2021;**120**:111787

[84] Ma S, Wang Z, Guo Y, Wang P, Yang Z, Han L, et al. Enhanced osteoinduction of electrospun scaffolds with assemblies of hematite nanoparticles as a bioactive interface. International Journal of Nanomedicine. 2019;**14**:1051

[85] Song H, Zhang Y, Zhang Z, Xiong S, Ma X, Li Y. Hydroxyapatite/NELL-1 nanoparticles electrospun fibers for Osteoinduction in bone tissue engineering application. International Journal of Nanomedicine. 2021;**16**:4321

[86] Wang C, Wang M. Electrospun multicomponent and multifunctional nanofibrous bone tissue engineering scaffolds. Journal of Materials Chemistry B. 2017;**5**(7):1388-1399

[87] Théry C, Witwer KW, Aikawa E, Alcaraz MJ, Anderson JD, Andriantsitohaina R, et al. Minimal information for studies of extracellular vesicles 2018 (MISEV2018): A position statement of the International Society for Extracellular Vesicles and update of the MISEV2014 guidelines. Journal of extracellular vesicles. 2018;**7**(1):1535750

[88] Man K, Brunet MY, Jones M-C, Cox SC. Engineered extracellular vesicles: Tailored-made nanomaterials for medical applications. Nanomaterials. 2020;**10**(9):1838

[89] Nieuwoudt M, Woods I, Eichholz KF, Martins C, McSweeney K, Shen N, et al. Functionalization of electrospun polycaprolactone scaffolds with matrix-binding osteocyte-derived extracellular vesicles promotes osteoblastic differentiation and mineralization. Annals of Biomedical Engineering. 2021;**49**(12):3621-3635

[90] Xing X, Han S, Ni Y, Cheng G, Cheng Y, Ni X, et al. Mussel-inspired functionalization of electrospun scaffolds with polydopamine-assisted immobilization of mesenchymal stem cells-derived small extracellular vesicles for enhanced bone regeneration. International Journal of Pharmaceutics. 2021;**609**:121136

[91] Pankongadisak P, Tsekoura E, Suwantong O, Uludağ H. Electrospun gelatin matrices with bioactive pDNA polyplexes. International journal of biological macromolecules. 2020;**149**:296-308

[92] Ren Z, Ma S, Jin L, Liu Z, Liu D, Zhang X, et al. Repairing a bone defect with a three-dimensional cellular construct composed of a multi-layered cell sheet on electrospun mesh. Biofabrication. 2017;**9**(2):025036

[93] Song J, Zhu G, Wang L, An G, Shi X, Wang Y. Assembling of electrospun meshes into three-dimensional porous scaffolds for bone repair. Biofabrication. 2017;**9**(1):015018

[94] Santana-Melo GF, Rodrigues BV, da Silva E, Ricci R, Marciano FR, Webster TJ, et al. Electrospun ultrathin PBAT/nHAp fibers influenced the in vitro and in vivo osteogenesis and improved the mechanical properties of neoformed bone. Colloids and Surfaces B: Biointerfaces. 2017;**155**:544-552

[95] Wu R, Gao G, Xu Y. Electrospun fibers immobilized with BMP-2 mediated by polydopamine combined with autogenous tendon to repair developmental dysplasia of the hip in a porcine model. International Journal of Nanomedicine. 2020;**15**:6563

[96] Ye K, Liu D, Kuang H, Cai J, Chen W, Sun B, et al. Three-dimensional electrospun nanofibrous scaffolds displaying bone morphogenetic protein-2-derived peptides for the promotion of osteogenic differentiation of stem cells and bone regeneration. Journal of colloid and interface science. 2019;**534**:625-636

[97] Yang J, Liu Y, He L, Wang Q, Wang L, Yuan T, et al. Icariin conjugated hyaluronic acid/collagen hydrogel for osteochondral interface restoration. Acta Biomaterialia. 2018;**74**:156-167

[98] Shirazi RN, Ronan W, Rochev Y, McHugh P. Modelling the degradation and elastic properties of poly (lactic-co-glycolic acid) films and regular open-cell tissue engineering scaffolds. Journal of the mechanical behavior of biomedical materials. 2016;**54**:48-59

[99] Zhao H, Tang J, Zhou D, Weng Y, Qin W, Liu C, et al. Electrospun icariin-loaded core-shell collagen, polycaprolactone, hydroxyapatite composite scaffolds for the repair of rabbit tibia bone defects. International Journal of Nanomedicine. 2020;**15**:3039

[100] Yan Y, Yao R, Zhao J, Chen K, Duan L, Wang T, et al. Implantable nerve guidance conduits: Material combinations, multi-functional strategies and advanced engineering innovations. Bioactive Materials. 2022;**11**:57-76

[101] Daly W, Yao L, Zeugolis D, Windebank A, Pandit A. A biomaterials approach to peripheral nerve regeneration: Bridging the peripheral nerve gap and enhancing functional recovery. Journal of the Royal Society Interface. 2012;**9**(67):202-221

[102] Moharrami Kasmaie F, Zamani F, Sayad-Fathi S, Zaminy A. Promotion of nerve regeneration by biodegradable nanofibrous scaffold following sciatic nerve transection in rats. Progress in Biomaterials. 2021;**10**(1):53-64

[103] Idini M, Wieringa P, Rocchiccioli S, Nieddu G, Ucciferri N, Formato M, et al. Glycosaminoglycan functionalization of electrospun scaffolds enhances Schwann cell activity. Acta Biomaterialia. 2019;**96**:188-202

[104] Gryshkov O, Al Halabi F, Kuhn AI, Leal-Marin S, Freund LJ, Förthmann M, et al. PVDF and P (VDF-TrFE) electrospun scaffolds for nerve graft engineering: A comparative study on piezoelectric and structural properties, and in vitro biocompatibility. International journal of molecular sciences. 2021;**22**(21):11373

[105] Ahmadi S, Shafiei SS, Sabouni F. Electrospun Nanofibrous scaffolds of Polycaprolactone/gelatin reinforced with layered double hydroxide Nanoclay for nerve tissue engineering applications. ACS Omega. 2022;**7**(32):28351-28360

[106] Chen C, Tang J, Gu Y, Liu L, Liu X, Deng L, et al. Bioinspired hydrogel electrospun fibers for spinal cord regeneration. Advanced Functional Materials. 2019;**29**(4):1806899

[107] Xi K, Gu Y, Tang J, Chen H, Xu Y, Wu L, et al. Microenvironment-responsive immunoregulatory electrospun fibers for promoting nerve function recovery. Nature Communications. 2020;**11**(1):1-18

[108] Mohamadi F, Ebrahimi-Barough S, Reza Nourani M, Ali Derakhshan M, Goodarzi V, Sadegh Nazockdast M, et al. Electrospun nerve guide scaffold of poly (ε-caprolactone)/collagen/nanobioglass: An in vitro study in peripheral nerve tissue engineering. Journal of biomedical materials research Part A. 2017;**105**(7):1960-1972

[109] Fasolino I, Bonadies I, Ambrosio L, Raucci MG, Carfagna C, Caso FM, et al. Eumelanin coated PLA electrospun micro fibers as bioinspired cradle for SH-SY5Y neuroblastoma cells growth and maturation. ACS applied materials & interfaces. 2017;**9**(46):40070-40076

[110] Dong R, Tian S, Bai J, Yu K, Liu C, Liu L, et al. Electrospun Polycaprolactone (PCL)-amnion Nanofibrous membrane promotes nerve repair after Neurolysis. Journal of Biomaterials Applications. 2022;**36**(8):1390-1399

[111] Quigley HA, Broman AT. The number of people with glaucoma worldwide in 2010 and 2020. British journal of ophthalmology. 2006;**90**(3):262-267

[112] Chen T-C, She P-Y, Chen DF, Lu J-H, Yang C-H, Huang D-S, et al. Polybenzyl

glutamate biocompatible scaffold promotes the efficiency of retinal differentiation toward retinal ganglion cell lineage from human-induced pluripotent stem cells. International journal of molecular sciences. 2019;**20**(1):178

[113] Carvalho CR, Oliveira JM, Reis RL. Modern trends for peripheral nerve repair and regeneration: Beyond the hollow nerve guidance conduit. Frontiers in bioengineering and biotechnology. 2019;**7**:337

[114] Yang Y, De Laporte L, Rives CB, Jang J-H, Lin W-C, Shull KR, et al. Neurotrophin releasing single and multiple lumen nerve conduits. Journal of Controlled Release. 2005;**104**(3):433-446

[115] Frost HK, Andersson T, Johansson S, Englund-Johansson U, Ekström P, Dahlin LB, et al. Electrospun nerve guide conduits have the potential to bridge peripheral nerve injuries in vivo. Scientific Reports. 2018;**8**(1):1-13

[116] Sun B, Zhou Z, Wu T, Chen W, Li D, Zheng H, et al. Development of nanofiber sponges-containing nerve guidance conduit for peripheral nerve regeneration in vivo. ACS applied materials & interfaces. 2017;**9**(32):26684-26696

[117] Zhang J, Zhang X, Wang C, Li F, Qiao Z, Zeng L, et al. Conductive composite fiber with optimized alignment guides neural regeneration under electrical stimulation. Advanced healthcare materials. 2021;**10**(3):2000604

[118] Wu T, Li D, Wang Y, Sun B, Li D, Morsi Y, et al. Laminin-coated nerve guidance conduits based on poly (l-lactide-co-glycolide) fibers and yarns for promoting Schwann cells' proliferation and migration. Journal of Materials Chemistry B. 2017;**5**(17):3186-3194

[119] Lee YS, Wu S, Arinzeh TL, Bunge MB. Enhanced noradrenergic axon regeneration into schwann cell-filled PVDF-TrFE conduits after complete spinal cord transection. Biotechnology and Bioengineering. 2017;**114**(2):444-456

[120] Cheong H, Kim J, Kim BJ, Kim E, Park HY, Choi B-H, et al. Multi-dimensional bioinspired tactics using an engineered mussel protein glue-based nanofiber conduit for accelerated functional nerve regeneration. Acta Biomaterialia. 2019;**90**:87-99

[121] Liu C, Wang Z, Yao X, Wang M, Huang Z, Li X. Sustained biochemical signaling and contact guidance by electrospun Bicomponents as promising scaffolds for nerve tissue regeneration. ACS Omega. 2021;**6**(48):33010-33017

[122] Afrash H, Nazeri N, Davoudi P, FaridiMajidi R, Ghanbari H. Development of a bioactive scaffold based on NGF containing PCL/chitosan nanofibers for nerve regeneration. Biointerface Res Appl Chem. 2021;**11**:12606-12617

[123] Aznar-Cervantes S, Pagán A, Martínez JG, Bernabeu-Esclapez A, Otero TF, Meseguer-Olmo L, et al. Electrospun silk fibroin scaffolds coated with reduced graphene promote neurite outgrowth of PC-12 cells under electrical stimulation. Materials Science and Engineering: C. 2017;**79**:315-325

[124] Heidari M, Bahrami SH, Ranjbar-Mohammadi M, Milan P. Smart electrospun nanofibers containing PCL/gelatin/graphene oxide for application in nerve tissue engineering. Materials science and engineering: C. 2019;**103**:109768

Chapter 7

Influence of Mechanical Properties of Biomaterials on the Reconstruction of Biomedical Parts via Additive Manufacturing Techniques: An Overview

Babatunde Olamide Omiyale, Akeem Abiodun Rasheed, Robinson Omoboyode Akinnusi and Temitope Olumide Olugbade

Abstract

Additive manufacturing (AM) is the opposite of conventional manufacturing technologies, creating an opportunity to fabricate parts using a layer upon layer approach to obtain 3D patterns. AM technology has provided an opportunity for biomaterials usage in the bio-fabrication of organs and scaffolds for tissues engineering. In recent times, AM has been well-utilized for the printing of organs, customized implants, anatomical models for surgery training kits, drug formulations, prosthetics, orthotics, dentistry, and scaffolds for tissue engineering with the use of metals, ceramics, polymers, and composites materials. Printing of biomaterial that has a suitable viscosity, enough strength, good biocompatibility, and degradability has been reported by many researchers to be an arduous task. Biomaterials printed with robust mechanical properties are considered highly essential for the fabrication of soft tissues such as cartilage and skin because the function of such tissues mainly relies on their mechanical properties that possess the capacity to support cell proliferation and extracellular matrix production. For repairing and regenerating organs or tissue, the implant must provide sufficient mechanical support to endure *in vivo* stresses and load-bearing cycles. This book chapter aims to document the mechanical properties of 3D printed biomaterials and provides a keys future research direction.

Keywords: additive manufacturing, biomaterial, mechanical properties, tissue engineering, bio-fabrication

1. Introduction

The unique material that can be used to fabricate biomedical parts inside a human body to treat, repair, replace any tissue is known as biomaterials. Biomaterials are referred to as any material that comes into contact with humans or animals to fulfill their intended purpose without causing any toxic reaction [1]. In the field of biomedical engineering, biomaterials can be used to replace or mimic part of an organ or a tissue while still maintaining interaction with living tissue. One major setback when working with biomaterial is the lack of mechanical strength [2]. The reports have suggested that this constraint can either be subdued by utilizing appropriate material and manufacturing techniques of processing the scaffolds to enhance the mechanical integrity of the fabricated parts. Additive manufacturing (AM) has been described as an emerging advanced technology in which materials are linked layer by layer to fabricate functional components from three-dimensional (3D) model data [3]. The ability of 3D printing to produce complex shapes at low cost placed it a higher advantage in the production of biomedical parts than other manufacturing methods [4–10]. **Figure 1** summarizes common synthetic polymers in 3D bioprinting applications [4–11].

3D bioprinting has become a successful technique used in fabricating biomaterial scaffolds such as customized implant, organ, drug delivery systems, prosthetics, orthotic, and tissues engineering [3, 12]. Bioprinting technology has emerged as a powerful bio-fabrication tool where biomaterials such as cells and growth factors are combined to fabricate biomedical parts with bio-ink [13]. The advantages of bioprinting include accurate control of cell distribution, high-resolution cell deposition, scalability, and cost-effectiveness. Bioprinting has become an effective fabrication tool to create complex micro-and macro-scale biomedical systems. Biomaterials, such as hydrogels, are currently extensively studied for their ability to reproduce both the ideal 3D extracellular environment for tissue growth and to have required mechanical properties for load-bearing. Microfabrication techniques such as electrospinning [14] and 3D printing have emerged as promising strategies for manufacturing complex hydrogel structures for tissue engineering applications. 3D bioprinting application has extensively found its potential use in biomedical applications such as tissue engineering, drug discovery, toxicology [15], regenerative medicine to generate a variety of transplantable tissues including skin, cartilage, and bone [16]. In 3D bioprinting, 3D fabrication techniques have been precisely used to dispense cell-laden biomaterials for the construction of complex 3D functional living

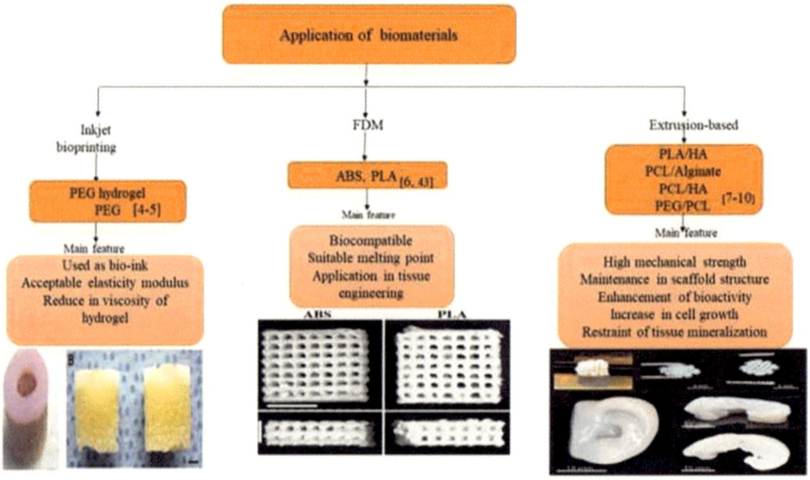

Figure 1.
Summarizes common synthetic polymers in 3D bioprinting applications.

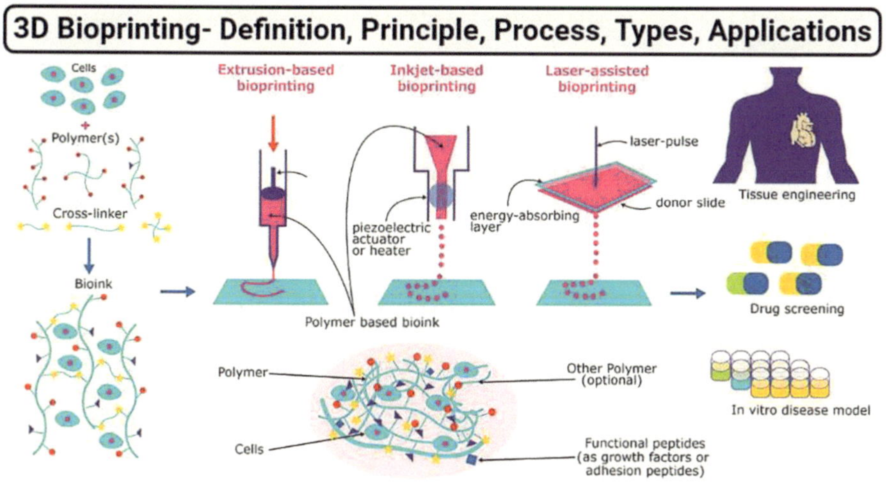

Figure 2.
Bioprinting process, techniques, and applications [23].

tissues or artificial organs using an additive manufacturing strategy by depositing substrates such as living cells, nucleic acids, drug particles, proteins, and other biological components [17, 18]. Nowadays, several bioprinting signs of progress have been demonstrated, such as bionic ears [19], multilayered skins [20], artificial bones [21], vascular tissues [22], and cartilaginous structures. Three major bioprinting techniques are being commonly used which include extrusion bioprinting, laser-assisted bioprinting, and inkjet bioprinting. **Figure 2** summarizes the bioprinting process, techniques, and applications.

According to Dinesh & Devaprakasam [24], the life span, reliability, bio-compatible and mechanical properties of the implant material are not up to the expectation but research has to be done to find out the most reliable, high strength, and long-lasting material.

Laasri et al. [25] studied the influence of powder manufacturing and sintering temperature on densification, microstructure, and mechanical properties of calcium phosphate biomaterials. It was reported that the manufacturing of β-TCP ceramic improved its densification, microstructure homogeneity, and mechanical properties. In another study, Wang et al. [26] discussed the influence of mechanical properties of biomaterials on degradability, cell behavior, and signaling pathways. It is concluded that both biomaterial degradability and signaling cascades of cell interactions with biomaterials are significantly influenced by mechanical properties of biomaterials, determining the final repair effect of bio-implants [26]. Lv et al. [27] designed biomaterials to mimic the mechanical properties of muscles. It is confirmed that the mechanical properties of biomaterials can be modified by fine-tuning the composition of the elastomeric proteins, providing the opportunity to develop biomaterials that are mimetic of different types of muscles [27]. Fabrication of the appropriate 3D scaffold with required mechanical properties is a crucial factor in tissue regenerative medicine in promoting in *vivo-like* cell behavior. This book chapter documents the influence of mechanical properties of additively manufactured biomaterials in reconstructing biomedical parts.

2. Mechanical properties of biomaterials manufactured by using 3D printing techniques

The tensile strength, yield strength, elastic modulus, corrosion, creep, and hardness are some of the most paramount properties of biomaterials that need to be carefully examined and improved before implantation [1]. Production of functional parts needs to demonstrate their stable mechanical properties. The importance of the mechanical control of cellular phenotype and fate are as critical as biochemical factors in regulating cellular function [28]. Biomaterials can experience several mechanical degradations during processing, storage, and use. Due to environmental conditions during the printing process, mechanical degradation can take place due to shear forces, tension, and compression [29–31]. In one study, Yang et al. [32] demonstrated that the complex stress state of maxillofacial bone tissues requires metal implants with matching mechanical properties to support mandibular functions. Soufivand et al. [33] also investigated the applicability of CAD-based FEM analysis to tune and predict the mechanical behaviors of the printed PCL scaffolds based on the inner geometries, as the inner geometries of the tissue-engineered scaffolds play an important role in biomechanical and biological aspects for tissue engineering applications. Their findings revealed that the CAD-based FEM prediction could be used for designing tissue-specific constructs to mimic the mechanical properties of targeted tissues or organs. Moroni et al. [34] also proved that the mechanical signaling from tissue-engineered scaffolds could impact cellular activities, including cell proliferation and differentiation, as well as tissue hemostasis and development. In one investigation, Ambrosio et al. [35] evaluated the mechanical and viscoelastic behavior of a soft composite material based on a hydrogel matrix reinforced with the range of polyethylene terephthalate (PET) fibers designed to mimic mechanical properties of soft tissue such as tendons, ligaments, and intervertebral discs. It is concluded that the control of the geometrical configuration of the wound fibers allows a wide range of mechanical properties to be secured [35]. Iannace et al. [36] determined the mechanical behavior of composite artificial tendons and ligaments. The results recommended that large variations in the mechanical behavior can be attained by altering the winding angle of the fibers in the composite which determines the extent of the 'toe' region and the sensitivity of the system to the rigidity of the fibers. In another study, Hippler et al. [37] informed that biophysical factors such as mechanical properties of the surrounding extracellular matrix (ECM) can significantly impact cellular reactions. As reported, modifications in the composition or the mechanical properties of the ECM are often connected to cancer progression and metastasis as well as pathological conditions [37]. Hukins et al. [38] also reviewed the difficulties which are encountered in defining the mechanical properties of natural tissues, and in replacing them with synthetic materials in the human body. In any support system like the frame in engineering, bone in the human body and others required suitable mechanical properties to make it function optimally. Mechanical properties such as tensile and compressive stresses, strain, torque, bending moment shear force, and others have a significant influence on the suitability of any materials [39]. Capurro and Barberis [40] discussed the mechanical properties of biomaterials and their influence in the field of medicine, surgery, and physiology applications. It was presented that the mechanical properties can be categorized into two main branches: elastic and viscoelastic properties, and ultimate properties (such as plasticity, fracture, fatigue damage, and others) [41]. A lot of previous research works have been carried out on biomaterials but little has been said on the evaluation

AM technology	Materials	Mechanical properties	Applications	Ref
Electron beam melting	Co–29Cr–6Mo alloy and a Ti–6Al–4Vtibial component	The UTS increased by 20%, and the elongation increased by 900%. In contrast to the ASTM-F75 Co–29Cr–6Mo transplant standard, the yield stress for the EBM-fabricated and HIP manufactured component increased by 30%, and a 200% increase in elongation	Knee replacement implants	[42]
3D bioprinting	Alginate hydrogel	Results demonstrated that alginate concentration, $CaCl_2$ cross-linking concentration and cross-linking ratios as well as gelling conditions, such as cross-linking reaction time and temperature which have a significant effect on its mechanical properties and printability of 3D alginate scaffolds network.	Tissue engineering	[32]
Micro-extrusion 3D bioprinting	Poly(ε-caprolactone) (PCL)	Results showed that the theoretical compressive elastic moduli of the designed constructs were 23.3, 56.5, 67.5, and 1.8 MPa, and the experimental compressive elastic moduli were 23.6 ± 0.6, 45.1 ± 1.4, 56.7 ± 1.7, and 1.6 ± 0.2 MPa for lattice, wavy, hexagonal, and shifted microstructures, respectively, while maintaining the same construct dimension and porosity.	Tissue engineering and regenerative medicine	[33]
3D bioprinting	Bio-ink	On day 20, all tissues showed an improved stiffness compared to day 10. P2 tissue exhibited the lowest variation of young's modulus with a final value of 103.3 Pa, the lowest of the three modalities. M1 and control tissues exhibited similar elastic moduli of 155.8 and 149.0 Pa.	Tissue engineering	[10]
3D bioprinting	Nanocellulose/ chitosan-based bio-ink	The addition of CNCs and cells (5 million cells) significantly improved the viscosity of bio-inks and the mechanical properties of chitosan scaffolds post-fabrication.	Bone tissue engineering and regeneration applications	[43]

Table 1.
Summary of biomaterials in 3D printing applications.

of mechanical properties of biomaterials in many fields of physiology, medicine, and Surgery applications. **Table 1** presented a summary of biomaterials in 3D printing applications.

3. Summary and future work

Loss of many lives associated with getting a replacement for defective or loss of human tissues and bones have become a great matter of concern. For implants and natural tissue materials, the biocompatibility and mechanical strength of biomaterials are critical for a variety of skeletal repair joint replacements and dental restorations. Printing of scaffolds with the use of design strategies could be an effective platform to fabricate functional tissue engineering that will provide significantly enhanced different mechanical properties. In recent times, materials with high concentration and high viscosity, including ceramics, poly(caprolactone) (PCL), polylactic acid (PLA), beta-tricalcium phosphate (β-TCP), have been widely utilized to produce bio-inks for the bioprinting of defective bone tissue. For future work, the following conclusions are drawn from the literature comprehensively reviewed:

Further research is required to investigate appropriate material and manufacturing techniques of processing the scaffolds to enhance the mechanical integrity of the fabricated parts.

There is a need to determine the effects of processing parameters on the biochemical and biophysical characteristics of biomaterials for the fabrication of tissue structure.

The tensile strength, yield strength, elastic modulus, corrosion, creep, and hardness are some of the most paramount properties of biomaterials that should be carefully examined and improved before implantation.

Conflict of interest

The authors declare no conflict of interest.

Abbreviations

AM	Additive manufacturing
ABS	Acrylonitrile butadiene styrene
ASTM	American Society for Testing and Materials
ALG	Alginate
CS	Chitosan
CAD	Computer aided design
HY	Hydrogel
HIP	Hot isostatic pressing
β-TCP	β-tricalcium phosphate
ECM	Extracellular matrix
PET	Polyethylene terephthalate
FDM	Fused deposition modeling
PCL	Poly(ε-caprolactone)
PLA	Polylactic acid
EBM	Electron beam melting
CNCs	Cellulose Nanocrystals
ABS	Acrylonitrile Butadiene Styrene
PEG	Polyethylene glycol
CaCl2	Calcium chloride
HA	Hydroxyapatite
FEM	Finite element modeling
UTS	Ultimate tensile strength

Author details

Babatunde Olamide Omiyale*, Akeem Abiodun Rasheed, Robinson Omoboyode Akinnusi and Temitope Olumide Olugbade
Industrial and Production Engineering, Federal University of Technology Akure, Ondo, Nigeria

*Address all correspondence to: boomiyale@futa.edu.ng

References

[1] An Introduction to Biomaterials Science and Engineering. Available from: www.worldscientific.com by 105.112.209.217. [Accessed: 02 April 2022]

[2] Billiet T, Vandenhaute M, Schelfhout J, Van Vlierberghe S, Dubruel P. A review of trends and limitations in hydrogel. Rapid Prototyping for Tissue Engineering. 2012;**33**(26):6020-6041. DOI: 10.1016/j.biomaterials.2012.04.050

[3] Lee VK, Dai G. Printing of three-dimensional tissue analogs for regenerative medicine. Annals of Biomedical Engineering. 2016;**45**(1): 115-131. DOI: 10.1007/s10439-016-1613-7

[4] Cui X, Breitenkamp K, Finn MG, Lotz MD'L, Darryl D. Direct human cartilage repair using three-dimensional bioprinting technology. Tissue Engineering Part A. 2012;**18**(11-12):1304-1312. DOI: 10.1089/ten.tea.2011.0543

[5] Gao G, Yonezawa T, Hubbell K, Dai G, Cui X. Inkjet -bioprinted acrylated pep- tides and PEG hydrogel with human mesenchymal stem cells promote robust bone and cartilage formation with minimal printhead clogging. Biotechnology Journal. 2015;**10**(10):1568-1577

[6] Rosenzweig DH, Carelli E, Steffen T, Jarzem P, Haglund L. 3D-printed ABS and PLA scaffolds for cartilage and nucleus pulposus tissue regeneration. International Journal of Molecular Sciences. 2015;**16**(7):15118-15135

[7] Shim J-H, Kim JY, Park M, Park J, Cho D-W. Development of a hybrid scaffold with synthetic biomaterials and hydrogel using solid freeform fabrication technology. Biofabrication. 2011;**3**(3):034102

[8] Kundu JH, Shim J, Jang SWK, Cho DW. An additive manufacturing -based PCL-alginate–chondrocyte bioprinted scaffold for cartilage tissue engineering. Journal of Tissue Engineering and Regenerative Medicine. 2015;**9**(11):1286-1297

[9] Gonçalves EM, Oliveira FJ, Silva RF, Neto MA, Fernandes MH, Amaral M, et al. Three-dimensional printed PCL-hydroxyapatite scaffolds filled with CNTs for bone cell growth stimulation. Journal of Biomedical Materials Research. Part B, Applied Biomaterials. 2016;**104**(6): 1210-1219. DOI: 10.1002/jbm.b.33432

[10] Fedore C, Tse L, Nam H, Barton K, Hatch N. Analysis of polycaprolactone scaffolds fabricated via precision extrusion deposition for control of craniofacial tissue mineralization. Orthodontics & Craniofacial Research. 2017;**20**:12-17

[11] Farayibi PK, Omiyale BO. Mechanical behaviour of Polylactic acid parts fabricated via material extrusion process: A Taguchi-Grey relational analysis approach. International Journal of Engineering Research in Africa. 2020;**46**:32-44. DOI: 10.4028/www.scientific.net/jera.46.32

[12] Gao G, Yonezawa T, Hubbell K, Dai G, Cui X. Inkjet-bioprinted acrylated peptides and PEG hydrogel with human mesenchymal stem cells promote robust bone and cartilage formation with minimal printhead clogging. Biotechnology Journal. 2015;**10**:1568

[13] Agarwal S, Saha S, Balla VK, Pal A, Barui A, Bodhak S. Current developments in 3D bioprinting for tissue and organ regeneration–a review. Frontiers in Mechanical Engineering. 2020;**6**:589171. DOI: 10.3389/fmech.2020.589171

[14] Xue J, Wu T, Dai Y, Xia Y. Electrospinning and electrospun nanofibers: Methods, materials, and applications. Chemical Reviews. 2019;**119**(8):5298-5415. DOI: 10.1021/acs.chemrev.8b00593

[15] Murphy SV, Atala A. 3D bioprinting of tissues and organs. Nature Biotechnology. 2014;**32**:773-785. DOI: 10.1038/nbt.2958

[16] 3D Printing.com, (2018), What Is 3D Printing? Available from: https://www.3dprinting.com/what-is-3d-printing. Accessed: 2018 November 14.

[17] Huang Y, Zhang XF, Gao G, et al. 3D bioprinting and the current applications in tissue engineering. Biotechnology Journal. 2017;**12**(8):1600734. DOI: 10.1002/biot.201600734

[18] Ulucan-Karnak F. 3D bioprinting in medicine. Global Journal of Biotechnology and Biomaterial Science. 2021;**7**:001-005

[19] Mannoor MS et al. 3D printed bionic ears. Nano Letters. 2013;**13**:2634-2639

[20] Lee V et al. Design and fabrication of human skin by three-dimensional bioprinting. Tissue Engineering. Part C, Methods. 2013;**20**:473-484

[21] Bose S, Vahabzadeh S, Bandyopadhyay A. Bone tissue engineering using 3D printing. Materials Today. 2013;**16**:496– 504

[22] Norotte C, Marga FS, Niklason LE, Forgacs G. Scaffold-free vascular tissue engineering using bioprinting. Biomaterials. 2009;**30**:5910-5917

[23] Mandrycky C, Wang Z, Kim K, Kim D. 3D bioprinting for engineering complex tissues. Biotechnology Advances. 2016;**34**:422-434

[24] Dinesh BR, Devaprakasam R. A review on additive manufacturing techniques in medical applications. Int. J. Adv. Sci. Eng. 2019;**5**(3):988-997

[25] Laasri S, Taha M, Hlil EK, Laghzizil A, Hajjaji A. Manufacturing and mechanical properties of calcium phosphate biomaterials. Comptes Rendus Mécanique. 2012;**340**(10):715-720. DOI: 10.1016/j.crme.2012.09.005

[26] Wang L, Wang C, Wu S, Fan Y, Li X. Influence of mechanical properties of biomaterials on degradability, cell behaviors and signaling pathways: Current progress and challenges. Biomaterials Science. 2020;**8**:2714-2733. DOI: 10.1039/d0bm00269k

[27] Murr LE, Amato KN, Li SJ, Tian YX, Cheng XY, Gaytan SM, et al. Microstructure and mechanical properties of open-cellular biomaterials prototypes for total knee replacement implants fabricated by electron beam melting. Journal of the Mechanical Behavior of Biomedical Materials. 2011;**4**(7):1396-1411. DOI: 10.1016/j.jmbbm.2011.05.010

[28] Liu WF. Mechanical regulation of cellular phenotype: Implication for vascular tissue regeneration. Cardiovascular Research. 2012;**95**(2):215-222. DOI: 10.1093/cvr/cvs168

[29] La Mantia FP, Morreale M, Botta L, Mistretta MC, Ceraulo M, Scaffaro R. Degradation of polymer blends: A brief review. Polymer Degradation and Stability. 2017;**145**:79-92

[30] Briassoulis D. Mechanical behaviour of biodegradable agricultural films under real field conditions. Polymer Degradation and Stability. 2006;**91**:1256-1272

[31] Briassoulis D. Analysis of the mechanical and degradation

performances of optimised agricultural biodegradable films. Polymer Degradation and Stability. 2007;**92**:1115-1132

[32] Abbas IA, Al-Darkazly. Optimization of mechanical properties of alginate hydrogel for 3D bioprinting self-standing scaffold architecture for tissue engineering applications. International Scholarly and Scientific Research & Innovation. 2020;**14**(12):419-427

[33] Soufivand AA, Abolfathi N, Hashemi SA, Lee SJ. Prediction of mechanical behavior of 3D bioprinted tissue-engineered scaffolds using finite element method (FEM) analysis. Additive Manufacturing. 2020;**33**:101181. DOI: 10.1016/j.addma.2020.101181

[34] Moroni L, de Wijn JR, van Blitterswijk CA. 3D fiber-deposited scaffolds for tissue engineering: Influence of pores geometry and architecture on dynamic mechanical properties. Biomaterials. 2006;**27**(7):974-985

[35] Ambrosio L, De Santis R, Nicolais L. Composite hydrogels for implants. Proceedings of the Institution of Mechanical Engineers, Part H: Journal of Engineering in Medicine. 1998;**212**(2):93-99

[36] Iannace S, Sabatini G, Ambrosio L, Nicolais L. Mechanical behaviour of composite artificial tendons and ligaments. Biomaterials. 1995;**16**(9):675-680. DOI: 10.1016/0142-9612(95)99693-g PMID: 7578769

[37] Hippler M, Lemma ED, Bertels S, Blasco E, Barner-Kowollik C, Wegener M, et al. 3D scaffolds to study basic cell biology. Advanced Materials. 2019;**31**:1808110. DOI: 10.1002/adma.201808110

[38] Hukins DWL, Leahy JC, Mathias KJJ. Materials Chemistry. 1999;**9**:629

[39] Amini M, Reisinger A, Pahr DH. Influence of processing parameters on mechanical properties of a 3D-printed trabecular bone microstructure. Journal of Biomedical Materials Research. Part B, Applied Biomaterials. 2020;**108**(1):38-47. DOI: 10.1002/jbm.b.34363

[40] Capurro M, Barberis F. Evaluating the mechanical properties of biomaterials. Biomaterials for Bone Regeneration. 1st ed. 2014:270-323. DOI: 10.1533/9780857098104.2.270

[41] Dussoyer M, Courtial EJ, Albouy M, Thépot A, Santos MD, Marquette C. Mechanical properties of 3D bioprinted dermis: Characterization and improvement. International Journal of Regenerative Medicine. 2019;**2**(1):1-6

[42] Murr LE, Amato KN, Li SJ, Tian YX, Cheng XY, Gaytan SM, et al. Microstructure and mechanical properties of open-cellular biomaterials prototypes for total knee replacement implants fabricated by electron beam melting. Journal of the Mechanical Behavior of Biomedical Materials. 2011;**4**(7):1396-1411. DOI: 10.1016/j.jmbbm.2011.05.010

[43] Maturavongsadit P, Narayanan LK, Chansoria P, Shirwaiker R, Benhabbour SR. Cell-laden Nanocellulose/chitosan-based bioinks for 3D bioprinting and enhanced osteogenic cell differentiation. ACS Applied Bio Materials. 2021;**4**(3):2342-2353. DOI: 10.1021/acsabm.0c01108

Section 4

Biosensors

Chapter 8

Challenges and Emerging Technologies in Biomanufacturing of Monoclonal Antibodies (mAbs)

Susan McDonnell, Raymon Floyd Principe, Maycou Soares Zamprognio and Jessica Whelan

Abstract

Therapeutic antibodies dominate the biopharmaceutical market with continual innovations being made to provide novel and improved antibody treatment strategies. Speed to-market and cost-efficiency are of increasing importance due to the changing landscape of the biopharmaceutical industry. The increasing levels of competition from biosimilars, the increase in small volume products and political and social pressure to reduce the cost of treatments are some of the challenges currently being faced. Chinese hamster ovary (CHO) cells have been the workhorse in the production of therapeutic antibodies over the last 36 years due to the robust nature and high productivity of these cell lines. However, there are many biomanufacturing challenges remaining. The aim of this review is to examine the current biological, and engineering challenges facing the biomanufacturing of antibodies and to identify the mitigations and emerging technologies that can be employed to overcome them. Developments in cell line engineering, intensified processing, continuous manufacturing, automation and innovations in process analytical technologies and single use technology will be discussed with regard to their ability to improve the current performance of mAb production processes.

Keywords: mAbs, biomanufacturing, challenges, emerging technologies, therapeutic antibodies

1. Introduction

Therapeutic monoclonal antibodies, referred to as mAbs throughout this chapter, have emerged as the dominant player in the Pharmaceutical/Biopharmaceutical sector and are extremely effective agents for the treatment of cancer, inflammatory disorders and infectious diseases [1]. The effectiveness of mAbs as therapeutics is due primarily to their specificity in recognizing and binding to specific antigens through the antigen binding site. The efficacy of full length mAbs as anti-cancer agents is due to their ability to activate both complement-dependent cytotoxicity (CDC) and antigen-dependent cell cytotoxicity (ADCC) [2].

Worldwide, the mAbs market represents approximately 50% of the biotherapeutics market and according to Global Market Insights Inc., the global antibody therapy market is projected to reach $445 billion by 2028 which represents a 13.2% compound annual growth rate (CAGR) [3]. The first mAb, Muromonab-CD3 (proprietary name rather than brand name will be used throughout this chapter) was approved in 1986 and in June 2021 a landmark achievement was reported with the approval of the 100th mAb by the FDA [4]. As of May 2022 there are currently over 111 mAbs approved by either the European Medicine Agency (EMA) or the US Food & Drug Administration (FDA) [5]. In addition, there are over 15 mAbs pending confirmation of approval by either one or both regulatory agencies. Interestingly, several of the products pending approval are targeting the Covid-19 spike protein [6]. Over the same time period marketing authorisation for several products including Daclizumab, a humanized mAb for the treatment of multiple sclerosis, has been withdrawn and several products including the first approved mAb Muromonab-CD3 have been discontinued. Biosimilars are generic versions of biologics and thus far, over 20 mAb biosimilar products have been approved by the EMA or FDA [7]. Humira, a human antibody targeting tumor necrosis factor alpha used to treat rheumatoid arthritis and related disorders, is the top selling biopharmaceutical drug with a market value of $20 billion in 2021 [8]. Currently at least 8 biosimilar versions of Humira have been approved in various markets [7].

Immunoglobulin G (IgG) the dominant type of immunoglobulin manufactured is composed of two heavy (H) chains and two light (L) chains and has a molecular weight ranging from 140 to 160 kDa depending on the type of IgG subtype. Each of the light chains contain a variable (VL) and constant (CH) domain and the heavy chains contain one variable (VH) and 3 constant domains (CH1, CH2 and CH3). mAbs contain a number of glycosylation sites in the CH3 region. There is a huge variety in the type of mAb products approved with the major mAb formats being: canonical (full-length antibodies), antibody drug conjugates (ADCs) and antigen-binding fragments (Fabs). **Figure 1A** shows the distribution of the major types of mAbs currently approved by EMA or FDA [5].

Currently there are 92 canonical antibodies approved (83% of total) which are typically of the IgG1 subtype, have a molecular weight of ~150 kDa and are subdivided into chimeric, humanized or human antibodies. The majority of approved

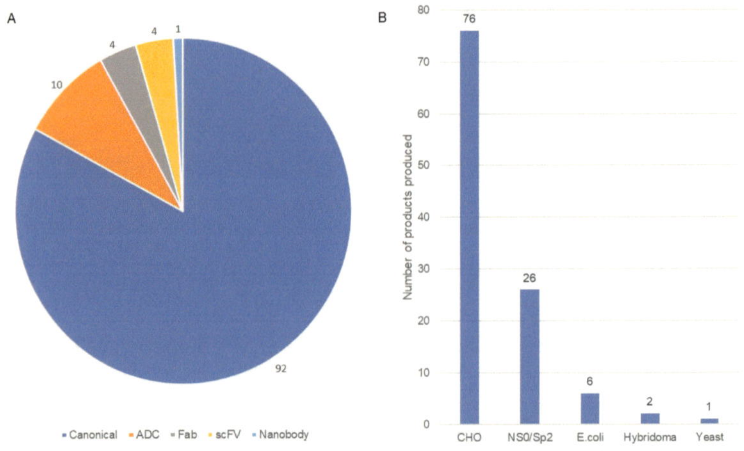

Figure 1.
A: Pie chart shows number of different mAbs format approved B: Bar chart shows the number of products produced in a range of host cells.

canonical mAbs target autoimmune diseases and many different types of cancers. Most canonical mAbs are monospecific with 3 bispecific antibodies (bsAbs) which recognize two different antigens (Emicizumab, Amivantamab and Faricimab) approved. ADCs are full-length antibodies with the addition of a highly toxic molecule usually attached to cysteine residues via a linker molecule. Upon arrival at specific cancer cells, targeted by the mAb component, the ADC is internalized by the cell and the toxin released through enzymatic cleavage [9]. ADCs represent highly specific and targeted therapies due to the specificity of the mAb component and currently 10 ADCs (9% of total) have been approved, of which two are targeting Her-2 positive breast cancer, ado-trastuzumab emtansine and trastuzumab deruxtecan [9]. Antibody fragments come in a variety of formats with molecular weights ranging from 12 to 50 kDa and contain a wide range of heavy and light variable and constant domains [10]. Currently, 4 Fab antibodies (4% of total) which contain the antigen binding site (i.e. one heavy and one light chain each with a variable and constant domain) have been approved. Fabs have a molecular weight between 40 and 50 kDa which improves tissue penetration due to their smaller size. The lack of the Fc domain means that these mAbs are not glycosylated so they can be produced in bacterial expression systems. There are 4 (4% of total) scFv products approved which are typically 25 kDa in molecular weight and are all produced in *E.coli* [10]. Interestingly, the scFv mAbs have the most diverse range of functionality: one is bispecific, one is a fusion-protein and the other is linked to an immunotoxin. The smallest of all therapeutic antibodies is the Nanobody with a molecular weight of 12 kDa. The first and currently only approved Nanobody is Caplacizumab which targets von Willebrand factor and is being used for the treatment of acquired thrombotic thrombocytopenic purpura [11].

Glycosylation plays an essential role in the biological efficacy of antibodies and is one of the critical quality attributes of mAbs [12]. N-linked glycosylation occurs in the Fc domain of the antibody. Canonical antibodies require glycosylation so production of full length mAbs is in mammalian cells, primarily CHO and murine myeloma cells (**Figure 1B**) as they have the biological capability to make these types of post-translational modifications (PTMs). CHO cells have been the workhorse in the production of antibody products over the last 36 years. CHO cells currently act as hosts for approximately 69% of all mAbs approved. Approximately 23% of approved products are produced in mouse myeloma cells lines with the predominant type being the non-secreting NS0 cell line followed by Sp2/0. Interestingly, the first mAb product approved Muromonab-CD3 was produced in a murine hybridoma cell line. A small number of Fab and scFv products that do not require glycosylation have been produced in bacterial cells, Escherichia Coli (*E.coli*). Thus far, just 1 product, Eptinezumab which is a humanized antibody has been produced in yeast cells Pichia *pastoris* [13]. Biomanufacturing of mAbs in CHO cells will be the focus of this chapter and looking at alternative expression systems is beyond the scope. Several comprehensive reviews of expression in bacterial, insect and yeast cells have been published [14–16].

Figure 2 shows a typical process flow diagram for the production of mAb drug substance, divided into upstream and downstream processing and formulation and fill-finish. Upstream processing encompasses the steps from vial thaw through inoculum and cell expansion with the aim of generating sufficient cells for the production stage bioreactor where the mAb is produced. Downstream processing encompasses all the operations from product capture, purification to formulation and fill finish.

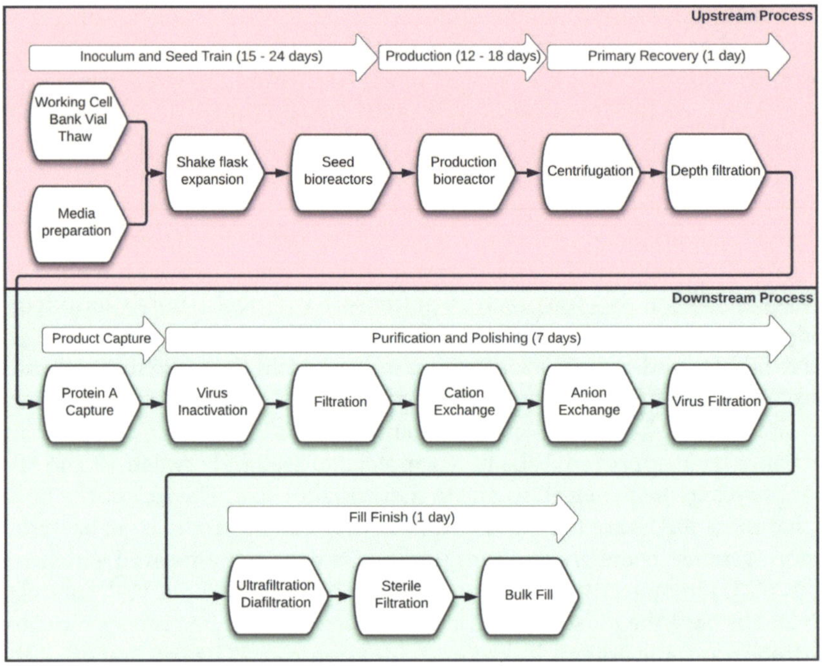

Figure 2.
Process flow diagram of mAb biomanufacturing process. Typical fed-batch process duration indicated.

The biomanufacturing process is initiated with a vial of cryopreserved cells from the working cell bank (WCB) which is thawed into a small volume of media and then expanded through a series of shake flasks and bioreactors of increasing size, in order to generate sufficient cell numbers to inoculate into the production scale bioreactor. The final production vessel referred to as the N-stage bioreactor, is in the order of between 1000 and 20,000 L depending requirements. A fed-batch upstream production bioreactor step typically takes 10–20 days [17]. The production stage bioreactor duration can increase up to approximately 60 days if operated in perfusion [18]. The upper limit is determined by the validated Limit of In Vitro Cell Age (LIVCA) which is the number of generations for which the cell line has been demonstrated to be genetically stable.

The downstream process begins with recovery of the product from the bioreactor. Since CHO based systems secrete the product into the media, the product is usually harvested through centrifugation and/or depth filtration. Purification generally involves several chromatography steps for capture, intermediate purification and polishing. Affinity chromatography using Protein A which specifically binds to the Fc domain of the mAb is used to capture full length antibodies. Protein L can be used for purification of Fab fragments as it binds specifically to kappa light chains [19]. Anion and cation-exchange chromatography are typically used for intermediate purification and hydrophobic interaction chromatography can be used for final polishing. In addition to purification to >99.9%, the downstream process must include a viral removal and inactivation step before final formulation and finishing [19].

mAb processing methods have become the model for the production of both therapeutic proteins and emerging products like cell and gene therapy-based products as these have matured and been optimized over the last 36 years. However, with the increasing focus on speed-to-market and cost efficiency, it is necessary to continue to innovate and improve mAb biomanufacturing. The aim of this chapter is to examine the current challenges facing the industry and to discuss the mitigations and emerging technologies that have the potential to address them. This chapter is presented in 2 sections and will focus on the *biological and engineering* aspects associated with the manufacture of mAbs in CHO cells. The main challenges and mitigations within these themes are discussed and future directions and innovations for the biopharmaceutical industry are presented.

2. Biological and bioprocessing challenges and mitigations involved in biomanufacture of mAbs

Mammalian expression systems are chosen for full-length mAb production as they have the necessary cell machinery required to facilitate the correct protein-folding and glycosylation. Advances in genetic engineering and cell line development methods have been used to improve cell productivity and glycosylation control [20]. However, cell-based expression systems are challenging due to their inherent biological nature which results in variations in product yields and protein quality inconsistencies which can lead to challenges in manufacturing and increased production costs.

2.1 Production cell lines

CHO cell lines are the preferred host expression system in mAb production due to their capacity for complex human-like post translational modification, ease of genetic manipulation, lack of susceptibility to human viral infection and regulatory approval [21]. In terms of bioprocessing, CHO cells are robust and are easily adapted to grow in suspension in chemically defined and serum-free media making them readily amenable to process scale-up. CHO cell line development has generated cells with specific productivities (Qp) in the range of 5-20 pg/cell/day [22] and advances in single cell sequencing could increase this further [23]. Other mammalian cell lines like baby hamster kidney (BHK) cells and murine lymphoma cells lines such as NS0 and Sp2/0 also provide human-like glycosylation patterns but productivity levels can be significantly lower [24]. In addition, both NS0 and Sp2/0 have been shown to express two predominant glycan epitopes, galactose-alpha 1,3-Gal (alpha-gal) and N-glycolylneuraminic acid (Neu5Gc), which are immunogenic in humans and can lead to deleterious side effects [24].

2.2 Genetic engineering of cells

The first step in developing a biomanufacturing process is to generate a cell line producing the desired product, referred to as a *'production cell line'*. Generation of production cell lines involves many steps as outlined in **Figure 3** and includes transfection of the gene of interest into the cells, followed by selection and screening to generate the optimal clonal production cell line for manufacturing. Once a production cell line has been generated there are regulatory guidelines in place governing the use of cells in production processes (Q5D ICH, 1998) that must be followed. The most common industrial approach for cell line development is transfection of the gene of interest using a non-targeted plasmid delivery system that contains a selection marker (Section 2.3) to enable the selection of stably transfected cells. Ideally, a high-quality

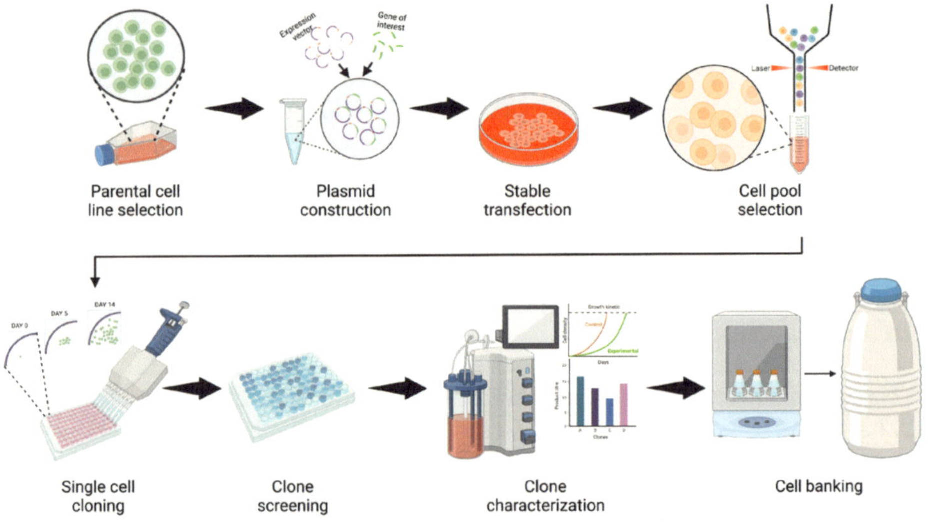

Figure 3.
Cell line development process showing the steps required for the generation of production cell lines. Created with BioRender.com.

production cell line should demonstrate high stability, scalability and high titer levels, in order to provide reproducible results with consistent product quality attributes.

CHO cell lines used commercially were derived from the original spontaneous immortalized culture established in 1956 by Dr. Theodore Puck [25]. Several variants referred to as subclones were generated using both chemical and radiation mutagenesis. The most frequently used variants are CHO-K1 (single copy of dihydrofolate reductase (DHFR)), CHO-DG44 (both copies of DHFR missing) and CHO-DUBX [26]. The plasmids typically used in the transfection of the genes encoding the product are not designed for integration at specific chromosomal sites so this step generates a population or pool of cells with a wide range of expression levels that reflects the gene copy number and the transcriptional activity of the locus based on the integration site [27]. The transfection efficiency can range between 15 and 80% depending on the system used [28]. This means that if 1 million cells were transfected, potentially between 10,000 and 100,000 potential new cell lines can be generated. This random integration can provide a diverse range of expression levels but can also result in integration into unstable areas of the genome [29], resulting in varying levels of expression and lack of genetic stability.

Although the original CHO cell line contained 22 chromosomes, the subclones used commercially contain a range of chromosome numbers, e.g. CHO-K1 typically contains between 18 and 21 chromosomes [30]. CHO cells are inherently genetically unstable which has resulted in a genetically and phenotypically diverse lineage, manifested by many single nucleotide polymorphisms (SNPs), copy number variations, and karyotypes [31]. A recent review by Wurm and Wurm described CHO cell lines as having a "quasi species" genome [32]. Studies have shown that over the course of passaging of CHO cells, the DNA is unstable. The notion that CHO-K1 cells were truly clonal in origin has also been questioned [32].

This random integration of plasmid DNA approach has remained relatively unchanged over the last 36 years but the availability of advanced gene-editing tools now makes transfection at targeted chromosomal sites more feasible. These include

CRISPR/Cas9, zinc finger nucleases (ZFNs), and transcription activator-like effector nucleases (TALENs), with alongside RNAi (RNA interference) and ribozyme engineering which facilitates "multiplexing gene-editing approach[es]" allowing precision and acceleration of genomic rearrangements for enhanced generation of successful clones with improved product yield and quality [33]. It is expected that future cell line development will include vectors targeted to stable areas of the CHO genome and will also be capable of expressing not only the transgene for the product required but additional genes involved in regulatory cell processes like apoptosis [34].

2.3 Selection, cloning and screening

In order to identify the transfected cells that have stably integrated the plasmid into the host genome, a selective marker is used for selection of cells which have acquired the highest expression of the associated product gene. First generation products were produced in CHO-K1 and CHO-DG44 cells that incorporated DHFR as the selection marker on the plasmid [35]. Transfected cells could be selected in media that did not contain nucleotides required for growth in the absence of DHFR activity. One advantage of the DHFR selection system was that gene expression and ultimately product production could be amplified using methotrexate (MTX). Antibiotic resistant genes like neomycin (selected with geneticin), are frequently used. The use of glutamine synthetase (GS) enzyme as a selection marker allows selection of cells growing in media without glutamine. The GS system has been successfully employed for the production of several mAbs. CHO cells produce low levels of GS activity and require selection in methionine sulfoximine (MSX). Interestingly, several studies have shown that amplification methods using DHFR and methotrexate may be susceptible to instability of the transgene that results in a decrease of recombinant gene copies in long-term culture [36]. For this reason, the MSX-GS selection system tends to be the most widely used alternative. NS0 cells do not express endogenous GS therefore can be selected without using MSX, thereby creating a simpler selection process [37].

The guidelines for using cells to manufacture products must be strictly adhered to and a key regulatory requirement is that production cells must originate from a single clone i.e. individual cell. Following the selection process, cells must be separated or cloned into single cells using limited dilution in suspension cultures or clonal rings if cells have been adapted to grow as adherent cultures. Single cell limited dilution relies strictly on a probabilistic approach which is time consuming and offers low throughput as a process platform. The use of plasmids incorporating green fluorescent protein (GFP) and the application of fluorescence-activated cell sorting (FACS) has helped simplify this process. Following transfection, GFP positive cells can be selected as pools of transfected cells and can be automatically separated into individual cells using single-cell isolation technologies such as FACS, magnetic activated cell sorting (MACS), microfluidic and manual cell picking [38]. Advancements in automated screening and selection systems such as Clonepix and CellCelector offer multitude cell screening through immobilization of cells within a semi-solid media matrix and relies with a fluorescent identifier in the vicinity of a resulting colony [39]. Additionally, these automated cloning systems are combined with imaging analysis to provide compelling visual evaluation that can be used in real-time.

Selection and cloning generate a large number of individual candidate cell lines that require screening and evaluation to select the optimal production cell line which is challenging with regard to throughput and creates a major bottleneck in upstream process development. A key requirement is a high specific productivity level and evaluation of mAb production by ELISA or ELISA spot assay is the first step in the screening process. A typical screening strategy will involve multiple evaluation stages in which a proportion of cell candidates are discarded after each stage to attain a small number of highly ranked possible cell line candidates [40]. A typical screening scenario involves initial screening of ~300 individual clones for productivity. Following elimination of low producing cell lines, the candidate cell line panel of ~50 is screened for proliferation capacity. Cells with high productivity (2-7 g/L) and specific growth rates (0.010-035 h-1) are then expanded into shake flasks and screened for their ability to adapt to production conditions [41]. The final step will generally involve ~5–10 cells lines grown under different bioreactor conditions in order to finally select the optimal cell line for commercial production. The introduction of miniature bioreactors such as the Ambr system allows for significant scale-down of this step and increased through-put significantly accelerating the final evaluation step.

2.4 Cell line characterization

A key requirement and challenge for manufacturing is to maintain optimal productivity and consistent quality between batches of product. As part of the regulatory requirements, a master cell bank (MCB) must be created from the initial cell line generated. A WCB is generated from the MCB and a new vial of cells is used to initiate each batch. As mentioned previously there are specific guidelines that must be followed when working with cell lines and these guidelines include specific tests that must be undertaken to characterize and authenticate cells within the MCB and WCB to ensure purity, identity and stability of the cells. In terms of purity the lack of microbial or viral contamination must be confirmed. The identity of the cells can be confirmed using STR profiling. Maintaining the genetic stability of the cells for more than 60 generations at high cell density must be demonstrated [42].

2.5 Media selection and optimization

Cell culture media are complex mixtures of nutrients, hormones, growth factors, salts, trace elements and buffers. Early media formulations used serum but regulatory concerns about possible prion and viral contamination led to the development of serum-free, protein-free and chemically defined media formulations that were free from animal-derived material. Serum-free media is widely used in CHO biomanufacturing and the basal media consists of 50–100 components. The optimal composition and amounts of these components must be determined to be capable of supporting cell growth and production of product with acceptable quality attributes. Overall, media development is an multiple process with iterative rounds of performance testing. Spent media analysis allows the development of feeding strategies for cells in fed-batch and nutrients like glucose and specific amino acids are routinely added to extend the lifetime and productivity of cells in culture [43]. Design of Experiments (DoE) has been used for development of cell culture media to evaluate both component concentrations and component interaction [44]. Many companies use a platform based approach and use the same media formulation for several different products.

Bioreactor systems like the Ambr systems have benefitted the media development process as it allows for much greater throughput, smaller volumes and less labour-intensive experimentation. Dynamic flux balance analysis (DFBA) is a new approach which elucidates the relationship between media supplementation with amino acids, feeding strategy, increased product yield, an extended growth phase and increased density [44]. DFBA illustrates that metabolic state varies more at the beginning of the culture and rather less in the middle of culture [45]. Other advancements are towards multiplexed at-line and operator-independent analytics in benchtop bioreactors such as use of multi-analyte analyzers including NIR, Raman and 2D-fluorescence spectrometry which provide useful process measurements for feed media optimization [44–45].

3. Engineering challenges and mitigations

Commercial manufacture of mAbs has evolved significantly since the first approval almost 40 years ago. The unit operations and technology routinely used are mature and well-understood. However, the landscape and business considerations associated with their production continue to evolve and drive new innovations and approaches. Increased expectations on the part of regulatory authorities, a burgeoning biosimilars market, ever-growing numbers of approved mAbs and wide ranging volume requirements have resulted in new challenges that must be addressed in order to successfully meet the patients' needs while maintaining a viable business. Efforts typically focus on accelerating the time to market and reducing CoGs. There are several engineering-related challenges to achieving this, namely: process variability, sub-optimal volumetric productivity, long cycle-times, and the complexity of managing multi-product facilities [46–47]. In this section, a number of technologies and strategies capable of addressing these challenges are presented alongside the opportunities that they provide the industry with as it moves towards the paradigm of Industry 4.0.

3.1 Bioprocessing considerations

The challenges facing mAb production include producing the product with a tight product quality profile, maintaining the biophysical properties of protein and reducing product and process-related impurities below the acceptable levels to meet quality specifications and ensure safety, efficacy and stability [48]. The first step is to produce a product of acceptable quality within the N stage bioreactor. Many of the critical quality attributes (CQAs) including glycosylation profile can be affected by operating conditions within the bioreactor including pH, temperature, dissolved oxygen (DO), media formulation and metabolites. Real time monitoring and control of CQAs is still aspirational and currently these attributes are at best managed through a QbD-based approach and automated control of basic parameters such as temperature, pH and DO as well as procedural controls such as raw material specifications which will be further discussed in later sections.

Impurities typically fall into two categories: product-related and process-related. Product-related impurities occur where the biophysical properties of the product are compromised and a portion of it degraded. This can occur by multiple mechanisms including aggregation, oxidation, fragmentation, deamidation and denaturation. These degradative mechanisms are the result of the conditions to which the protein is exposed such as high mechanical or physical stress due to agitation or gassing in vessels or fluid flow through piping, filtration and chromatography skids. Enzymatic reactions during cell cultivation can result in cleavage of the protein backbone and the generation of mAb fragments [49]. Aggregation where protein is converted from the desired monomers to dimers, trimers or even larger structures is one of the most common product-related impurities that results from physical stress [50]. Chemical degradation occurs predominantly during downstream processing as a result of changes in pH or buffer solutions. This generally occurs during the viral clearance, chromatography and ultrafiltration/diafiltration (UF/DF) steps. Mitigation strategies are focussed on reducing the residence time at low pH conditions and gentle pH shifts in UF/DF can circumvent the inherent critical effects of these processes in the final mAb product [51].

Process-related impurities are introduced as a result of the materials used in the process and include growth selection agents, surfactants, purification column binding agents and viral inactivation agents. Cell lysis within the bioreactor and harvest equipment results in the release of host cell proteins (HCP) and DNA. The final product must contain less than 100 pg of cellular DNA per dose, and ppm or ng quantities of HCP per mg of antibody product [52]. Conventionally, intact cellular components are removed during the harvest step by centrifugation and depth filtration, but additional purification steps are required to remove contaminants resulting from cell lysis. Whilst Protein A affinity chromatography is the gold standard for the capture step in the mAb purification process, it can leach from the column matrix and contaminate the product and so must be removed at a later stage with additional chromatography steps like ion exchange (IEX) and hydrophobic interaction to levels less than 1 ppm or 1 ng Protein A per mg product [52].

Effective purification is challenging due to the similarities between many of the contaminants and the product of interest in conjunction with the extremely low acceptable levels of contaminants, due to the parenteral nature of mAb products. Therefore, the best approach is to minimize the generation or introduction of these species. QbD, process monitoring and control and process modeling discussed in Section 3.2 and 3.3 can be leveraged to support this goal.

3.2 Quality-by-design (QbD), process monitoring and control

Much greater process variability is observed for mAb biomanufacturing processes as compared to traditional small molecule pharmaceutical processes. The main sources of variability fall into four main categories: biological factors, raw materials and consumables, operational inputs (measurements, methods, personnel and equipment) and environmental conditions [53]. Ultimately, the result of this is variable productivity and product quality, both of which must be controlled and optimized in order to minimize the CoGs. Additionally, from a regulatory perspective, the FDA's guidance document on process validation recommends that manufacturers understand and control the sources of variation [54]. Understanding and reducing variability can be achieved by effective application of QbD, process monitoring and control.

Traditionally, biomanufacturing companies have maintained that "the process is the product" or the quality-by-testing (QbT) approach. Under QbT, the product quality attributes are empirically linked with the manufacturing process and its inputs, both materials and process parameters, during the clinical phases of development, with little or no mechanistic understanding. One of the many disadvantages to this approach is that Proven Acceptable Ranges (PARs) for process parameters are extremely narrow, limiting both the opportunity for post-approval optimization and the capacity for the process to be adjusted in response to process variability [55].

QbD overcomes the limitations of QbT by taking a science and risk-based approach to drug development in order to ensure process and product understanding and to implement effective control strategies. According to the European Medicine

Agency (EMA): *"Quality by design (QbD) is an approach that aims to ensure the quality of medicines by employing statistical, analytical and risk-management procedures in the design, development and manufacturing of medicines"*. Summarily, this methodology is focused on the 1) identification of each source of process variability, 2) understanding of their effects on product's critical quality attributes and 3) control strategies by applying statistical inferences, such as multivariate analysis. Additionally, according to EMA, process analytical technology (PAT) is commonly part of QbD methodologies and it is defined as the system of integrated technologies and methods for control of critical quality attributes of raw and in-process materials [56]. An excellent overview of the QbD approach is provided by Yu et al. [57].

The result of using QbD during drug development is a strong understanding of the link between process inputs and product quality within the design space mapped. The benefits of this are manifold and include:

- Product defects, rejections and recalls are reduced due to an increase in process capability and decrease in process variability.
- The increased process understanding allows for enhanced root cause analysis, resulting in reduced deviations and batch failures.
- Post-approval process optimization is allowed to occur without regulatory notification if it lies within the design space filed.

Once a mAb moves from development into production, it is necessary to implement a control strategy. There are three levels of control possible [57]:

- Level 1 uses automated control in order to maintain the CQAs of the product at the desired value in real-time. Raw material attributes and process conditions are monitored and the process parameters are automatically adjusted in order to control the CQAs. Level 1 represents a highly adaptive form of control which requires process analytical technology (PAT), process models and advanced control algorithms to be implemented and can enable real time release (RTR).
- Level 2 is a mixture of automated and procedural controls which leverage the QbD design space in order to reduce end-product testing and allow for more flexibility in raw materials and process parameters.
- Level 3 is the traditional QbT-based control strategy where raw materials and process parameters are tightly controlled and there is extensive end-product testing.

Moving from Level 3 to 1 is desirable, due to regulatory pressure and the business benefit of reducing the CoGs through reduced variability, batch failures, process optimization and the removal of the time and cost associated with product quality testing. In order to reach Level 1, it is essential to be able to quantify both the control targets and the associated process responses. This may be done by direct measurement or with the use of predictive models coupled with indirect measurements.

Currently, a few variables (e.g. DO, pH and temperature) are measured routinely with in-line probes, however, most key process parameters (e.g. cell density and viability etc.) are measured off-line and the delay between sample extraction and analytical results can hinder the process productivity and the ability to implement adaptive closed loop control [58]. Much work has been done to address the gap in process measurement through the development of various sensors.

Sensors can be classified by their principle and their structure:

- Structure - hard vs. soft sensors: A hard sensor is based on equipment capable of direct measurement of the required attribute whilst a soft sensor is a combination of equipment and a mathematical or data-driven model which infers the measurement required [58]. For instance, Narayanan et al. [59] investigated the use of a hybrid model coupled with a Kalman filter (EKF) for real-time monitoring and control of mammalian cell culture processing. The hybrid model coupled with EKF was applied over a data set with 15% added Gaussian noise and the approach improved predictive accuracy for process variables (e.g. cell density, titer, lactate concentration and glucose processes. concentration) by about 35%, when compared to partial least square (PLS) algorithm.

- Principle - refers to the scientific principle on which the measurement is based. For instance, biosensors (ligand base), sensors based on optical absorption (e.g. spectroscopy – UV/VIS, fluorescence, infrared, near-infrared, mid-infrared), sensors based on light scattering and optical density, etc. [60]. As an example, the work of Whelan et al. [61] highlighted the application of Raman spectroscopy for the measurement of cell density and metabolite concentration in a CHO production bioreactor. They found an acceptable agreement between the in-line Raman measurement and off-line samples across different process scales.

The availability of real time process measurements is a prerequisite for the implementation of automated closed-loop feedback control which can adapt and respond to process variability in order to maintain a steady CQA profile in the final product. Automated control also reduces the risk of human error in the process. There have been reports in the literature demonstrating the capability and benefit of such control. Craven et al. [62] applied a nonlinear model predictive controller NMPC to a 15 L CHO fed-batch bioreactor to control glucose concentration at a defined set-point (11 mM) by adjusting the feed rate to the bioreactor. The substrates (glucose and glutamine) and byproducts (lactate and ammonia) were measured by in-situ Raman spectroscopy and the concentration values were determined by a partial least squares (PLS) calibration model analysis. The determined metabolite concentrations were inputted to the NMPC algorithm which used a first principles mechanistic process model in conjunction with an optimization algorithm to determine the optimal control output or feed rate. Both simulated and real-time application of the NMPC showed similar performance and the results highlighted the feasibility and capability of NMPC for bioreactor control. While it was shown that process model inaccuracy could hinder NMPC performance, the controller showed good ability to function with both noisy and non-continuous process measurements. While there are multiple reports of the benefits of this type of control in literature, currently, it is not implemented in commercial manufacture due to the conservative nature of the biopharmaceutical industry. It is however routinely employed in other sectors such as commodity chemicals and petrochemicals e.g. Quin & Badgell [63] reported 93 industrial applications of NMPC due to the improved safety, quality and efficiency that it enables [64]. As such, it presents a future opportunity for mAb production.

In summary, innovations in and implementation of QbD, process monitoring and control in mAb biomanufacturing can address the key challenge of reduction of CoGs by increasing process capability and robustness, reducing process variability and failures and facilitating post-approval optimization.

3.3 Process modeling

Process modeling or process simulation can be applied in order to address the challenges of speed-to-market and the reduction in CoGs. There are numerous types of models that may be deployed depending on the objective. These range from mechanistic to data-driven models with reports of hybrid model approaches for bioprocesses increasing in the literature [65–66].

Mechanistic models, also known as mathematical, first principle or white-box models depend on the laws of nature to describe a specific phenomenon. They require fundamental understanding of the phenomenon being studied. This is often difficult for bioprocesses due to the complexity of the system and the strong level of interaction and dependency between multiple parameters. They have high extrapolation capacity but are limited by the degree of accuracy of the model equations describing the behavior. Conversely, data-driven or black-box models rely on larger data sets and greater computational efforts in order to predict process responses, with no reference to the underlying mechanisms. This limits the ability to extrapolate predictions to unseen scenarios but significantly reduces the complexity of the model [65]. Hybrid models have elements of both mechanistic and data-driven models, circumventing the challenges of both by reducing the amount of data and level of process knowledge required. It should be noted that regardless of the type of model used, it is crucial to verify and validate the model outputs using independent or unseen datasets before using them.

Process models can be used to accelerate process development. Typically bioprocess development has a heavy reliance on experimentation under a traditional Design of Experiment (DoE) framework, augmented with prior expert knowledge. Moller et al. [67] reported the use of model-assisted DoE for bioprocess development. A mathematical model was developed that described cell growth, metabolism and antibody production for a CHO DP-12 cell line under both a batch and fed-batch mode of operation. The model was then used to reduce the boundary values for the experimental DoE. It was found that the same optimal conditions were identified for both the traditional and model-assisted approaches with a reduction in the number of experiments required from 16 to 4 in the case of batch and 29 to 4 in the case of fed-batch. Given the time-consuming, expensive nature of bioprocess experiments, this represents a significant potential to accelerate development timelines.

A key enabler for commercial production is the scale up of the bioprocess from small laboratory scale equipment sets through an intermediate scale required to supply the clinical trials to large scale commercial manufacturing plants. As discussed in Section 3.1, both the cell and the mAb product can be affected by factors such as shear stresses, extremes of temperature and pH among other things. As an example, as the production scale increases, it becomes more difficult to maintain a fully homogenous environment and so a balance needs to be identified and maintained between effective mixing and exposure of the cell and product to damaging conditions. This may be achieved empirically, however, the conditions identified by such means are potentially sub-optimal and the opportunity to develop fundamental process understanding for use subsequently for troubleshooting and tech transfer is missed. As an alternative, computational fluid dynamics (CFD) can be used to support and enhance process scale up.

CFD mathematically models fluid flow and its interactions with solid bodies by numerically solving systems of partial differential equations governing fluid dynamics problems (e.g. Navier-Stokes equations) [68, 69]. CFD can be used to derisk and support process scale up by predicting the conditions created under a range of agitation and aeration rates. For example, Mishra et al. [70] studied the effect of agitation and aeration rates on the mass transfer of oxygen and shear stress in the liquid phase of a 10 L single-use bioreactor. Their approach combined computational fluid dynamics with species transport and population balance models in order to predict the maximum total stress and energy dissipation rates that the cell culture would be exposed to. The simulations were performed for stirring speeds between 50 and 300 rpm and aeration rates between 0.1 and 0.2 LPM, considering 45% fill volume. The results indicated a maximum total stress of 34.17 Pa and energy dissipation rate of 1.352 m^2/s^3 (at 300 rpm and 0.2 LPM), which are unlikely to affect mammalian cells according to the author's consulted literature. The simulations were validated by experimental determination of the oxygen's mass transfer coefficient (stirring speed range: 50 rpm - 200 rpm, 0.2 LPM and 6.75 L of liquid volume) and the observed errors (predicted *vs* experimental) were between 0.47% and 10%.

Process modeling can support a reduction in CoGs via a number of approaches. One such approach is the use of multivariate data analysis to generate a "golden batch trajectory", a form of data-driven process model, against which real time production plant data is compared. The process insight gained can be used to rapidly identify root causes for batch failures and deviations and to prevent them reoccurring, hence reducing the rate of batch failures and process variability experienced. The knowledge gained can also be used to fine-tune the Normal Operating Ranges (NORs) for a process within its design space and maximize the productivity or particular CQA of the product. Sokolov et al. [71] applied multivariate analysis in the form of partial least square regression - PLSR to predict mAb-based product quality attributes (aggregates, fragments, charge variants and glycan profile) from process data (media supplements, pH and temperature shifts). The data set was obtained from a 91 run DoE at milliliter scale and the model performance was evaluated using the root mean square error in cross-validation (RMSECV). They firstly used principal component analysis (PCA) to analyze the correlation among 14 product quality attributes (QAs) and, since their findings indicated a strong correlation among QAs, they concluded that the variables should be treated as one characteristic. Additionally, PLSR1 and PLSR2 were used to predict product quality attributes and provided comparable prediction accuracy. The PLSR2 models were further investigated by the addition of genetic algorithm (GA), in which the results became more accurate and, in complex cases, the GA was able to remove noise, inconsistency and redundancy in the data set. Availability of such models can also enable statistical model predictive control the benefits of which were be discussed further in Section 3.2.

Digital twins are another approach to supporting a reduction in CoGs. Essentially, a digital twin is a digital representation of a physical process. The application is similar to that of the golden batch trajectory, however, digital twins are typically hybrid process models which may leverage PAT and other process monitoring (see Section 3.2). They are used to identify process bottlenecks, key engineering targets and identify operational strategies that improve the reliability and productivity of their physical twin. As an example, a digital twin of a production bioreactor could represent the physiology and metabolism of the cell culture by applying genome-scale metabolic models (e.g. Flux Balance Analysis - FBA, with appropriate objective function and adequate constraints) alongside the process kinetics obtained from in-line monitoring (e.g. Raman-based monitoring system) [72, 73]. Digital twins are key enablers of Industry 4.0, which seeks to revolutionize how industry operates through the use of smart, autonomous systems running on data and machine learning and have huge potential to improve the biomanufacturing of mAbs [66].

3.4 Process intensification

An intensified process can be defined as one that increases productivity (e.g. per batch, per facility) and/or reduces environmental impact (energy, waste, materials), facility footprint (smaller equipment, shorter process streams), manufacturing costs, process times, or process bottlenecks. It should be noted that having a genetically stable cell line with good growth and productivity characteristics and media systems capable of supporting the increased nutrient demands are prerequisites for upstream process intensification.

The pre-production stage encompasses all process steps before the production bioreactor, i.e. cell revival from the working cell bank (WCB) and the inoculum and seed train (**Figure 2**). These steps represent a significant portion of the process cycle time and typically require highly skilled labour and expensive equipment and infrastructure in order to be reliably executed. Two main approaches to circumventing the time and resources required at this stage have been reported in the literature, namely the modification of the cell banking approach to provide a greater number of cells upon thaw by either increasing volume or cell density or a combination of both and the use of the perfusion mode of operation at the N-1 seed bioreactor stage.

Intensification of cell banking eliminates multiple expansion steps which reduces the resources and time required to inoculate the production reactor, an example of which is described by the work of Seth et al. [74]. Their work investigated the cryopreservation of CHO cells at low density (LD, ~$30x10^6$ cell/mL), mid density (MD, ~$70x10^6$ cells/mL) and high density (HD, ~$110x10^6$ cells/mL) in single-use cryopreservation bags. They named the strategy the Frozen Accelerated Seed Train for Execution of a Campaign (FASTEC) which allowed the seed train to be bypassed. Seth et al. [74] evaluated the FASTEC approach in an 80 L seed bioreactor by comparing two processes: i) an inoculum seed train with 3 passages; ii) an inoculum seed train with only 1 passage. The processes were able to produce final cell densities of approximately $7x10^6$ viable cells/mL in 10 days and 4 days, respectively. Additionally, after inoculating the production bioreactor (400 L) operated in fed-batch, product titres (1.1 g/L & 1.2 g/L), triple-light chain impurities (3LC, 1.7% & 2.6%) and aggregates (3.1% & 2.2%) were comparable to the seed train control (1.1 ± 0.2 g/L; 3.7 ± 1.0%; 4.3 ± 1.0%, respectively). Seth et al. [74] concluded that, despite cells displaying lower growth rate and viability immediately post-inoculation, the FASTEC process was able to produce comparable titer and quality of mAbs than the standard process whilst significantly reducing the duration of the upstream process.

The second strategy commonly reported in the literature is the use of a perfusion N-1 seed bioreactor. The benefit of this approach is that a significantly higher cell density is achieved in the seed train which allows a fed-batch production reactor to be inoculated at a higher initial cell density. As a result the duration of the production reactor can be reduced by a number of days and/or higher titres can be achieved due to the increased cell time within the reactor. Xu et al. [75] applied an N-1 perfusion seed step in conjunction with media enrichment to a CHO fed-batch production reactor for four processes producing four different mAb products. They reported that increasing the initial seed density from 0.3–1.2 x 10^6 cells/ml to 10–20 x 10^6 cell/ml resulted in an up to 10 fold increase in titer in addition to up to a 4 day decrease in production reactor duration while maintaining product quality.

The benefits of continuous processing have been discussed extensively in the literature for a range of industries including the small molecule pharmaceutical sector [76]. They are generally smaller, faster and cheaper with greater levels of flexibility and quality assurance achievable. As such there is much interest in the application to mAb production. However, fully integrated end-to-end continuous processes in this space are not yet a reality in commercial manufacture although research and development of such systems is ongoing and has demonstrated feasibility [77]. As a first step, continuous unit operations, primarily perfusion bioreactor steps and continuous chromatography have been implemented.

Many authors have discussed the advantages and disadvantages of perfusion and fed-batch operations for mAb production. The discussion is complex as the best choice is dependent on the specific scenario i.e. the quantity of material required, the productivity of the cell line, the nature of the protein, the cycle times of the processes etc. Direct comparison is often hampered by differing scales and definitions of productivity [78]. For instance, Lee et al. [79] studied the change from perfusion to fed-batch for expression of biosimilar monoclonal antibody A (CR-mAb-A) by recombinant Sp2/0 mouse myeloma cells. The fed-batch operation was evaluated in 8 runs at the following bioreactor scales: 3 L (3×), 100 L (2×) and 12,500 L (3×). The results indicated that, although the perfusion mode provided higher volumetric productivity, the fed-batch operation showed increased total productivity (7.5 fold increase) due to its higher volume capacity. Lee et al. [79] also investigated the mAb-based product quality by measuring oligosaccharide profiles and charge variants of mAbs expressed by fed-batch and perfusion. They observed slight differences in heavy chain glycoforms (G0, G1 and G2) between fed-batch and perfusion, while different scales of fed-batch provided comparable proportions. Additionally, by performing capillary electrophoresis sodium dodecyl sulfate (CE-SDS), they observed a slightly lower amount of intact IgG (4%, normalized) obtained in fed-batch in comparison to perfusion.

It is important to highlight that, in order to benefit from upstream intensification, one must guarantee that downstream operations are capable of handling high volumes and titres while maintaining the product quality. To this end, efforts to improve resin capture and capacity in chromatography operations have been documented in addition to continuous processing. Gerstweiler et al. [80] highlighted applications to enable continuous processing, such as: periodic countercurrent chromatography (PCC), simulated moving bed chromatography (SMB), continuous flow-through chromatography and multi-column designs (e.g. continuous multicolumn countercurrent solvent gradient purification). Moreover, in regard to Protein A ligand-based columns, Somasundaram et al. [81] stated that the dynamic binding capacity and resin reusability are important aspects to be considered in continuous processing.

3.5 Platform processes

Producers of mAbs typically have platform production processes that span the majority of their portfolio. A platform process comprises the expression system, typically a suspension CHO cell line, the associated basal and feed media formulations and the series of unit operations used to produce and purify the mAb. The platform may be fine-tuned for each product, for example, the media formulation may be slightly modified for the particular nutrient requirements of a specific clone in order to boost productivity or ensure product quality, but there are no major fundamental changes unless essential for a given product. The benefits of utilizing a platform approach are: faster process development, easier facility fit for scale up and tech transfer, greater process knowledge and more robust processes, all of which contribute to delivering a faster time to market and lower CoGs.

The majority of manufacturers rely on a CHO fed-batch upstream process platform with serum-free media although a significant number have opted for perfusion. There is also a general move away from using undefined media components such as hydrolysates to fully chemically defined media formulations in order to reduce raw material and hence, process variability. The downstream process generally follows a platform process flow of one or two harvest steps (centrifugation/depth filtration) followed by a Protein-A chromatography step for product capture, a low pH hold for viral inactivation, two polishing chromatography steps, a virus filtration and a final ultrafiltration/diafiltration step to the final concentration in the formulation buffer [82].

Efforts to improve the existing platform processes focus on process intensification strategies such as high cell density cell banks and N-1 perfusion steps as discussed in Section 3.4 as well as the potential to replace the Protein-A capture step. Protein-A is the single biggest contributor to the Op-Ex costs associated with mAb production. It is widely used as it is extremely effective and reliable. Both chromatographic and non-chromatographic options have been explored. These include: precipitation, crystallization, cation exchange chromatography and multimodal chromatography [82–83]. Other approaches seek to improve the efficiency of Protein-A usage. Typically, Protein-A is used in a single product, packed bed format. There is work currently underway exploring resin use across multiple products, particularly useful for small volume products, as well as alternative formats such as membrane chromatography and monolithic chromatography which allow for higher flow rates and hence, throughput [84].

3.6 Facility design and single use technology (SUT)

The portfolio of mAb products currently approved ranges from large volume blockbuster therapies to low volume products that address orphan indications. Therefore, the scale of commercial manufacture varies considerably across products. For a given product, the annual requirement can also vary significantly. It may increase as a new product gains market share or is approved for additional indications or may decrease if a competing product, a new alternative therapy or a biosimilar version of the product in question, is launched. Biosimilar manufacturers target high volume products coming off patent, supplying a comparable product at a reduced cost. This results in a significant drop in the volume required for the original product, as observed for granulocyte colony stimulating factor (G-CSF). According to an IMS report [85], in 2016 within the EU, biosimilars accounted for 88% of the market as compared with the reference product. This resulted in a 37% reduction in price as compared with the year prior to entry of the biosimilars into the market.

As a result, where once a dedicated high volume production plant using stainless steel equipment was the norm, there has been a shift towards flexible multi-product facilities in order to accommodate the ever-changing numbers and volumes of mAbs to be supplied to meet patient needs. SUT and modular or ballroom style facilities help to satisfy these requirements.

In SUT, all surfaces that come in contact with the process are disposable and are replaced after a single batch. This includes the vessel itself which is typically a bag supported externally by a metal exoskeleton and fabricated from FDA (Food and Drug Administration) approved polymers such as polyethylene (PE), polytetrafluoroethylene (PTFE) and polypropylene (PP) supplemented with additives to enhance performance and/or extend useable life as well as impellers, probes, resins, filter cartridges etc. SUT eliminates the need for the validation requirements associated with cleaning and sterilization of equipment, reduces the turn-around time between batches and reduces the risk of both microbial contamination and cross-product contamination in multi-product facilities [86]. The cap-ex investment required to establish a single use facility is significantly lower than the stainless steel equivalent and the utility requirements particularly for steam and water for injection (WFI) are massively decreased [87]. Studies have also reported that despite the increased plastic waste produced from SUT, overall they are environmentally less impactful than stainless steel [87]. Currently, there are commercially marketed SUT solutions for each unit operation typically used to produce a mAb [86].

There are however some disadvantages and challenges that remain to be overcome. Firstly, leachables and extractables are a concern due to the material of construction. These substances may be detrimental to process performance and/or human health [86]. Typically this is overcome by performing studies on the material to prove suitability. Secondly, scale is limited. There are mechanical challenges in producing what are essentially plastic bags with sufficient strength to withstand the loads associated with large volumes. The largest volume routinely seen at commercial scale is 2000 L as opposed to 25,000 L in stainless steel [86, 88–89]. There is one system currently on the market at 6000 L made by ABEC [90]. Other challenges include extremely long lead times of 9–12 months currently for the sterile consumables required in addition to high op-ex costs associated with them. The SUT available for downstream is less mature than in the upstream space and as such is less likely to be adopted. This is evolving over time especially when considering the increase of demand by industry. According to American Pharmaceutical Review [91], 46.9% of a survey's respondents (12th Annual Report and Survey of Biopharmaceutical Manufacturing) had claimed to investigate single-use technologies in downstream bioprocessing to improve purification operations, in contrast to 36.8% in 2012. Despite these challenges however, many manufacturers have adopted either fully SU equipment trains or a hybrid approach where the upstream process up to a volume of 2000 L is SU and the remainder of the upstream and the downstream processes are stainless steel.

Facility design and construction takes an average of 1 year to design and 3–4 years to build and costs several hundred million to over one billion euros depending on the size of the facility. To maintain strategic relevance in the current market, biopharmaceutical companies must design these facilities to be flexible and multi-product while still maintaining a high standard of product safety and efficacy. Modified-ballroom or dance-floor facility design integrated with closed systems and SUT is the most common approach to achieve these objectives while managing the risks associated with large integrated production spaces. In this type of design, a series of rooms that meet the Clean-Not-Classified (CNC) criteria are interconnected through wall panels. Within each room, single use, closed systems are operated. This equipment can be based on modular skids that can be changed if requirements change in the future. This approach reduces the footprint of the facility by removing the need for personnel and material airlocks to a large extent as well as decreasing the both cap-ex and op-ex costs associated with graded cleanroom environments. Time for construction is also reduced. It eliminates the need for gowning and simplifies installation, maintenance and operation of equipment as there are less restrictions on activities in the production space [92].

4. Conclusion

mAbs represent the largest category of biopharmaceuticals on the market and the number of approved products continues to grow. The commercial production of these products has evolved significantly since the first approval in 1986. Expression systems, once only capable of titres in the mg/L range, are now routinely producing 5–10 g/L due to advances in cell line development. Efforts have moved from establishing reliable robust platform processes to optimization of product quantity and quality. New technologies and approaches are being adopted in order to achieve this despite the mAb specific challenges associated with processing protein molecules and controlling biological processes. QbD, process monitoring, and control can be harnessed to manage the inherent variability associated with the raw materials and biological expression system. Process modeling, particularly hybrid models, can mitigate the expensive, time-consuming nature of experimental approaches and the empirical approach taken to process development historically in order to accelerate time to market and optimize and troubleshoot the manufacturing process. There are currently multiple strategies for process intensification being adopted in order to reduce cycle time and increase productivity. New approaches to facility design coupled with SUT reaching greater levels of maturity have reduced the risk and complexity associated with multi-product facilities. Alternative technologies on the horizon such as greater offerings for SUT in the downstream space and cheaper alternatives to Protein-A packed bed chromatography are opening up new avenues for significant cost reduction. Over the coming decades, mAb production will continue to evolve. There are many promising technologies and approaches to address the existing challenges. While adoption is slow due to the regulated, conservative nature of the biopharmaceutical industry, where strong business drivers exist, this will be overcome and, in the future, integration of these technologies will become widespread. It is exciting to consider the next evolution of mAb production.

Acknowledgements

The authors acknowledge receipt of funding to Raymon Floyd Principe From the UCD Advance PhD Core Scheme and funding for Maycou Soares Zamprognio from SSPC, the Science Foundation Ireland Research Centre for Pharmaceuticals (12/RC/2275_P2).

Conflict of interest

The authors declare no conflict of interest.

Author details

Susan McDonnell[1*], Raymon Floyd Principe[1], Maycou Soares Zamprognio[2] and Jessica Whelan[2]

1 Biomanufacturing Research Group, UCD School of Chemical and Bioprocess Engineering, UCD, Ireland

2 SSPC – The Science Foundation Ireland Research Centre for Pharmaceuticals, Bernal Institute, University of Limerick, Ireland

*Address all correspondence to: susan.mcdonnell@ucd.ie

References

[1] Ecker D, Jones SD, Levine HL. The Therapeutic Monoclonal Antibody Market. mAbs. 2015;7:9-14. DOI: 10.4161/19420862.2015.989042

[2] Halimi V, Daci A, Netkovska KA, Suturkova L, Babar ZUD, Grozdanova A. Clinical and regulatory concerns of Biosimilars: A review of literature. International Journal of Environmental Research and Public Health. 2020;**17**(16):5800. DOI: 10.3390/ijerph17165800

[3] Global Market Insights. Antibody Therapy Market size to exceed $445bn by 2028 [Internet]. 2022. Available from: https://www.gminsights.com/pressrelease/antibody-therapy-market. [Accessed: April 26, 2022]

[4] Mullard A. FDA approves the 100th monoclonal antibody product. Nature Reviews: Drug Discovery. 2021;**20**:491-495. DOI: 10.1038/d41573-021-00079-7

[5] The Antibody Society. Therapeutic monoclonal antibodies approved or in review in the EU or US [Internet]. 2022. www.antibodysociety.org/resources/approved-antibodies. [Accessed: May 2022]

[6] Chavda VP, Prajapati R, Lathigara D, Nagar Bhumi N, Kukadiya J, Redwan EM, Uversky VN, Kher MN, Rajvi P. Therapeutic monoclonal antibodies for COVID-19 management: An update Expert Opinion on Biological Therapy. 2022;**22**:763-780. DOI: 10.1080/14712598.2022.2078160

[7] Gherghescu I, Begoña D-CM. The biosimilar landscape: An overview of regulatory approvals by the EMA and FDA. Pharmaceutics. 2021;**13**:48. DOI: 10:3390/pharmaceutics13010048

[8] Fierce Pharma. Humira rings up $20.7B in 2021, but AbbVie still mum on post-biosimilar expectations [Internet]. 2022. Available from: https://www.fiercepharma.com/pharma/humira-rings-up-20-7-billion-sales-but-abbvie-still-mum-a-projection-for-2023-when-it-faces

[9] Fu Z, Li S, Han S, Shi C, Zhang Y. Antibody drug conjugate: The "biological missile" for targeted cancer therapy. Signal Transduction and Targeted Therapy. 2022;7:93. DOI: 10.1038/s41392-022-00947-7

[10] Kang TH, Seong BL. Solubility, stability, and avidity of recombinant antibody fragments expressed in microorganisms. Frontiers in Microbiology. 2020;**11**:1927. DOI: 10.3389/fmicb.2020.01927

[11] Palanques-Pastor T, Megias-Vericat JE, Boso RV, Gomez SI, Poveda AJL. Effectiveness of Caplacizumab Nanobody in acquired thrombotic thrombocytopenic purpura refractory to conventional treatment. Acta Haematologica. 2022;**145**:75-76. DOI: 10.1159/000517813

[12] Delobel A. Glycosylation of therapeutic proteins: A critical quality attribute. In: Delobel A, editor. Mass Spectrometry of Glycoproteins. Methods in Molecular Biology. Vol. 2271. New York, NY: Humana; 2021. DOI: 10.1007/978-1-0716-1241-5_1

[13] Datta A, Maryala S, John R. A review of Eptinezumab use in migraine. Cureus. 2021;**13**(9):e18032. DOI: 10.7759/cureus.18032

[14] Tripathi NK, Shrivastava A. Recent developments in bioprocessing of

recombinant proteins: Expression hosts and process development. Frontiers in Bioengineering and Biotechnology. 2019;7(420):1-35. DOI: 10.3389/fbioe.2019.00420

[15] Madhavan A, Arun KB, Sindhu R, Krishnamoorthy J, Reshmy R, Sirohi R, et al. Customized yeast cell factories for biopharmaceuticals: From cell engineering to process scale up. Microbial Cell Factories. 2021;**20**(124): 1-17. DOI: 10.1186/s12934-021-01617-z

[16] Cox MMJ. Innovations in the insect cell expression system for industrial recombinant vaccine antigen production. Vaccine. 2021;**9**(1504):1-12. DOI: 10.3390/vaccines9121504

[17] Yongky A, Xu J, Tian J, Oliveira C, Zhao J, McFarland K, et al. Process intensification in fed-batch production bioreactors using non-perfusion seed cultures. MAbs. 2019;**11**(8):1502-1514. DOI: 10.1080/19420862.2019.1652075

[18] Bunnak P, Allmendinger R, Ramasamy SV, Lettieri P, Titchener-Hooker NJ. Life-cycle and cost of goods assessment of fed-batch and perfusion-based manufacturing processes for mAbs. Biotechnology Progress. 2016;**32**(5):1324-1335. DOI: 10.1002/btpr.2323

[19] Rodrigo G, Gruvegard M, Alstine JMV. Antibody fragments and their purification by protein L affinity chromatography. Antibodies. 2015;**4**:259-277. DOI: 10.3390/antib4030259

[20] Dangi AK, Sinha DS, Gupta SK, Shukla P. Cell line techniques and gene editing tools for antibody production: A review. Frontiers in Pharmacology. 2018;**9**(630):1-12. DOI: 10.3389/fphar.2018.00630

[21] Fisher S, Handrick R, Otte K. The art of CHO cell engineering: A comprehensive retrospect and future perspectives. Biotechnology Advances. 2015;**3**:1878-1896. DOI: 10.1016/j.biotechadv.2015.10.015

[22] Tihanyi B, Nyitray L. Recent advances in CHO cell line development for recombinant protein production. Drug Discovery Today: Technologies. 2021;**38**:25-34. DOI: 10.1016/j.ddtec.2021.02.003

[23] Diep J, Le H, Le K, Zasadzinska E, Tat J, Yam P, et al. Microfluidic chip-based single-cell cloning to accelerate biologic production timelines. Biotechnology Progress. 2021;**37**(e3192):1-9. DOI: 10.1002/btpr.3192

[24] Kunert R, Reinhart D. Advances in recombinant antibody manufacturing. Applied Microbiology and Biotechnology. 2016;**100**:3451-3461. DOI: 10.1007/s00253-016-7388-9

[25] Puck TT, Cieciura SJ, Robinson A. Genetics of somatic mammalian cells: III. Long-term cultivation of euploid cells from human and animal subjects. The Journal of Experimental Medicine. 1958;**108**(6):945-956. DOI: 10.1084/jem.108.6.945

[26] Drug Target Review. Cell line development for therapeutic proteins – current perspectives and future opportunities [Internet]. 2020. Available from: https://www.drugtargetreview.com/article/57784/cell-line-development-for-therapeutic-proteins-current-perspectives-and-future-opportunities/. [Accessed: May, 2022]

[27] Fus-Kujawa A, Prus P, Bajdak-Rusinek K, Teper P, Gawron K, Kowalczuk A, et al. An overview of methods and tools for transfection of eukaryotic cells in vitro. Frontiers in

Bioengineering and Biotechnology. 2021;**9**(701031):1-15. DOI: 10.3389/fbioe.2021.701031

[28] Elshereef AA, Jochums A, Lavrentieva A, Stuckenberg L, Scheper T, Solle D. High cell density transient transfection of CHO cells for TGF-β1 expression. Engineering in Life Sciences. 2019;**19**:730-740. DOI: 10.1002/elsc.201800174

[29] Kim JY, Han SK, Yoon S, Kim CW. Rich production media as a platform for CHO cell line development. AMB Expr. 2020;**10**(93):1-13. DOI: 10.1186/s13568-020-01025-3

[30] Xu X, Nagarajan H, Lewis NE, Pan S, Cai Z, Lui X, et al. The genomic sequence of the Chinese hamster ovary (CHO)-K1 cell line. Nature Biotechnol. 2011;**29**:735-741. DOI: 10.1038/nbt.1932

[31] Welch JT, Arden S. Considering "clonality": A regulatory perspective on the importance of the clonal derivation of mammalian cell banks in biopharmaceutical development. Biologicals. 2019;**62**:16-21. DOI: 10.1016/j.biologicals.2019.09.006

[32] Wurm FM, Wurm MJ. Cloning of CHO cells, productivity and genetic stability—A discussion. PRO. 2017;**5**(20):1-13. DOI: 10.3390/pr5020020

[33] Gupta SK, Shukla P. Gene editing for cell engineering: Trends and applications. Critical Reviews in Biotechnology. 2016;**37**:672-684. DOI: 10.1080/07388551.2016.1214557

[34] Hong JK, Lakshmanan M, Goudar C, Lee DY. Towards next generation CHO cell line development and engineering by systems approaches. Current Opinion in Chem Eng. 2018;**22**:1-10. DOI: 10.1016/j.coche.2018.08.002

[35] Urlaub G, Kas E, Carothers AM, Chasin LA. Deletion of the diploid Dihydrofolate reductase locus from cultured mammalian cells. Cell. 1983;**33**:405-412. DOI: 10.1016/0092-8674(83)90422-1

[36] Ainavarapu SRK, Li L, Badilla CL, Fernandez JM. Ligand binding modulates the mechanical stability of Dihydrofolate reductase. Biophysical Journal. 2005;**89**:3337-3344. DOI: 10.1529/biophysj.105.062034

[37] Barnes LM, Bentley CM, Moy N, Dickson AJ. Molecular analysis of successful cell line selection in transfected GS-NS0 myeloma cells. Biotechnology and Bioengineering. 2007;**96**(2):337-348. DOI: 10.1002/bit.21119

[38] Hu P, Zhang W, Xin H, Deng G. Sing cell isolation and analysis. Frontiers in Cell and Development Biology. 2016;**4**(116):1-12. DOI: 10.3389/fcell.2016.00116

[39] Sifniotis V, Cruz E, Eroglu B, Kayser V. Current advancements in addressing key challenges of therapeutic antibody design, manufacture, and formulation. Antibodies. 2019;**8**(36):1-23. DOI: 10.3390/antib8020036

[40] Porter AJ, Racher AJ, Preziosi R, Dickson AJ. Strategies for selecting recombinant CHO cell lines for cGMP manufacturing: Improving the efficiency of cell line generation. Biotechnology Progress. 2010;**26**(5):1455-1464. DOI: 10.1002/btpr.443

[41] Lopez-Mez J, Araiz-Hernandez D, Carrillo-Cocom LM, Lopez-Pacheco F, Rocha-Pizana MDR, Alvarez MM. Using simple models to describe the kinetics of growth, glucose consumption, and monoclonal antibody formation in naive and infliximab producer CHO cells.

Cytotechnology. 2016;**68**:1287-1300. DOI: 10.1007/s10616-015-9889-2

[42] BioPharm International. Best Practices for Selecting a Top-Quality Cell Line [Internet]. 2019. Available from: http://www.processdevelopmentforum.com/articles/best-practices-for-selecting-a-top-quality-cell-line/. [Accessed: May, 2022]

[43] Ling WL. Development of protein-free medium for therapeutic protein production in mammalian cells: Recent advances and perspectives. Pharm Bioprocess. 2015;**3**(3):215-226. DOI: 10.4155/PBP.15.8

[44] Cell Culture Dish. Fed-batch culture – Optimizing feed strategies now and in the future [Internet]. 2017. Available from: https://cellculturedish.com/fed-batch-culture-optimizing-feed-strategies-now-and-in-the-future/

[45] Reimonn TM, Park S, Agarabi CD, Brorson KA, Yoon S. Effect of amino acid supplementation on titer and glycosylation distribution in Hybridoma cell cultures—Systems biology-based interpretation using genome-scale metabolic flux balance model and multivariate data analysis. Biotechnology Progress. 2016;**32**(5):1163-1173. DOI: 10.1002/btpr.2335

[46] Kelley B. Industrialization of mAb production technology: The bioprocessing industry at a crossroads. MAbs. 2009;**1**(5):443-452. DOI: 10.4161/mabs.1.5.9448

[47] Papathanasiou MM, Kontoravdi C. Engineering challenges in therapeutic protein product and process design. Current Opinion in Chemical Engineering. 2020;**27**:81-88. DOI: 10.1016/j.coche.2019.11.010

[48] Basle YL, Chennell P, Tokhadze N, Astier A, Sautou V. Physicochemical stability of monoclonal antibodies: A review. Journal of Pharmaceutical Sciences. 2020;**109**(1):169-190. DOI: 10.1016/j.xphs.2019.08.009

[49] Vlasak J, Ionescu R. Fragmentation of monoclonal antibodies. Landes Bioscience. 2011;**3**(3):253-263. DOI: 10.4161/mabs.3.3.15608

[50] Goswami S, Wang W, Arakawa T, Ohtake S. Developments and challenges for mAb-based therapeutics. Antibodies. 2013;**2**:452-500. DOI: 10.3390/antib2030452

[51] Shukla AA, Hubbard B, Tressel T, Guhan S, Low D. Downstream processing of monoclonal antibodies—Application of platform approaches. Journal of Chromatography B. 2007;**848**:28-39. DOI: 10.1016/j.jchromb.2006.09.026

[52] Flatman S, Alam I, Gerard J, Mussa N. Process analytics for purification of monoclonal antibodies. Journal of Chromatography B. 2007;**848**:79-87. DOI: 10.1016/j.jchromb.2006.11.018

[53] Hutchinson N. Understanding and controlling sources of process variation: Risks to achieving product critical quality attributes. Bioprocess International. 2014;**12**:24-29

[54] Guidance for Industry. Process Validation, General Principles and Practices. Silver Spring: US Food and Drug Administration; 2011

[55] Val IJD, Jedrzejewski PM, Exley K, Sou SN, Kyriakopoulos S, Polizzi KM, et al. Application of quality by design paradigm to the manufacture of protein therapeutics. In: Petrescu S, editor. Glycosylation. London: IntechOpen; 2012. DOI: 10.5772/50261

[56] Maruthamuthu MK, Rudge SR, Ardekani AM, Ladisch MR,

Mohit S, Verma MS. Process analytical technologies and data analytics for the manufacture of monoclonal antibodies. Trends in Biotechnology. 2020;**38**:1169-1186. DOI: 10.1016/j.tibtech.2020.07.004

[57] Yu LX, Amidon G, Khan MA, Hoag SW, Polli J, Raju GK, et al. Understanding pharmaceutical quality by design. AAPS Journal. 2014;**16**:771-783. DOI: 10.1208/s12248-014-9598-3

[58] Luo Y, Kurian V, Ogunnaike BA. Bioprocess systems analysis, modeling, estimation, and control. Current Opinion in Chemical Engineering. 2021;**33**:00705. DOI: 10.1016/j.coche.2021.100705

[59] Narayanan H, Behle L, Luna MF, Sokolov M, Guillén-Gosálbez G, Morbidelli M, et al. Hybrid-EKF: Hybrid model coupled with extended Kalman filter for real-time monitoring and control of mammalian cell culture. Biotechnology and Bioengineering. 2020;**117**:2703-2714. DOI: 10.1002/bit.27437

[60] Roch P, Mandenius CF. On-line monitoring of downstream bioprocesses. Current Opinion in Chemical Engineering. 2016;**14**:112-120. DOI:10.1016/J.COCHE.2016.09.007

[61] Whelan J, Craven S, Glennon B. In situ Raman spectroscopy for simultaneous monitoring of multiple process parameters in mammalian cell culture bioreactors. Biotechnology Progress. 2012;**28**(5):1355-1362. DOI: 10.1002/btpr.1590

[62] Craven S, Whelan J, Glennon B. Glucose concentration control of a fed-batch mammalian cell bioprocess using a nonlinear model predictive controller. Journal of Process Control. 2014;**24**(4):344-357. DOI: 10.1016/j.jprocont.2014.02.007

[63] Qin SJ, Badgwell TA. A survey of industrial model predictive control technology. Control Engineering Practice. 2003;**11**(7):733-764. DOI: 10.1016/S0967-0661(02)00186-7

[64] Schwenzer M, Ay M, Bergs T, Abel D. Review on model predictive control: An engineering perspective. The International Journal of Advanced Manufacturing Technology. 2021;**117**:1327-1349. DOI: 10.1007/s00170-021-07682-3

[65] Zendehboudi S, Rezaei N, Lohi A. Applications of hybrid models in chemical, petroleum, and energy systems: A systematic review. Applied Energy. 2018;**228**:2539-2566. DOI: 10.1016/j.apenergy.2018.06.051

[66] Narayanan H, Luna MF, Von M, Cruz MN, Polotti G, Morbidelli M, et al. Bioprocessing in the digital age: The role of process models. Biotechnology Journal. 2020;**15**:1900172. DOI: 10.1002/biot.201900172

[67] Möller J, Kuchemüller KB, Steinmetz T, Koopmann KS, Pörtner R. Model-assisted Design of Experiments as a concept for knowledge-based bioprocess development. Bioprocess and Biosystems Engineering. 2019;**42**:867-882. DOI: 10.1007/s00449-019-02089-7

[68] Blazek J. Computational Fluid Dynamics: Principles and Applications. 3rd ed. Oxford: Butterworth-Heinemann - Elsevier; 2005. p. 3. DOI: 10.1016/B978-0-08-099995-1.00001-4

[69] Jamshed S. Using HPC for Computational Fluid Dynamics. 1st ed. Oxford: Academic Press, Elsevier; 2015. p. 4. DOI: 10.1016/B978-0-12-801567-4.00001-5

[70] Mishra S, Kumar V, Sarkar J, Rathore AS. CFD based mass transfer modeling of a single use bioreactor for production of monoclonal antibody

biotherapeutics. Chemical Engineering Journal. 2021;**412**:128592. DOI: 10.1016/j.cej.2021.128592

[71] Sokolov M, Ritscher J, MacKinnon N, Souquet J, Broly H, Morbidelli M, et al. Enhanced process understanding and multivariate prediction of the relationship between cell culture process and monoclonal antibody quality. Biotechnology Progress. 2017;**33**:1368-1380. DOI: 10.1002/btpr.2502

[72] Park S-Y, Park C-H, Choi D-H, Hong JK, Lee D-Y. Bioprocess digital twins of mammalian cell culture for advanced biomanufacturing. Current Opinion in Chemical Engineering. 2021;**33**:100702. DOI: 10.1016/j.coche.2021.100702

[73] Cinar ZM, Nuhu AA, Zeeshan Q, Korhan O. Digital Twins for Industry 4.0: A Review. In: Calisir F, Korhan O, editors. Industrial Engineering in the Digital Disruption Era. GJCIE 2019. Lecture Notes in Management and Industrial Engineering. Cham: Springer; 2020. DOI: 10.1007/978-3-030-42416-9_18

[74] Seth G, Hamilton RW, Stapp TR, Zheng L, Meier A, Petty K, et al. Development of a new bioprocess scheme using frozen seed train intermediates to initiate CHO cell culture manufacturing campaigns. Biotechnology and Bioengineering. 2013;**110**:1376-1385. DOI: 10.1002/bit.24808

[75] Xu J, Rehmann MS, Xu M, Zheng S, Hill C, Qin H, et al. Development of an intensified fed-batch production platform with doubled titers using N-1 perfusion seed for cell culture manufacturing. Bioresources and Bioprocessing. 2020;7:17. DOI: 10.1186/s40643-020-00304-y

[76] Domokos A, Nagy B, Szilágyi B, Marosi G, Nagy ZK. Integrated continuous pharmaceutical technologies - a review. Organic Process Research & Development. 2021;**25**(4):721-739. DOI: 10.1021/acs.oprd.0c00504

[77] Gomis-Fons J, Schwarz H, Zhang L, Andersson N, Nilsson B, Castan A, et al. Model-based design and control of a small-scale integrated continuous end-to-end mAb platform. Biotechnology Progress. 2020;**36**:e2995. DOI: 10.1002/btpr.2995

[78] Bausch M, Schultheiss C, Sieck JB. Recommendations for comparison of productivity between fed-batch and perfusion processes. Biotechnology Journal. 2019;**14**:1700721. DOI: 10.1002/biot.201700721

[79] Lee SY, Kwon YB, Cho JM, Park KH, Chang SJ, Kim DL. Effect of process change from perfusion to fed-batch on product comparability for biosimilar monoclonal antibody. Process Biochemistry. 2012;**47**:1411-1418. DOI: 10.1016/j.procbio.2012.05.017

[80] Gerstweiler L, Bi J, Middelberg APJ. Continuous downstream bioprocessing for intensified manufacture of biopharmaceuticals and antibodies. Chemical Engineering Science. 2021;**231**:116272. DOI: 10.1016/j.ces.2020.116272

[81] Somasundaram B, Pleitt K, Shave E, Baker K, Lua LHL. Progression of continuous downstream processing of monoclonal antibodies: Current trends and challenges. Biotechnology and Bioengineering. 2018;**115**:2893-2907. DOI: 10.1002/bit.26812

[82] Grilo AL, Mateus M, Aires-Barros MR, Azevedo AM. Monoclonal antibodies production platforms: An opportunity study of a non-protein-a chromatographic platform based on process economics. Biotechnology

Journal. 2017;**12**:1700260. DOI: 10.1002/biot.201700260

[83] Kateja N, Kumar D, Sethi S, Rathore AS. Non-protein a purification platform for continuous processing of monoclonal antibody therapeutics. Journal of Chromatography A. 2018;**1579**:60-72. DOI: 10.1016/j.chroma.2018.10.031

[84] Gagnon P. Technology trends in antibody purification. Journal Chromatography. 2012;**1221**:57-70. DOI: 10.1016/j.chroma.2011.10.034

[85] IMS. The Impact of Biosimilar Competition in Europe [Internet]. 2017. Available from: https://www.medicinesforeurope.com/wp-content/uploads/2017/05/IMS-Biosimilar-2017_V9.pdf [Accessed: May 31, 2022]

[86] Galliher PM. Chapter 29 - single use technology and equipment. In: Jagschies G, Lindskog E, Łącki K, Galliher P, editors. Biopharmaceutical Processing. Oxford: Elsevier; 2018. pp. 557-577. DOI: 10.1016/B978-0-08-100623-8.00029-3

[87] Guldager N. Cost advantages of single-use technologies. Pharmaceutical Technology. 2010;**S26**(S28):S30-S31

[88] Bioprocess International. Upstream Single-Use Bioprocessing Systems [Internet]. 2012. Available from: https://bioprocessintl.com/upstream-processing/upstream-single-use-technologies/upstream-single-use-bioprocessing-systems-326675/. [Accessed: May 31, 2022]

[89] Allison N, Richards J. Current status and future trends for disposable technology in the biopharmaceutical industry. Journal of Chemical Technology and Biotechnology. 2014;**89**:1283-1287. DOI: 10.1002/jctb.4277

[90] Bioprocess International. ABEC breaks plastic ceiling again with 6,000 L single-use bioreactor [Internet]. 2019. Available from: https://bioprocessintl.com/bioprocess-insider/upstream-downstream-processing/abec-breaks-plastic-ceiling-again-with-6000-l-single-use-bioreactor/. [Accessed: May 31, 2022]

[91] American Pharmaceutical Review. Single-Use Systems Not Helping Downstream Bioprocessing, Yet... Alternatives to Chromatography Still Slow in Adoption [Internet]. 2015. Available from: https://www.americanpharmaceuticalreview.com/Featured-Articles/177871-Single-Use-Systems-Not-Helping-Downstream-Bioprocessing-Yet-Alternatives-to-Chromatography-Still-Slow-in-Adoption/. [Accessed: June 1, 2022]

[92] Jones S. BioPhorum Operations Group Technology Roadmapping, Part 4: Efficiency, Modularity, and Flexibility As Hallmarks for Future Key Technologies. BioProcess Int. 2017;**15**:14-19

Chapter 9

Nanomaterials as Novel Biomarkers for Cancer Nanotheranostics: State of the Art

Hao Yu, Zhihai Han, Cunrong Chen and Leisheng Zhang

Abstract

Cancers including hematological malignancies and metastatic solid tumors are one of the life-threatening diseases to the general population, which have become a heavy burden for patients and their caregivers physically and mentally. Despite the great progression in preclinical and clinical studies, effective implementation strategies are urgently needed to optimize the advancements in cancer diagnosis and treatment. State-of-the-art updates have indicated the application of multifunctional nanotheranostics as an emerging diagnostic and therapeutic tool for cancer management. Herein, this chapter displayed the literature and description of various nanomaterial-based noninvasive diagnostic and therapeutic approaches for cancer administration from the view of nanomaterial classification and nanomaterial-based application in nanotheranostics as well as the promising perspectives and grand challenges in nanomedicine. Collectively, this review will provide overwhelming new references for cancer supervision and benefit the medical and pharmaceutical practice in the field of nanotheranostics.

Keywords: nanomaterials, nanotheranostics, chemoradiotherapy, cancer immunotherapy, nanomedicine

1. Introduction

Cancers with high heterogeneity and uncontrolled cell division are notoriously hard to conquer and have emerged as one of the leading causes of death worldwide with a prevalence of over 10 million mortalities annually [1, 2]. Over the years, a certain number of investigations have been accomplished to figure out the fundamental pathogenesis and the concomitant treatment regimens including surgery, oncolytic virotherapy, radiotherapy, chemotherapy, photothermal therapy, RNA vaccine, peptide-based neoantigen vaccine, hormone therapy, and immunotherapy [3–6]. Generally, surgery (e.g., robotic surgery, laparoscopic rectal surgery) has been considered the best option for localized cancers without transfer and diffusion, which usually works in combination with chemoradiotherapy for the eradication of residual cancerous cells [7, 8]. Chemoradiotherapy has become a notable and synergistic anticancer treatment for a variety of locally advanced solid tumors through a rationale of two concepts (chemotherapy, radiotherapy) of in-field cooperation

IntechOpen

and spatial cooperation but inevitably increases microbiota resistance and damage to normal tissues [9–11]. Current progresses have also highlighted the potential of anticancer immunotherapy including immune cells and checkpoint inhibitors for the significant clinical benefit [3, 4, 9, 12–14]. Meanwhile, despite new insights into RNA vaccine-derived immunity in melanoma treatment, those cancer vaccine trials in the late-stage patients with various treatment-refractory tumors have not been successful [6, 15–17]. Therefore, in overall consideration of the shortcomings (e.g., off-target effects, severe toxicity, drug delivery barriers, and graft-versus-host disease), the aforementioned treatment regimens fell short of expectation in cancer administration [1, 3, 12, 18–20].

State-of-the-art updates have highlighted the feasibility of nanomaterials as promising agents for cancer diagnosis and therapy based on the rapid progress of nanobiotechnology and clinical biomedicine [21–23]. To date, multidisciplinary research has further highlighted the superiority of the newly emerging bidimensional (2D) nanomaterials in multiple physicochemical properties and ultrathin layer-structured topology for theragnostic nanomedicine such as graphene and its derivatives, transition metal carbides (MXenes), hexagonal boron nitrides (h-BN), black phosphorus (BP), transition metal dichalcogenides (TMDCs), palladium (Pd) nanosheets, and transition metal oxides (TMOs) [24–27].

Therefore, this chapter principally focused on the current progress in nanomaterials for cancer nanotheranostics including the classification of nanomaterials (e.g., inorganic nanomaterials, organic nanomaterials, organic-inorganic hybrid nanomaterials), nanomaterials in cancer diagnostics (e.g., contrast agents for in vivo imaging, signal modes for *in vitro* diagnostics) and cancer treatment (e.g., cancer phototherapeutics, photothermal therapy, photodynamic therapy, cancer immunotherapy, combined therapy), and ultimately summarized the opportunities and challenges of nanomaterial-based cancer nanotheranostics. Collectively, the nanomaterial-mediated nanotheranostics had constituted a promising area of oncology theragnostics.

2. Nanomaterials and classification

Nanoparticles, with a size ranging from 1 nm to 100 nm, reveal many unique properties in terms of light, heat, electricity, magnetism, sound and chemistry, and in particular, the "hobby" of lodging with tumor cells endow themselves with enhanced permeability and retention (EPR) effect [28, 29].

Generally, nanomaterials are categorized as inorganic nanomaterials, organic nanomaterials, and organic-inorganic hybrid nanomaterials, which have been extensively developed for tumor diagnosis and treatment based on their unique biofunctions and biomedical characteristics [30, 31]. Among them, inorganic nanomaterials are the earliest studied and most widely used biomaterials in clinical oncology treatment including noble metal nanoparticles, metal chlorocarbon nanomaterials, magnetic nanoparticles, and quantum dots. These inorganic nanomaterials usually possess a series of excellent properties such as strong near-infrared light absorption capacity, high photothermal conversion efficiency, easy preparation, and modification, which thus enable the applications in fluorescence imaging, photoacoustic imaging, or nuclear magnetic resonance imaging [32]. Organic nanomaterials can be divided into organic small molecule nanomaterials and organic polymer (polymeric) nanomaterials, which are employed in the area of bioluminescent probes, photothermal therapy, and drug carriers due to their unique properties (e.g., diverse structure, easy to cut, low assembly cost) [33]. Organic-inorganic hybrid nanomaterials not only possess improved stability and biocompatibility of inorganic nanoparticles but also reveal enhanced hardness and strength of organic matrix materials, which thus have a wider range of applications over organic and inorganic materials [34].

Of note, the pathological structure of tumor tissue has the characteristics of low pH value, hypoxia, new blood vessels, and lymphatic vessels in the microenvironment owing to the anatomical structure and physiological function are quite different from normal tissues [35, 36]. Owing to the aforementioned characteristics, tumor tissues can be specifically targeted by nanoparticles in order to achieve an efficient and accurate diagnosis and treatment.

3. Contrast agents for *in vivo* imaging

As one of the most life-threatening diseases worldwide, the morbidity and mortality of cancer are increasing year by year [37]. Traditional diagnostics mainly focus on pathological examinations and endoscopic examinations, which often cause certain trauma to the patient's body. In recent years, clinical imaging diagnoses of solid tumors have mainly relied on computed tomography (CT), magnetic resonance imaging (MRI), positron emission tomography (PET), and ultrasound (**Figure 1**). However, these methodologies are not safe and efficient enough for monitoring the changing microstructure of tumors due to the ionizing radiation damage, insufficient resolution, and lack of targeting. Therefore, it is urgently needed to develop a new

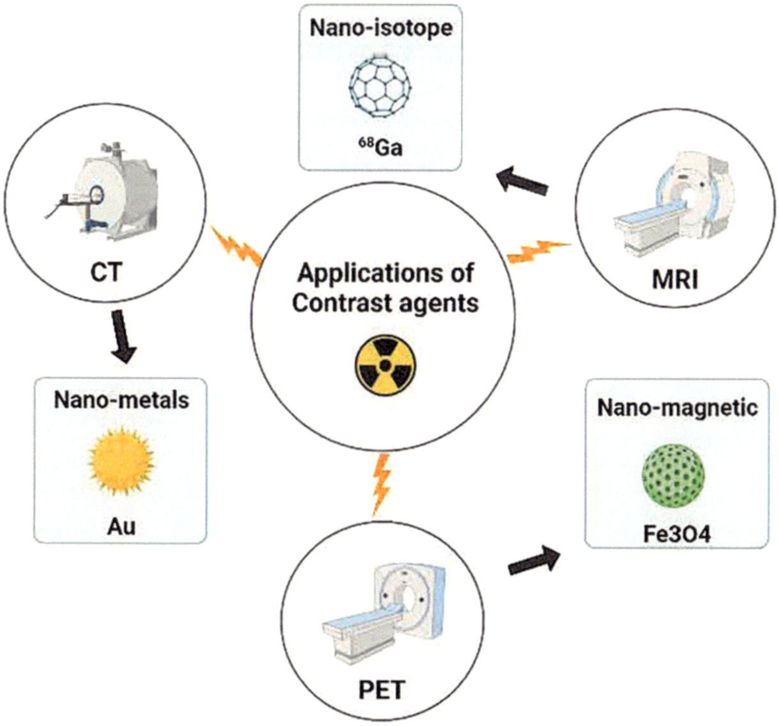

Figure 1.
Schematic illustration of the biofunction of nanomaterials.

type of nanomaterial contrast agent with high efficiency, accuracy, low toxicity, and side effects for clinical tumor diagnosis and treatment, which thereby benefits the enhancement of sensitivity and accuracy of tumor diagnosis [38, 39].

3.1 Metal nanomaterials and CT

CT is a noninvasive imaging technique that uses X-rays, γ-rays, or ultrasound to scan a certain area of the human body in order to achieve differential diagnosis via the variations in signal absorption among different types of cells [40]. In recent years, the application of based-metal elements (e.g., gold, bismuth, tantalum, and ytterbium) of nanomaterials as contrast agents has been extensively reported (**Table 1**).

Generally, differing from the iodine-containing nanoparticle imaging system (denoted as "soft particles"), metal-based compound nanoparticles (denoted as "hard particles") manifest more reliable stability in the body and are easier to bypass the body's immunity system barrier to reach the tumor lesions and improve imaging efficiency. Meanwhile, these metal elements exhibit higher density and atomic numbers for effective absorption of X-rays, which makes up for the insufficient contrast ability of iodine as a contrast agent and sharply reduces the X-ray radiation dose of patients during imaging. In addition, the surface of metal-based nanoparticles is easy to be modified by physically or chemically methods, which enhances the targeting and versatility of CT imaging probes in clinical applications [22]. For instance, Luo et al. reported the accumulation of prostate-specific membrane antigen (PSMA) targeted AuNPs in prostatic cancer revealed a size-dependent pattern [41]. Shao et al. developed a novel Bi2S3 nanoparticle coated with a hyaluronic acid (HA)-modified tantalum oxide (TaOx) nanoshell (Bi2S3@TaOx-HA) for multimodality breast cancer diagnosis, which manifested excellent biocompatibility, photothermal transducing performance and computed tomography imaging capacity [42]. Instead, the carboxybetaine zwitterionic-coated tantalum oxide (TaCZ) nanoparticle CT contrast agent was reported with greater contrast enhancement compared with a conventional iodinated contrast agent in swine models [43]. Notably, thrombocytopenia and neutropenia in patients could be predicted

Metal	Contrast agent	Characteristics	Application
Gold	Based on gold nanoparticles and its derivatives	Stability, biocompatibility, enhanced CT contrast, safety, high X-ray absorption, Expensive	Solid tumors, blood-pool imaging capability
bismuth	Bi_2S_3 nanoparticles	Biocompatibility, high X-ray absorption, safety, long blood half-life, cost-effective	Solid tumors; lymphoscintigraphy, cartilage imaging
tantalum	Tantalum oxide nanoparticles and its derivatives	Biocompatibility, enhanced CT contrast, safety, cost-effective	Lymphoscintigraphy, cartilage imaging
ytterbium	Based on ytterbium nanoparticles and its derivatives	Biocompatibility, X-ray extinction ability, enhanced CT contrast	Blood-pool imaging capability
lutetium	Based on lutetium nanoparticles and its derivatives	Biocompatibility, enhanced CT contrast	Metabolic contrast imaging

Table 1.
CT imaging system based on metal nanomaterials.

after 177Lutetium-lilotomab satetraxetan treatment based on the SPECT/CT-derived absorbed dose [44].

3.2 Magnetic nanomaterials and MRI

MRI is a type of noninvasive tomographic imaging, which is used to obtain electromagnetic signals from the body and reconstruct human body information. The combination of nanomaterials and MRI technology can improve the sensitivity and accuracy of MRI, and in particular, the iron-based magnetic nanomaterials with various shapes and sizes are extensively explored. For example, superparamagnetic iron oxide (SPIO) nanoparticles serve as an ideal MRI contrast agent and have been approved for clinical application attributed to their unique properties such as dual-function angiography, longer half-life in blood, specificity reticuloendothelial system, venography effect, and *in vivo* tracking of cell labeling [45].

Meanwhile, various raw materials with outstanding characteristics have also been reported such as high magnetic torque, saturation, and coercivity. For example, Wang et al. synthesized the Au-Fe_3O_4@PDA-PEG-DTPA-Gd hetero-nanostructure with reasonable biocompatibility and high photothermal conversion efficiency, which was adequate to completely inhibit the growth of MDA-MB-231 tumor *in vivo* [46]. Xu et al. generated the tumor-targeted NPs (DOX@Gd-MFe_3O_4 NPs) by combining Gd-doped mesoporous Fe_3O_4 nanoparticles with doxorubicin (DOX), which exhibited good colloidal dispersity, superior magnetic properties, superior NIR photothermal conversion, and NIR-triggered DOX release [47]. Additionally, amine-functionalized $CuFeSe_2$-NH_2 nanoparticles were reported with specificity against 4 T1 and $HepG_2$ cells due to the multifaceted signatures including water solubility, cytocompatibility, hemocompatibility, and biosafety [47].

3.3 Isotope nanomaterials and PET

Fluorodeoxyglucose is the main medium in PET, which functions as a critical element in various metabolisms and accumulates in high-metabolic tumor tissues rather than in low-metabolic normal tissues [48]. In recent years, radionuclide-labeled nanomaterials in PET have become a research hotspot for cancer diagnosis and monitoring due to their preferable properties such as high sensitivity and precise spatial quantification capabilities (**Table 2**). For example, Song *et al.* took advantage

Methods	Labeling	Characteristics
Chelating reactions	DOTA/NOTA/NODA	High costs, low efficiency, poor labeling stability
Direct bombardment of nanoparticles by proton or neutron beam	16O (p, α) 13 N18O (p, n)13 N165Ho (n, γ)166Ho	Rapid preparation, short-half-life radionuclide-labeled nanomaterials, unstable bioactive molecules on the surface
Direct synthesis with radioactive and non-radioactive materials	^{64}Cu, ^{68}Ga, ^{18}F	Highly stable radiolabeled nanomaterials
Radiolabeling without chelating agent	Superparamagnetic iron oxide nanoparticles (SPION)	Rapid preparation, high specificity and labeling rate, limitation in combining radionuclides with nanoparticles

Table 2.
Methodologies of radionuclide-labeled nanomaterials for PET.

of the ^{131}I-labeled copper sulfide-loaded microspheres for the treatment of hepatic tumors via hepatic artery embolization [49]. Peng *et al.* confirmed the excellent safety profile and favorable pharmacokinetics of a self-assembling [^{68}Ga] Ga-NOTA supramolecular dendrimer nanosystem for PET imaging, which was more competent for the detection of imaging-refractory low-glucose-uptake tumors compared to the clinical ^{18}F FDG [50]. Co-injection of CBT-NODA-^{68}Ga with CBT-NODA or CBT-NODA-Ga has been reported for the enhanced micro-PET tumor imaging in mice via accelerating the synthesis of hybrid gallium-68 nanoparticles in furin-overexpressing cancer cells [51].

4. Application of different signal modes in diagnostics *in vitro*

Nanomaterials can be used to generate different types of detection signals, amplify the intensity of detection signals, and simplify the detection process attributed to their unique optical properties (e.g., magnetic, electrical, and thermal), which thus have broad application prospects *in vitro* diagnosis upon nucleic acids, proteins, small molecules, bacteria and viruses (**Figure 2**). Currently, the applications of fluorescent signals, surface-Raman signals, magnetic signals, electrochemical signals, color signals, and thermal signals of nanomaterials are the most representative signal detection modes for diagnostics *in vitro* (**Table 3**). For instance, Liu et al. generated a versatile nanoprobe based on reduced graphene oxide (rGO) and nucleic acid (DNA) nanoprobe, which provided a general sensing platform for highly sensitive imaging of dual miRNAs in living cells [52]. Lin et al. developed a microfluidic biosensor for Salmonella detection based on viscoelastic inertial microfluidics for separating magnetic bacteria from unbound magnetic nanoparticles (MNPs) and enzyme catalytic colorimetry for amplifying biological signals [53]. Compared with the unmodified electrode, a glassy carbon electrode (GCE)-based ultrasensitive electrochemical biosensor modified by a unique sandwich-like nano-Au/ZnO sol-gel/nano-Au compound revealed high absorbability and surface activity, good electro-conductivity, and biocompatibility [54].

5. Nanomaterials in cancer treatment

For decades, multifaceted treatment options for cancer such as surgery, chemotherapy, radiation therapy, pharmacotherapy, targeted therapy, cellular therapy, and combined therapy have been developed, yet the clinic prognosis of tumor patients is still unsatisfactory [4, 55, 56]. For instance, despite the great efforts focused on cancer drug discovery pipeline (e.g., PD1/PDL1 axis), the undesirable outcomes and burdensome expenditures of pharmacotherapy alone or in combination with other strategies including nanomaterials still need to be overcome [57, 58].

5.1 Nanomaterials in cancer chemoradiotherapy

Radiotherapy, including external radiation and internal radiation therapy, is one of the main treatments and adjuvant therapy for oncologic treatment, which can efficiently reduce the misery and pressure as well as affect the tumor environment (TME) but may cause a severe untoward effect upon patients [59, 60]. Chemotherapy is treatment with specific drugs to obliterate or shrink the metastatic cancer cells

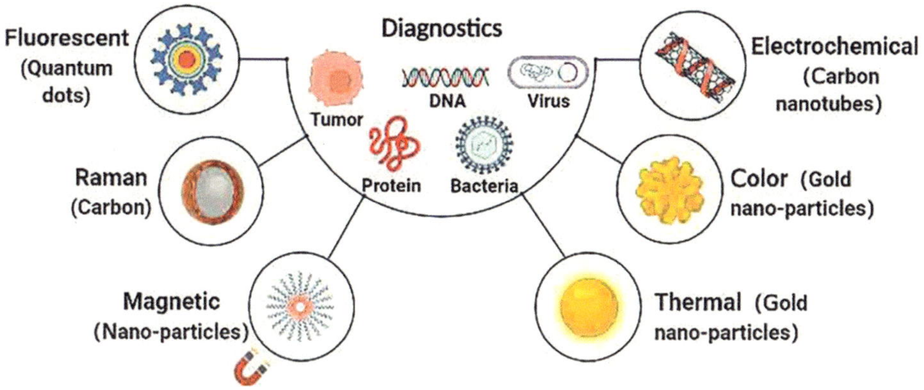

Figure 2.
Nanomaterial-based tumor diagnostics.

Signal types	Carriers	Target objects	Applications	Principles
Fluorescent signals	Quantum dots, graphene oxide, gold nanoparticles	Nucleic acid, protein, influenza virus antigen, intracellular virus titer	ELISA immunoassay, DNA microarray	Fluorescence encoding capability, energy transfer (FRET) detection
Raman signals	Carbon	Nucleic acid, protein, tumor cells, bacteria, virus and tracer of small molecule drugs in living cells	Analysis method (molecular structure)	The surface plasmon resonance effect of metal nanoparticles to enhance Raman signals
Magnetic signals	Magnetic nano-particles	Proteins, nucleic acids, tumor cells and bacteria detection	Bio-assay reagent	The surface of microspheres is modified by magnetic nanoparticles
Electrochemical signals	Gold nanoparticles or carbon nanotubes	DNA, protein, virus, bacteria	Bio-affinity electrochemical sensor	Sandwich-like structure of using nanoparticle-labeled surface-binding analyte
Color signals	Gold nanoparticles	DNA	Nucleic acid-colorimetric detection	Coupling of DNase and gold nanoparticles
Thermal signals	Gold nanoparticles	Influenza A, malaria, clostridium	Thermal contrast signal detection	Near-infrared detector

Table 3.
Application of representative signal detection modes for diagnosis.

before or after surgery. The chemotherapy drugs can be divided into antimetabolites (e.g., 6-mercaptopurine), alkylating agents (e.g., cyclophosphamide), topoisomerase inhibitors (e.g., Topotecan), and anticancer antibiotics (e.g., Bleomycin), which mainly function by suppressing cell division of both cancer cells and normal cells in the body (e.g., bone marrow, gastrointestinal mucosa) and thus cause adverse effects.

State-of-the-art updates have reported the involvement of unidimensional (1D) and bidimensional nanomaterials (2D) with aromatic ring carbon particles in cancer chemoradiotherapy and device fabrication based on the unique nanosheet structures, tunable chemical composition, the large surface areas, surface functionalization, minimal thickness, and other extraordinary physicochemical properties (**Figure 3**) [61, 62]. For drug delivery via encapsulation or covalent linking or surface adsorption, nanoparticles are loaded with biomolecules and chemotherapeutic drugs based on noncovalent bonding (e.g., van der Waal's force, hydrophobic interaction, π–π stacking) [63, 64].

To date, a variety of 2D nanomaterials with potential of acting as drug delivery nanoplatforms have been synthesized by different methodologies, which attract the tremendous interest of investigators in the field such as transition metal dichalcogenides (TMDC), layered double hydroxides, transition metal dichalcogenides, nitrides and carbonitrides, metal-organic framework nanosheets, graphene and its derivatives, and black phosphorus nanosheets [61, 65]. In particular, those with unique X-ray attenuation and easily tunable properties such as graphene and TMDCs are adequate to be harnessed for radiotherapy or phototherapy of cancer.

5.2 Nanomaterials in cancer phototherapeutics

Phototherapeutics, a next-generation therapeutic modality, is a type of photoresponsive regulation of biological function and relative stimuli-responsive features, which thus supplies promising prospective for promoting the accuracy and efficacy of cancer treatments via producing reactive oxygen species (ROS) by photosensitizers and eliminating cancer cells by specific wavelength light irradiation [66, 67]. Generally, phototherapeutics can be divided into three typical categories including photobiomodulation (PBM), photodynamic therapy (PDT), and photothermal therapy, which are widely applied to cancer administration such as colorectal cancer, head and neck cancer, breast cancer, and colorectal cancer [68, 69]. However, ineffective treatment of cancers by PDT can be caused by specific tumor environments and even hindered by the deep tumor cells [70].

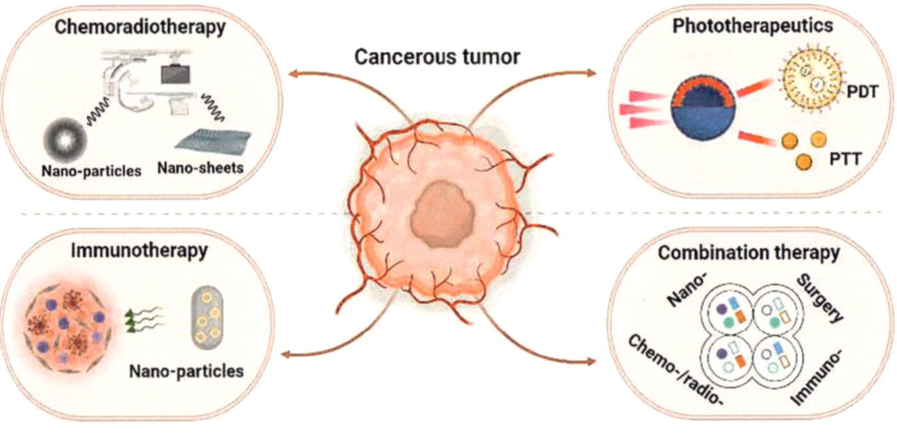

Figure 3.
Categories of nanomaterial-based tumor therapeutics.

To date, increasing literatures in the cutting-edge research area have turned to phototherapy combined with various nanomaterials in cancer therapy. Of them, carbon-based materials such as graphene and carbon nanotubes have attracted attention in the field of cancer phototherapy worldwide attribute to their unique physical and chemical properties including large surface area, thermal conductivity, and electrical properties [68]. Additionally, several kinds of nontoxic photosensitizers involved in phototherapy are also functionalized on the aforementioned carbon-based nanomaterials. Current research has also highlighted the potential role of stimuli-responsive nanomaterials (PNMs) with characteristics of responding to endogenous pathological changes for smart tumor-specific phototherapeutics [66]. For instance, Fu et al. took advantage of the porous shuttle-shape platinum methylene blue (IV-Mb) coordination polymer nanotheranostics-loaded 10-hydroxycamptothecin (CPT) for synergistically enhancing the in situ mitochondrial reactive oxygen species (ROS) and highly efficient tumor ablation by phototherapy, which was regarded as a promising method for synergistic oncotherapy [70]. Furthermore, a TME-sensitive oxygen-dual-generating nanosystems named MnO_2@Chitosan-CyI (MCC) has been developed to decrease the level of glutathione (GSH) and relieve environmental tumor hypoxia, which reveals synergistic effects with PDT in cancer treatment by triggering an acute immune response and reducing tumor metastasis [67].

Of note, the chemodynamic therapy (CDT) based on photothermal-enhanced Fenton has also attracted considerable research attention in the field, and the nanocatalyst-based strategy with high specificity and limited side effects has also emerged as a promising therapeutic option for the in-situ treatment of various cancers [71]. For instance, a number of multifunctional nanomaterials (e.g., metal oxide- or metal-sulfide-based nanocatalysts) have been manufactured to trigger the reaction within the TME and generate highly cytotoxic hydroxyl radicals as well [71].

5.2.1 Nanomaterials in cancer photothermal therapy (PTT)

Nanomaterial-based PTT has been recognized as a promising therapeutic modality for whole-body anti-tumor immune response and tumor ablation in the tumor microenvironment [72, 73]. Recently, Yang et al. took advantage of the magnetite nanomedicine in the administration of lung cancer and reported the synergistic effect with hyperthermia and chemotherapy, which collectively suggested the designed SPIO@PSS/CDDP/HSA-MTX nanoparticles with good biocompatibility and stability as powerful candidate nanoplatform for future antitumor treatment strategies [55]. Meanwhile, gold-nanobranched-shell-based vehicles and near-infrared nanomaterial-liposome hybrid nanocarriers (NIRN-Lips) with dual superiorities such as higher tumor permeability, enhanced photoluminescence, stimulus-responsive drug release, better tumor-targeted drug delivery, and anti-tumor efficacy have been applied in cancer PTT and chemo-photothermal therapy as well [23, 74, 75]. Collectively, it is of paramount importance for the future improvement of photothermal therapy (PTT) via incorporating drug conjugates and polymer linkers with the surface of nanomaterials, which will further enhance the multiplexing capability and surface functionalization of nanomaterials as well as the advanced cancer imaging and therapies [61].

5.2.2 Nanomaterials in cancer photodynamic therapy (PDT)

Photodynamic therapy (PDT) is a noninvasive form of therapy that combines both photophysical and photochemical processes, which has emerged as a promising therapeutic modality for cancer and nononcological diseases of various types and locations [59]. Differ from the aforementioned chemoradiotherapy, the third-generation photosensitizers of PDT are more affordable and dispense with hospitalization. PDT mainly functions via the activation of photosensitizers with an applicable wavelength of light and the upregulation of transient concentration of reactive oxygen species (ROS) accumulated at tumor sites, which has emerged as an important therapeutic option in oncology [76]. In recent years, PDT has attracted widespread attention as a highly selective and noninvasive approach for various cancer treatments, and in particular, the carrier nanoparticles with additional active supplementary and complementary roles [77]. However, PDT has inherent defects in treating deep tumors due to the insufficient luminous flux and limitation in approved drugs as well as the inevitable occurrence of peripheral tissue damage [59]. Additionally, due to the unique tumor microenvironment, the PDT-induced immune responses upon cancers are generally mild and thus not sufficient to ultimately eradicate metastatic cells as well [67].

The combination of nanomaterials with photosensitizers can further potentiate the efficiency and selectivity of PDT and help eliminate the side effects [78]. Current investigations have illuminated the practicality of utilizing the persistent or scintillation luminescence nanoparticles (e.g., porphyrins) with conjunctive photosensitizers for photodynamic therapy, which is adequate to enhance the effectiveness of X-ray-based ionizing radiation and minimalize the potential damage to healthy cells [79]. For example, Wang et al. developed novel biphasic and bimetallic Rh-based core-shell Au@Rh-ICG-CM nanostructures with good biocompatibility and photoacoustic imaging properties for the treatment of hypoxic tumors in combination with PDT and verified the synergistic enhancement upon oxygen generation from the endogenous hydrogen peroxide in cancer [80].

Generally, nanoparticles as delivery vehicles in PDT can be functional and classified into active participants and passive carriers during photosensitizer excitation [76]. Meanwhile, a series of oxygen-evolving agents (e.g., perfluorocarbon, catalase, HbO_2) for self-supplying oxygen and Manganese dioxide (MnO_2)-based nanoparticles with high reactivity toward H_2O_2 have been incorporated into the PDT nanosystems [67]. Distinguish from the nonbiodegradable carriers with extraneous functions, active nanoparticles can be mechanistically subclassified and divided into self-illuminating nanoparticles, upconverting nanoparticles, and photosensitizer nanoparticles [81]. Nevertheless, the cancer regions deep in the body and the deficiency of the second-generation PDT nanoparticles still remain the major obscure challenges before the adoption in large-scale clinical application [67]. In consequence, there is an urgent need for the development of intelligent nanosystems capable of functioning in the TME and enhancing the therapeutic efficacy of PDT for deep cancers.

5.3 Nanomaterials in cancer immunotherapy

The complex orchestration of cancer cells with tumor immune microenvironment results in the emergence of novel immunotherapy-based treatment regimens in patients [3, 14]. Immunotherapy such as immune checkpoint blockade and adoptive cell infusion has turned into a powerful clinical alternative for cancer administration attributes to their durable responses in hematologic malignancies and multiple metastatic solid tumors [82–85]. Generally, cancer immunotherapy functions mainly via stimulating or training the inherent immunological systems and thus benefits the recognition, attack, and eradication of cancer cells with minimal damage to normal cells as well [13, 83]. Notably, cancer immunotherapy (e.g., natural killer cells, chimeric antigen receptor transduced T cells, cytotoxic T-lymphocyte antigen 4, programmed cell death-1) might cause unique toxicity profiles or an insignificant spectrum of immune-related adverse events (irAEs) differ from the toxicities of chemoradiotherapy and phototherapeutics depending on their mode of action [82, 86]. Worse still, despite the potentially favorable outcomes for advanced-stage patients such as complete cures and long-term survival, it is reported that cancer immunotherapy only works well in a relatively small subset of patients [87]. For example, Gong et al. recently reported the prominent challenges to the further broad implementation of T-cell-based immunotherapies including insufficient expansion, decreased cellular vitality *in vitro*, and trafficking of T cells into solid tumors [88].

In recent years, nanoparticle-based nanomedicine has revealed dramatic progress in the fast-rising field of cancer immunotherapy and has boosted therapeutic outcomes. Nanomaterials with unique chemical and physical features offer advantaged therapeutic platforms for photo-induced hyperthermia and cancer immunotherapy by turning the "cold" nonimmunoresponsive cancers and metastases into the "hot" immuno-responsive lesions [87, 89]. Moreover, nanomaterial-based nanomedicines can also be employed to target the tumor immune microenvironment, potentiate antigen presentation, trigger the release of danger-associated molecular patterns, inhibit immunosuppressive cells, and thus boost the therapeutic outcomes of cancer immunotherapy [87]. Of note, the fourth generation of biomaterials including nanomaterials is expected to stimulate a more specific cellular response and a more accurate control of sophisticated immunomodulation to the implants or cancers [90]. Collectively, nanomaterials after rational designation are uniquely suited to overcome the aforementioned challenges in cancer immunotherapy.

5.4 Nanomaterials in combined therapy of cancer

Due to the aforementioned deficiency in cancer treatment, investigators have turned to exploring the feasibility of combining nanomaterials with other strategies (e.g., surgery, chemotherapy, radiotherapy, immunotherapy) for increasing the coordination of treatment effects as well as reducing the side effects [77, 91, 92]. Different from monotherapy, combination therapy for cancer patients usually provokes a good response to tumor surveillance and clearance [93]. Of note, Wang et al. recently summarized the ferroptosis-inducing nanomedicine by combining ferroptosis with nanomaterials, the conventional treatment, and emerging therapy for cancer therapy, yet most of the ferroptosis inducers such as system Xc-inhibitors (e.g., erastin and sorafenib) and GPX4 inhibitors (e.g., RSL3 and altretamine) had not been clinically approved due to nonspecific distribution, poor solubility, and unpredictable side effects [1, 94].

In the last years, a variety of 2D nanomaterials (e.g., ceramic-based biomaterials and 2D MXenes) with prominent physiochemical properties and specific surface properties (e.g., protein corona formation, unique planar structure, and chemical modification) have been explored in cancer management in combination with surgical treatment, radiotherapy, chemotherapy, photothermal therapy, photodynamic therapy, chemodynamic therapy, radiodynamic therapy, and immunotherapy [84, 95]. For example, current achievements in the combination therapy of glioblastoma with nanocarriers have demonstrated increasing benefits against the disappointing clinical outcomes and existing challenges such as blood-brain barrier (BBB), tumor heterogeneity, glioma stem cells, drug efflux pumps, and toxicity, which are particularly formidable challenges in developing cancer therapeutics [63, 64]. In addition, nanocarrier-based combination therapy has been supposed to ensure the targeted colocalization of drugs into the tumor sites and facilitate sequential drug exposures and the synergistic drug ratio [64]. As reviewed by Zhao et al., numerous nanoformulations (e.g., Doxil, Abraxane, and DaunoXome) were not only adequate to load hydrophobic and hydrophilic drugs and prolong the half-life of diagnostic or theranostic agents, but also could reduce the toxicity of the parent compound and thereby ameliorate its therapeutic index [64].

6. Discussion and conclusions

Cancers of various kinds remain a core challenge and life-threatening disease taking millions of peoples' lives as well as exacerbating the quality of life of the survivors [2, 13]. Despite the inspiring advances in the cancer treatment paradigm, high mortality and the concomitant toxicities of traditional therapies result in a significant challenge to adherence and tolerability of patients, and in particular, the severe adverse effects and toxic effects of chemoradiotherapy and phototherapeutics on patients cannot be neglected [96, 97]. Therewith, pioneering clinicians and researchers turned to alternate treatment regimens with a complete response and minimum side effects during cancer treatment. Among them, anticancer nanomedicine has been studied for over 30 years and a handful of formulations have been approved for clinical purposes, which has revolutionized the remedy of several advanced-stage tumors [30, 98]. For example, Chen et al. reported a novel and low-cost modality for augmented efficacy upon cancers *via* a combination of *in vivo* luminescent nanoparticle agent-based radiation and photodynamic therapies [79]. Meanwhile, chemodynamic therapy (CDT) in combination with photothermal therapy (PTT) and multifunctional nanomaterials has also been utilized to enhance therapeutic efficacy in cancer theragnostic, which also provides more effective efficacy when compared with monotherapy [71]. Despite the unique physical and chemical properties including targeting specificity and profound stability, the application of nontoxic nanomaterials coated with appropriate structures and biocompatibility for *in vivo* imaging is of great importance for clinical purposes. Moreover, considering the influence of the large size of nanomaterials for localization in tissues, nanoparticles should be degraded into the essential components before they can be excreted *via* metabolism or the kidneys.

Nevertheless, the tumor microenvironment as well as the toxic and side effects of current therapeutic regimens still remain the major obstacle to be overcome [3, 58, 60, 72]. For this purpose, great efforts have been expended on the modification of the physicochemical surface properties of nanomaterials with increased complexity and adaptability for the more sophisticated immunomodulation against various tumor cells in the past decades [90, 93]. For instance, a series of novel nanomaterials based on the surface modification of MXenes for combination therapy with magnetic resonance (MR), magnetic resonance imaging (MRI), or computed tomography (CT) have been manufactured such as Nb_2C nanosheets with polyvinylpyrrolidone (PVP) decoration, PEGylation assembled into Ti_3C_2 nanosheets, Ta_4C_3 nanosheets modified with the soybean phospholipid (SP), Ta_4C_3 nanosheets coupled with Fe_3O_4 nanoparticles, Ti_3C_2 MXene attached to mesoporous silica nanoparticles (MSNs) [99].

To date, topographical modification of nanomaterials has become an attractive and expanding field aiming to dissect the sophisticated diversity of synergistic interactions between surface nanotopography and cancer cells, and thus holds the promising prospect for solving the long-lasting challenges in cancer nanotheranostics [84, 90]. Notably, self-assembled nanomedicines with unique and versatile features have been extensively explored for dealing with the malignancy and heterogeneity of tumors, which are designed to enhance antitumor immune responses *via* a series of immuno-potentiating biofunctions and controlled pharmacokinetics in the tumor regions [98]. However, considering the heterogeneity of tumors and inefficiency of nanoparticle loading and releasing, it remains challenging to ensure agents specifically targeting cancer cells and alleviating collateral toxicity to healthy tissue. Most of all, despite the plethora of information on cell-surface interaction and nanofabrication at the research level, there is still a long way to obtain more advanced nanopatterning techniques and transform the academic knowledge into commercial technologies or clinical practice [90, 100]. Nanomaterials are acknowledged as advantaged sources for tumor surveillance and elimination. Distinguish from our previously reported biomaterials and various counterparts of immune cells such as T cells, dendritic cells, natural killer cells, and Treg cells, the nanomaterial-based nanomedicine efficaciously fulfills the function of combating transformed hematological malignancies and metastatic solid tumors. Moreover, considering the inherent properties, nanomaterials are "off-the-shelf" products satisfying the clinical demand for large-scale manufacture for cancer diagnosis and treatment.

Acknowledgements

The coauthors thank the members of the Key Laboratory of Molecular Diagnostics and Precision Medicine for Surgical Oncology in Gansu Province & NHC Key Laboratory of Diagnosis and Therapy of Gastrointestinal Tumor, Gansu Provincial Hospital, The First Affiliated Hospital of Shandong First Medical University, Hefei Institute of Physical Science in Chinese Academy of Sciences, Tianjin Key Laboratory of Engineering Technologies for Cell Pharmaceutical, and National Engineering Research Center of Cell Products for their technical support. This work was supported by grants from the project Youth Fund supported by Shandong Provincial Natural Science Foundation (ZR2020QC097), Fujian Provincial Ministerial Finance Special Project (2021XH018), Jiangxi Provincial Key New Product Incubation Program Funded by Technical Innovation Guidance Program of Shangrao City (2020G002), Science and technology projects of Guizhou Province (QKH-J-ZK[2021]-107), The 2021 Central-Guided Local Science and Technology Development Fund (ZYYDDFFZZJ-1), the Non-profit Central Research Institute Fund of Chinese Academy of Medical Sciences (2019PT320005), Medical Innovation Project of Fujian Provincial Health and Health Commission (2019-CX-21), Natural Science Foundation of Jiangxi Province (20212BAB216073), Key project funded by Department of Science and Technology of Shangrao City (2020AB002, 2020 K003, 2021F013), the Non-profit Central Research Institute Fund of Chinese Academy of Medical Sciences (2019PT320005), Gansu Key Laboratory of Molecular Diagnosis and Precision treatment of surgical tumors (18JR2RA033), and Horizontal Project upon Retrospective Analyses of COVID-19 (2020XH001).

Conflict of interest

The authors declare no conflict of interest.

Notes/thanks/other declarations

Not applicable.

Appendices and nomenclature

h-BN	hexagonal boron nitrides
BP	black phosphorus
TMDCs	palladium
Pd	Low-glucose DMEM with GlutaMax™ supplemented with 15% human serum
TMOs	transition metal oxides
EPR	enhanced permeability and retention
CT	computed tomography
MRI	magnetic resonance imaging
PET	positron emission tomography
PSMA	prostate-specific membrane antigen
HA	hyaluronic acid
TaOx	tantalum oxide
TaCZ	carboxybetaine zwitterionic-coated tantalum oxide
SPIO	superparamagnetic iron oxide
rGO	reduced graphene oxide
MNPs	magnetic nanoparticles
GCE	glassy carbon electrode
TME	tumor environment
TMDC	transition metal dichalcogenides
ROS	reactive oxygen species
PBM	photobiomodulation
(PDT)	photodynamic therapy
GSH	glutathione
PTT	photothermal therapy
irAEs	immune-related adverse events
CDT	chemodynamic therapy
PVP	polyvinylpyrrolidone
MSNs	mesoporous silica nanoparticles
MR	magnetic resonance (MR).

Author details

Hao Yu[1,2,3], Zhihai Han[4], Cunrong Chen[5]* and Leisheng Zhang[6,7,8,9]*

1 School of Medicine, Nankai University, China

2 National Engineering Research Center of Cell Products, AmCellGene Engineering Co., Ltd, China

3 Tianjin Key Laboratory of Engineering Technologies for Cell Pharmaceutical, China

4 Jiangxi Research Center of Stem Cell Engineering, Jiangxi Health-Biotech Stem Cell Technology Co., Ltd., China

5 Department of Critical Care Medicine, Union Hospital of Fujian Medical University, China

6 Key Laboratory of Molecular Diagnostics and Precision Medicine for Surgical Oncology in Gansu Province and NHC Key Laboratory of Diagnosis and Therapy of Gastrointestinal Tumor, Gansu Provincial Hospital, China

7 Key Laboratory of Radiation Technology and Biophysics, Hefei Institutes of Physical Science, Chinese Academy of Sciences, China

8 Institute of Health-Biotech, Health-Biotech (Tianjin) Stem Cell Research Institute Co., Ltd., China

9 Center for Cellular Therapies, The First Affiliated Hospital of Shandong First Medical University, China

*Address all correspondence to: 13705056799@139.com and leisheng_zhang@163.com

References

[1] Wang Y, Liu T, Li X, Sheng H, Ma X, Hao L. Ferroptosis-inducing nanomedicine for cancer therapy. Frontiers in Pharmacology. 2021;**12**:735965

[2] Siegel RL, Miller KD, Fuchs HE, Jemal A. Cancer statistics, 2021. CA: A Cancer Journal for Clinicians. 2021;**71**:7-33

[3] Zhang L, Liu M, Yang S, Wang J, Feng X, Han Z. Natural killer cells: Of-the-shelf cytotherapy for cancer immunosurveillance. American Journal of Cancer Research. 2021;**11**:1770-1791

[4] Depil S, Duchateau P, Grupp SA, Mufti G, Poirot L. 'Off-the-shelf' allogeneic CAR T cells: Development and challenges. Nature Reviews Drug Discovery. 2020;**19**:185-199

[5] Chen M, Hu S, Li Y, Jiang TT, Jin H, Feng L. Targeting nuclear acid-mediated immunity in cancer immune checkpoint inhibitor therapies. Signal Transduction and Targeted Therapy. 2020;**5**:270

[6] Sahin U, Oehm P, Derhovanessian E, Jabulowsky RA, Vormehr M, et al. An RNA vaccine drives immunity in checkpoint-inhibitor-treated melanoma. Nature. 2020;**585**:107-112

[7] Holyoake DLP, Smyth EC. Chemoradiotherapy or surgery for very early esophageal squamous cancer: Can a nonrandomized trial give us the answer? Gastroenterology. 2021;**161**:1793-1795

[8] Solaini L, Perna F, Cavaliere D, Vaccaro C, Avanzolini A, Cucchetti A, et al. Average treatment effect of robotic versus laparoscopic rectal surgery for rectal cancer. The International Journal of Medical Robotics. 2021;**17**:e2210

[9] Rallis KS, Lai Yau TH, Sideris M. Chemoradiotherapy in cancer treatment: Rationale and clinical applications. Anticancer Research. 2021;**41**:1-7

[10] He Y, Liu QW, Liao HX, Xu WW. Microbiota in cancer chemoradiotherapy resistance. Clinical and Translational Medicine. 2021;**11**:e250

[11] Conibear J, Astra Zeneca UKL. Rationale for concurrent chemoradiotherapy for patients with stage III non-small-cell lung cancer. British Journal of Cancer. 2020;**123**:10-17

[12] Basar R, Daher M, Rezvani K. Next-generation cell therapies: The emerging role of car-nk cells. Hematology American Society of Hematology Education Program. 2020;**2020**:570-578

[13] Daher M, Melo Garcia L, Li Y, Rezvani K. CAR-NK cells: The next wave of cellular therapy for cancer. Clinical & Translational Immunology. 2021;**10**:e1274

[14] Zhang L, Meng Y, Feng X, Han Z. CAR-NK cells for cancer immunotherapy: From bench to bedside. Biomarker Research. 2022;**10**:12

[15] Bordon Y. An RNA vaccine for advanced melanoma. Nature Reviews Immunology. 2020;**20**:517

[16] Rohatgi A, Kirkwood JM. Cancer vaccine induces potent T cell responses - but is it enough? Nature Reviews Clinical Oncology. 2020;**17**:721-722

[17] Sahin U, Derhovanessian E, Miller M, Kloke BP, Simon P, et al. Personalized RNA mutanome vaccines mobilize

poly-specific therapeutic immunity against cancer. Nature. 2017;**547**:222-226

[18] Liu E, Marin D, Banerjee P, Macapinlac HA, Thompson P, et al. Use of car-transduced natural killer cells in CD19-positive lymphoid tumors. New England Journal of Medicine. 2020;**382**:545-553

[19] Zhang Y, Li Y, Cao W, Wang F, Xie X, Li Y, et al. Single-cell analysis of target antigens of car-t reveals a potential landscape of "on-target, off-tumor toxicity". Frontiers in Immunology. 2021;**12**:799206

[20] Pan J, Tan Y, Wang G, Deng B, Ling Z, Song W, et al. Donor-derived CD7 chimeric antigen receptor t cells for t-cell acute lymphoblastic leukemia: First-in-human, phase I trial. Journal of Clinical Oncology. 2021;**39**:3340-3351

[21] Ning L, Zhu B, Gao T. Gold nanoparticles: Promising agent to improve the diagnosis and therapy of cancer. Current Drug Metabolism. 2017;**18**:1055-1067

[22] Nazir S, Hussain T, Ayub A, Rashid U, MacRobert AJ. Nanomaterials in combating cancer: Therapeutic applications and developments. Nanomedicine. 2014;**10**:19-34

[23] Liang P, Mao L, Dong Y, Zhao Z, Sun Q, Mazhar M, et al. Design and application of near-infrared nanomaterial-liposome hybrid nanocarriers for cancer photothermal therapy. Pharmaceutics. 2021;**13**:2070

[24] Huang X, Tang S, Mu X, Dai Y, Chen G, Zhou Z, et al. Freestanding palladium nanosheets with plasmonic and catalytic properties. Nature Nanotechnology. 2011;**6**:28-32

[25] Wyatt BC, Nemani SK, Anasori B. 2D transition metal carbides (MXenes) in metal and ceramic matrix composites. Nano Convergence. 2021;**8**:16

[26] Wu Z, Shang T, Deng Y, Tao Y, Yang QH. The assembly of mxenes from 2d to 3d. Advanced Science (Weinh). 2020;7:1903077

[27] Hong YL, Liu Z, Wang L, Zhou T, Ma W, Xu C, et al. Chemical vapor deposition of layered two-dimensional MoSi2N4 materials. Science. 2020;**369**:670-674

[28] Stater EP, Sonay AY, Hart C, Grimm J. The ancillary effects of nanoparticles and their implications for nanomedicine. Nature Nanotechnology. 2021;**16**:1180-1194

[29] Bayda S, Adeel M, Tuccinardi T, Cordani M, Rizzolio F. The history of nanoscience and nanotechnology: From chemical-physical applications to nanomedicine. Molecules. 2019;**25**:112

[30] Chiang CL, Cheng MH, Lin CH. From nanoparticles to cancer nanomedicine: Old problems with new solutions. Nanomaterials (Basel). 2021;**11**:1727

[31] Ang MJY, Chan SY, Goh YY, Luo Z, Lau JW, Liu X. Emerging strategies in developing multifunctional nanomaterials for cancer nanotheranostics. Advanced Drug Delivery Reviews. 2021;**178**:113907

[32] Tang L, Zhang A, Zhang Z, Zhao Q, Li J, Mei Y, et al. Multifunctional inorganic nanomaterials for cancer photoimmunotherapy. Cancer Communication (Lond). 2022;**42**:141-163

[33] Cheng Z, Li M, Dey R, Chen Y. Nanomaterials for cancer therapy: Current progress and perspectives. Journal of Hematology & Oncology. 2021;**14**:85

[34] Yang C, Lin ZI, Chen JA, Xu Z, Gu J, Law WC, et al. Organic/inorganic self-assembled hybrid nano-architectures for cancer therapy applications. Macromolecular Bioscience. 2022;**22**:e2100349

[35] Zhang M, Gao S, Yang D, Fang Y, Lin X, Jin X, et al. Influencing factors and strategies of enhancing nanoparticles into tumors in vivo. Acta Pharmaceutica Sinica B. 2021;**11**:2265-2285

[36] Huang D, Sun L, Huang L, Chen Y. Nanodrug delivery systems modulate tumor vessels to increase the enhanced permeability and retention effect. Journal of Personalized Medicine. 2021;**11**:124

[37] Siegel RL, Miller KD, Jemal A. Cancer statistics, 2020. CA-A Cancer Journal for Clinicians. 2020;**70**:7-30

[38] Chaturvedi VK, Singh A, Singh VK, Singh MP. Cancer nanotechnology: A new revolution for cancer diagnosis and therapy. Current Drug Metabolism. 2019;**20**:416-429

[39] Han X, Xu K, Taratula O, Farsad K. Applications of nanoparticles in biomedical imaging. Nanoscale. 2019;**11**:799-819

[40] Reuveni T, Motiei M, Romman Z, Popovtzer A, Popovtzer R. Targeted gold nanoparticles enable molecular CT imaging of cancer: An in vivo study. International Journal of Nanomedicine. 2011;**6**:2859-2864

[41] Luo D, Wang X, Zeng S, Ramamurthy G, Burda C, Basilion JP. Prostate-specific membrane antigen targeted gold nanoparticles for prostate cancer radiotherapy: Does size matter for targeted particles? Chemical Science. 2019;**10**:8119-8128

[42] Jin Y, Tang C, Tian J, Shao B. Integration of taox with bi2s3 for targeted multimodality breast cancer theranostics. Bioconjugate Chemistry. 2021;**32**:161-171

[43] Lambert JW, Sun Y, Stillson C, Li Z, Kumar R, Wang S, et al. An intravascular tantalum oxide-based ct contrast agent: Preclinical evaluation emulating overweight and obese patient size. Radiology. 2018;**289**:103-110

[44] Blakkisrud J, Londalen A, Dahle J, Martinsen AC, Kolstad A, Stokke C. Myelosuppression in patients treated with (177)lutetium-lilotomab satetraxetan can be predicted with absorbed dose to the red marrow as the only variable. Acta Oncologica. 2021;**60**:1481-1488

[45] Cheng FY, Su CH, Yang YS, Yeh CS, Tsai CY, et al. Characterization of aqueous dispersions of Fe(3)O(4) nanoparticles and their biomedical applications. Biomaterials. 2005;**26**:729-738

[46] Xu F, Li X, Chen H, Jian M, Sun Y, Liu G, et al. Synthesis of heteronanostructures for multimodality molecular imaging-guided photothermal therapy. Journal of Materials Chemistry B. 2020;**8**:10136-10145

[47] Zheng S, Jin S, Jiao M, Wang W, Zhou X, Xu J, et al. Tumor-targeted Gd-doped mesoporous Fe3O4 nanoparticles for T1/T2 MR imaging guided synergistic cancer therapy. Drug Delivery. 2021;**28**:787-799

[48] Ni D, Jiang D, Ehlerding EB, Huang P, Cai W. Radiolabeling silica-based nanoparticles via coordination chemistry: Basic principles, strategies, and applications. Accounts of Chemical Research. 2018;**51**:778-788

[49] Liu Q, Qian Y, Li P, Zhang S, Liu J, Sun X, et al. (131)I-labeled copper sulfide-loaded microspheres to treat hepatic tumors via hepatic artery embolization. Theranostics. 2018;**8**:785-799

[50] Garrigue P, Tang J, Ding L, Bouhlel A, Tintaru A, Laurini E, et al. Self-assembling supramolecular dendrimer nanosystem for PET imaging of tumors. Proceedings of the National Academy of Sciences of the United States of America. 2018;**115**:11454-11459

[51] Chen P, Wang H, Wu H, Zou P, Wang C, Liu X, et al. Intracellular synthesis of hybrid gallium-68 nanoparticle enhances micropet tumor imaging. Analytical Chemistry. 2021;**93**:6329-6334

[52] Xiong X, Dang W, Luo R, Long Y, Tong C, Yuan L, et al. A graphene-based fluorescent nanoprobe for simultaneous imaging of dual miRNAs in living cells. Talanta. 2021;**225**:121947

[53] Yao L, Zheng L, Cai G, Wang S, Wang L, Lin J. A rapid and sensitive salmonella biosensor based on viscoelastic inertial microfluidics. Sensors (Basel). 2020;**20**:2737

[54] Zhang L, Wang Y, Wang J, Shi J, Deng K, Fu W. rhEPO/EPO discrimination with ultrasensitive electrochemical biosensor based on sandwich-type nano-Au/ZnO sol-gel/nano-Au signal amplification. Biosensors & Bioelectronics. 2013;**50**:217-223

[55] Yang SJ, Huang CH, Wang CH, Shieh MJ, Chen KC. The synergistic effect of hyperthermia and chemotherapy in magnetite nanomedicine-based lung cancer treatment. International Journal of Nanomedicine. 2020;**15**:10331-10347

[56] Siegler EL, Zhu Y, Wang P, Yang L. Off-the-shelf CAR-NK cells for cancer immunotherapy. Cell Stem Cell. 2018;**23**:160-161

[57] Mullard A. Addressing cancer's grand challenges. Nature Reviews Drug Discovery. 2020;**19**:825-826

[58] Elkin EB, Bach PB. Cancer's next frontier: Addressing high and increasing costs. Journal of the American Medical Association. 2010;**303**:1086-1087

[59] Chilakamarthi U, Giribabu L. Photodynamic therapy: Past, present and future. Chemical Record. 2017;**17**:775-802

[60] Jarosz-Biej M, Smolarczyk R, Cichon T, Kulach N. Tumor microenvironment as a "game changer" in cancer radiotherapy. International Journal of Molecular Sciences. 2019;**20**:3212

[61] Murugan C, Sharma V, Murugan RK, Malaimegu G, Sundaramurthy A, et al. Two-dimensional cancer theranostic nanomaterials: Synthesis, surface functionalization and applications in photothermal therapy. Journal of Controlled Release. 2019;**299**:1-20

[62] Da Silva GH, Franqui LS, Petry R, Maia MT, Fonseca LC, Fazzio A, et al. Recent advances in immunosafety and nanoinformatics of two-dimensional materials applied to nano-imaging. Frontiers in Immunology. 2021;**12**:689519

[63] Karim R, Palazzo C, Evrard B, Piel G. Nanocarriers for the treatment of glioblastoma multiforme: Current state-of-the-art. Journal of Controlled Release. 2016;**227**:23-37

[64] Zhao M, van Straten D, Broekman MLD, Preat V, Schiffelers RM.

Nanocarrier-based drug combination therapy for glioblastoma. Theranostics. 2020;**10**:1355-1372

[65] Cheng L, Wang X, Gong F, Liu T, Liu Z. 2D nanomaterials for cancer theranostic applications. Advanced Materials. 2020;**32**:e1902333

[66] Zhou B, Guo Z, Lin Z, Jiang BP, Shen XC. Stimuli-responsive nanomaterials for smart tumor-specific phototherapeutics. Chem Med Chem. 2021;**16**:919-931

[67] Shen Z, Xia J, Ma Q, Zhu W, Gao Z, Han S, et al. Tumor microenvironment-triggered nanosystems as dual-relief tumor hypoxia immunomodulators for enhanced phototherapy. Theranostics. 2020;**10**:9132-9152

[68] Sundaram P, Abrahamse H. Phototherapy combined with carbon nanomaterials (1d and 2d) and their applications in cancer therapy. Materials (Basel). 2020;**13**:4830

[69] Antunes HS, Herchenhorn D, Small IA, Araujo CMM, Viegas CMP, de Assis RG, et al. Long-term survival of a randomized phase III trial of head and neck cancer patients receiving concurrent chemoradiation therapy with or without low-level laser therapy (LLLT) to prevent oral mucositis. Oral Oncology. 2017;**71**:11-15

[70] Fu X, Yin W, Shi D, Yang Y, He S, Hai J, et al. Shuttle-shape carrier-free platinum-coordinated nanoreactors with O2 self-supply and ros augment for enhanced phototherapy of hypoxic tumor. ACS Applied Materials & Interfaces. 2021;**13**:32690-32702

[71] Manivasagan P, Joe A, Han HW, Thambi T, Selvaraj M, Chidambaram K, et al. Recent advances in multifunctional nanomaterials for photothermal-enhanced Fenton-based chemodynamic tumor therapy. Materials Today Bio. 2022;**13**:100197

[72] Xu P, Liang F. Nanomaterial-based tumor photothermal immunotherapy. International Journal of Nanomedicine. 2020;**15**:9159-9180

[73] Darrigues E, Nima ZA, Griffin RJ, Anderson JM, Biris AS, Rodriguez A. 3D cultures for modeling nanomaterial-based photothermal therapy. Nanoscale Horizons. 2020;**5**:400-430

[74] Liu Y, Zhang X, Luo L, Li L, Zhu RY, et al. Gold-nanobranched-shell based drug vehicles with ultrahigh photothermal efficiency for chemo-photothermal therapy. Nanomedicine. 2019;**18**:303-314

[75] Lee SY, Shieh MJ. Platinum(ii) drug-loaded gold nanoshells for chemo-photothermal therapy in colorectal cancer. ACS Applied Materials & Interfaces. 2020;**12**:4254-4264

[76] Chatterjee DK, Fong LS, Zhang Y. Nanoparticles in photodynamic therapy: An emerging paradigm. Advanced Drug Delivery Reviews. 2008;**60**:1627-1637

[77] Zhang Q, Li L. Photodynamic combinational therapy in cancer treatment. Journal of BUON. 2018;**23**:561-567

[78] Kwiatkowski S, Knap B, Przystupski D, Saczko J, Kedzierska E, Knap-Czop K, et al. Photodynamic therapy - mechanisms, photosensitizers and combinations. Biomedicine & Pharmacotherapy. 2018;**106**:1098-1107

[79] Chen W, Zhang J. Using nanoparticles to enable simultaneous radiation and photodynamic therapies for cancer treatment. Journal of

Nanoscience and Nanotechnology. 2006;**6**:1159-1166

[80] Wang J, Sun J, Hu W, Wang Y, Chou T, Zhang B, et al. A porous au@rh bimetallic core-shell nanostructure as an h2o2-driven oxygenerator to alleviate tumor hypoxia for simultaneous bimodal imaging and enhanced photodynamic therapy. Advanced Materials. 2020;**32**:e2001862

[81] Bakalova R, Ohba H, Zhelev Z, Ishikawa M, Baba Y. Quantum dots as photosensitizers? Nature Biotechnology. 2004;**22**:1360-1361

[82] Kennedy LB, Salama AKS. A review of cancer immunotherapy toxicity. CA-A Cancer Journal for Clinicians. 2020;**70**:86-104

[83] Riley RS, June CH, Langer R, Mitchell MJ. Delivery technologies for cancer immunotherapy. Nature Reviews Drug Discovery. 2019;**18**:175-196

[84] Ma B, Bianco A. Recent advances in 2d material-mediated immuno-combined cancer therapy. Small. 2021;**17**:e2102557

[85] Wang Q, Su X, He Y, Wang M, Yang D, Zhang R, et al. CD11c participates in triggering acute graft-versus-host disease during bone marrow transplantation. Immunology. 2021;**164**:148-160

[86] Wang G, Sun X, Zuo S, Li C, Niu Q, Xia Y, et al. Homogeneously high expression of CD32b makes it a potential target for CAR-T therapy for chronic lymphocytic leukemia. Journal of Hematology & Oncology. 2021;**14**:149

[87] Shi Y, Lammers T. Combining nanomedicine and immunotherapy. Accounts of Chemical Research. 2019;**52**:1543-1554

[88] Gong N, Sheppard NC, Billingsley MM, June CH, Mitchell MJ. Nanomaterials for T-cell cancer immunotherapy. Nature Nanotechnology. 2021;**16**:25-36

[89] Liu Y, Crawford BM, Vo-Dinh T. Gold nanoparticles-mediated photothermal therapy and immunotherapy. Immunotherapy. 2018;**10**:1175-1188

[90] Harawaza K, Cousins B, Roach P, Fernandez A. Modification of the surface nanotopography of implant devices: A translational perspective. Materials Today Bio. 2021;**12**:100152

[91] Chang M, Hou Z, Wang M, Li C, Lin J. Recent advances in hyperthermia therapy-based synergistic immunotherapy. Advanced Materials. 2021;**33**:e2004788

[92] Pan J, Zuo S, Deng B, Xu X, Li C, Zheng Q, et al. Sequential CD19-22 CAR T therapy induces sustained remission in children with r/r B-ALL. Blood. 2020;**135**:387-391

[93] Anand S, Chan TA, Hasan T, Maytin EV. Current prospects for treatment of solid tumors via photodynamic, photothermal, or ionizing radiation therapies combined with immune checkpoint inhibition (a review). Pharmaceuticals (Basel). 10 May 2021;**14**(5):447

[94] Wang R, Su Q, Yin H, Wu D, Lv C, Yan Z. Inhibition of SRSF9 enhances the sensitivity of colorectal cancer to erastin-induced ferroptosis by reducing glutathione peroxidase 4 expression. International Journal of Biochemistry & Cell Biology. 2021;**134**:105948

[95] Lin H, Chen Y, Shi J. Insights into 2d mxenes for versatile biomedical applications: Current advances and challenges ahead. Advanced Science (Weinh). 2018;**5**:1800518

[96] Mun EJ, Babiker HM, Weinberg U, Kirson ED, Von Hoff DD. Tumor-treating fields: A fourth modality in cancer treatment. Clinical Cancer Research. 2018;**24**:266-275

[97] Tsimberidou AM, Fountzilas E, Nikanjam M, Kurzrock R. Review of precision cancer medicine: Evolution of the treatment paradigm. Cancer Treatment Reviews. 2020;**86**:102019

[98] Cabral H, Kinoh H, Kataoka K. Tumor-targeted nanomedicine for immunotherapy. Accounts of Chemical Research. 2020;**53**:2765-2776

[99] Liu Z, Lin H, Zhao M, Dai C, Zhang S, Peng W, et al. 2D superparamagnetic tantalum carbide composite mxenes for efficient breast-cancer theranostics. Theranostics. 2018;**8**:1648-1664

[100] Ye Z, Sang T, Li K, Fischer NG, Mutreja I, Echeverria C, et al. Hybrid nanocoatings of self-assembled organic-inorganic amphiphiles for prevention of implant infections. Acta Biomaterialia. 1 Mar 2022;**140**:338-349

Chapter 10

Biosensor for the Detection of Cyanobacterial Toxin Microcystin-LR

Rasmus Rohtla, Kairi Kivirand, Eerik Jõgi and Toonika Rinken

Abstract

Cyanobacteria are found everywhere in the environment, and their growth accelerates significantly with rising amounts of sunlight and temperatures. The proliferation of cyanobacteria begins when the average temperatures rise above 15°C. The proliferation can lead to high amounts of secondary metabolites, such as cyanotoxins, in surrounding waters. The most common cyanotoxin is microcystin-LR (MC-LR). MC-LR can cause rashes, abdominal cramps, and liver damage in humans and animals, so continuous monitoring of its content in water is of great importance. MC-LR is commonly detected with high-performance liquid chromatography, but phosphatase inhibition-based bioassays and enzyme-linked immunosorbent tests are also available. However, these are all lab-based methods and require sample transport and preparation for analytical procedures, not allowing for obtaining quick results. Therefore, there is a need for a rapid and field-based analysis method, and one promising option is to use biosensors. The present study aimed to design and construct an aptamer/antibody-based biosensor to detect MC-LR and test its applicability to detect MC-LR in cyanobacteria culture (*Microcystis aeruginosa*).

Keywords: cyanobacteria, microcystin-LR, biosensor, aptamer, antibody, field-based method, environment, monitoring

1. Introduction

Cyanotoxins are metabolites produced by cyanobacteria, a group of photosynthetic prokaryotes found in freshwater. The intake of contaminated water, skin contact, or swallowing water during swimming are among the most common reasons for poisoning caused by cyanotoxins [1]. An increase in temperature causes the cyanobacteria to grow faster. In addition, the spread of cyanobacteria is also affected by the pH of the environment, salinity, the presence of necessary nutrients (e.g., nitrogen and phosphorus), and light. It has been observed that global warming may increase the frequency and extent of cyanobacterial proliferation [2, 3]. Cyanobacteria can release toxins into the environment during the mass spread of microorganisms, that is, the water blooming.

1.1 Cyanotoxins

Microcystins are the most widespread cyanobacterial toxins produced by *Microcystis aeruginosa* in freshwater lakes and rivers worldwide [4]. Microcystins are hepatotoxins that substantially affect serine/threonine protein phosphatases (PPs), which can remove phosphate from the protein in many biochemical pathways [5]. They are cyclic heptapeptides with a molecular weight of 800–1100 Da, and more than 250 different microcystins have been described [6]. The general structure of microcystins is cyclo-D-Ala^1-X^2-D-MeAsp^3-Z^4-Adda^5-D-Glu^6-Mdha^7 (superscript number indicates the position number, **Figure 1**), where X and Z are variable L-amino acids, D-MeAsp is D-erythro-β-methylaspartic acid, Adda is 3-amino-9-methoxy-2,6,8-trimethyl-10-phenyldeca-4,6-dienoic acid, and Mdha is N-methyldehydroalanine [5, 7]. Structure variations occur in all seven amino acid residues; most common are the replacement of L-amino acids in positions 2 and 4, replacement of Mdha by dehydrobutyrine (Dhb) or by serine in position 7, and a lack of methylation of amino acids in positions 3 and/or 7 [7]. The variations in Adda are essential because they may affect analytical test results, which use Adda as a marker, and in addition, Adda moiety is critical to microcystin activity [5]. The hydrophobicity of the amino acids at positions two and four influences the overall hydrophobicity. Hydrophobicity of the microcystin congener determines how the toxin interacts with cell membranes, and therefore, affects its specific toxicity [8].

The most common and most toxic is microcystin-LR (MC-LR), with leucine (Leu = L) in the second position and arginine (Arg = R) in the fourth position (**Figure 1**). Modeling of the MC-LR molecule has shown that the alanine and leucine residues in positions 1 and 2 extend beyond the ring plane; thus, allow selective binding to receptor molecules, which causes high toxicity of MC-LR and another metabolite, a cyclic non-ribosomal pentapeptide nodularin [7]. Also, three-dimensional structure studies of the MC-LR have shown that Adda and Arg side chains protrude from the ring distal from one another caused by the repulsion between the guanidino function of Arg and the hydrophobic Adda [9]. Microcystin-LR inhibits protein phosphatase type 1 and type 2A (PP1 and PP2A) activities in the cytoplasm of liver cells [10, 11], which leads to an increase in the phosphorylation of proteins in liver cells. The Adda

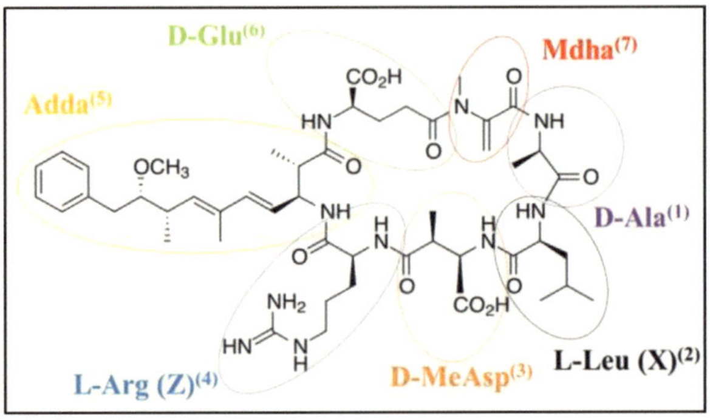

Figure 1.
Generic structure of microcystin-LR: Superscript numbers indicate the position numbers, and X and Z are the variable L-amino acids in different microcystins. The two specific L-amino acids of MC-LR are shown in black (leucine, L) and blue (arginine, R). Abbreviations: Ala is alanine; Leu is leucine; MeAsp is erythro-β-methylaspartic acid; Arg is arginine; Adda is 3-amino-9-methoxy-2,6,8-trimethyl-10-phenyldeca-4,6-dienoic acid; Glu is glutamic acid, and Mdha is N-methyldehydroalanine.

side-chain is accommodated to the hydrophobic channel [7]. The carboxylic D-Glu site makes hydrogen bonds to metal-bound water molecules [7]. The carboxyl group of the MeAsp site makes hydrogen bonds to conserved arginine and tyrosine residues in the PPP enzyme [7]. Finally, the methylene group at the Mdha site binds to an S-atom of a cysteine residue, and the leucine residue folds closely to another well-maintained tyrosine residue [12].

The degradation of microcystins is slow in most water environments [13]. Most mycotoxins are heat-resistant [14], and the water treatment process cannot altogether remove them. Still, they can be degraded, when using UV treatment close to their absorption peak (UV lambda max for microcystin-LR is 238 nm) [4, 15]. Due to carboxyl, amino, and acylamino groups in the structure, mycotoxins have different ionization propensities at different pH values.

Limits for MC-LR in natural waters have been set in only a few countries. In Hungary and some US states, such as Indiana and New York, the MC-LR limit in water is 4 μg/l [16]. The World Health Organization (WHO) has set a limit of 1 μg/l for MC-LR in drinking water [6, 17].

1.2 Methods to detect cyanotoxins

For the detection of cyanotoxins in water, the following methodologies are used: high-performance liquid chromatography (HPLC) combined with mass spectrometry (MS, MS/MS) or ultraviolet/photodiode array detectors (UV/PDA), enzyme-linked immunosorbent assays (ELISA), and protein phosphatase inhibition assay (PPIA).

HPLC is a selective and sensitive method that allows the simultaneous determination of different microcystins at very low concentrations (0.02 μg/l). Still, the determination is technically complex and time-consuming, and the cost of the apparatus and analysis is high [18]. In addition, pretreatment of samples is required [19]. Chromatographic methods do not allow on-site monitoring, and given the need to transport samples, results can be obtained in a minimum of 4–6 hours [19]. It is also important to consider matrix effects in chromatographic analysis, and prior calibration with the matrix is required [20].

ELISA and PPIA are the other technologies often used to detect microcystins. ELISA assays are based on antigen-antibody interactions, and analytes are detected by the color change resulting from the reaction. The ELISA assay is highly sensitive and relatively straightforward [19]. For commercial ELISA rapid tests, the limit of determination for microcystins is 0.06 μg/l, and the test time varies between 4 and 6 h [21]. A significant disadvantage of many commercial ELISAs is that they are based on anti-Adda antibodies and do not measure the specific microcystin, but the total microcystin and nodularin content [22] and there is cross-reactivity [20]. ELISA tests based on a monoclonal antibody against arginine at position 4 limit detection of as low as 0.002–0.006 μg/l [18, 23].

PPIA allows to perform assays efficiently and quickly (approx. 2 h), and is based on a protein phosphatase-catalyzed protein dephosphorylation reaction in which the presence of a chromogenic substrate (e.g., p-nitrophenol phosphate) releases p-nitrophenol, which is detected at 410 nm [24]. The enzymes used, such as protein phosphatase 1 (PP1), are readily available, and this method has a medium sensitivity of 0.1 μg/l for MC-LR [25]. The main disadvantage of PPIA is the low selectivity because cyanobacteria contain phosphatases, and it is impossible to identify different microcystins [26]. The interaction of microcystins with PP1 is thought to be related to non-coding amino acid residues: Adda, D-Glu, and Mdha at positions 5, 6, and 7

Methods	LOD (μg/l)	Analysis time (h)	Pretreatment of the sample	On-site possibility	Approximate cost	Ref.
Chromatography methods (HPLC-UV ja LC-MS/MS)	0.02	4–6	+	—	100–200€	[18, 20]
ELISA	0.002	4	+/–	+		[18]
Commercial ELISA tests	0.06	5–6	+/–	+	50€	[21]
PPIA	0.1	2	+	—	20–50€	[25]

LOD: limit of detection; HPLC-UV: high-performance liquid chromatography with UV detector; LC-MS/MS: liquid chromatography combined with mass spectrometry; ELISA: enzyme-linked immunosorbent assay; and PPIA: protein phosphatase inhibitors.

Table 1.
Overview of MC-LR determination methods.

of the microcystin molecules, respectively [27]. The essential analytical parameters characterizing the above-described microcystin determination methods are summarized in **Table 1**.

Each technique has some limitations in sensitivity, reliability, detection limit, or speed and cost. The selection of a suitable method is based on the information they provide and the technical expertise needed. The cost of analytical equipment, long-lasting measurements, and the need for qualified personnel to perform the analysis are a challenge for routine monitoring. Nowadays, methods suitable for the end-user that can be validated and accepted worldwide continue to be an objective for regulators and the industry. The variety of commercially available assays or testing kits for marine toxin analysis remains limited. The list of currently available point-on-site marine toxin end-product testing technologies is provided in ref. [28].

1.3 MC-LR biosensors

Cyanotoxins are not monitored regularly in most countries due to technical complications in the detection and quantification. Biosensors for freshwater monitoring and safety applications are prospective alternatives to traditional methods. Biosensors are analytical devices that include a bio-recognition element linked to a transducer that transforms the chemical information produced into a readable signal followed by a detector. Most biosensors used to detect MC-LR are immunosensors that use antibodies or aptamers to recognize the analyte. The detection limit of biosensors ranges from 0.00003 to 0.37 μg/l, and the assay time is from 0.8 to 2.3 hours [29–34]. The most important parameters characterizing the biosensors used to determine microcystins are summarized in **Table 2**. High selectivity of bio-recognition is assured by using specific antibodies or aptamers.

1.3.1 Bio-recognition elements

Antibodies or immunoglobulins (Ig) are glycoproteins used in nature to detect and neutralize foreign objects. They have a characteristic basic structure consisting

Methods	LOD (μg/l)	Analysis time (h)	Pretreatment of the sample	On-site possibility	Ref.
Aptasensor	0.37	0.7	Yes	+	[29]
Aptasensor (fluorescence)	0.002	~1	Not needed	+/−	[30]
Aptasensor (Raman)	0.002	2.3	Not needed	+/−	[31]
Aptasensor (square wave voltammetry)	0.002	~1	Yes	—	[32]
Sensor (differential cyclic voltammetry)	0.0003	~1	Not needed	—	[34]

LOD: limit of detection.

Table 2.
An overview of MC-LR biosensors.

of a protein chain linked by a disulfide bridge and a very high affinity for the antigen detected, described by the dissociation constant of the antigen/antibody complex. The values of this constant are usually between 10^{-12} and 10^{-8} M [35]. The MC-LR monoclonal antibody used in the present work has an affinity toward myco-toxins, which have arginine in the 4th position, with the dissociation constant of $1.4 \cdot 10^{-11}$ M [36].

Aptamers are synthetic single-stranded oligonucleotides capable of binding various molecules with high affinity and specificity. Aptamers are considered artificial antibodies and can adapt through intermolecular interactions [37, 38]. Nevertheless, compared with antibodies, they are more stable. When interacting with its target, the "lock key" is formed by matching the spatial conformation with the aptamer molecules [39]. Aptamers are produced using SELEX (systematic evolution of ligands by exponential enrichment), and once the aptamer sequence is developed, it can be reproduced with high precision. The characteristics of the aptamers selected for microcystin, identifying modifications at 3′ or 5′ ends to label or link the aptamers to the sensor platform, and their affinity to the target toxin are summarized in refs. [40, 41]. The MC-LR aptamer used in the present work, AN6 (5′ GGC GCC AAA CAG GAC CAC CAT GAC AAT TAC CCA TAC CAC CTC ATT ATG CCC CAT CTC CGC 3′), is a microcystin-LR specific aptamer with the affinity (K_d) of 5×10^{-8} M. It can also bind to microcystin-LA but with 3-fold reduced affinity (approx. $15.8 \cdot 10^{-8}$ M), and no binding to microcystin-YR has been observed [40, 41]. AN6 is a synthetic 60-base DNA aptamer with a molecular weight of 18167.79 Da.

1.4 Bead injection analysis

It is promising to use measurements in analyte-containing flows for continuous monitoring of toxins for continuous monitoring of on-site analyses. One option for designing in-flow sensor systems is to use the principle of BIA (Bead Injection Analysis) [42, 43]. This microgranule insertion assay uses microgranule transport in a flowing solution to form microcolumns required for the assay. A selective component recognizing an analyte is immobilized on the surface of the microcolumn-forming granules allowing it to pre-concentrate and bind the targeted compound. After removing the sample matrix, selective detection of the bound analyte occurs, for example, using an antigen/antibody interaction. The signal of the recognition

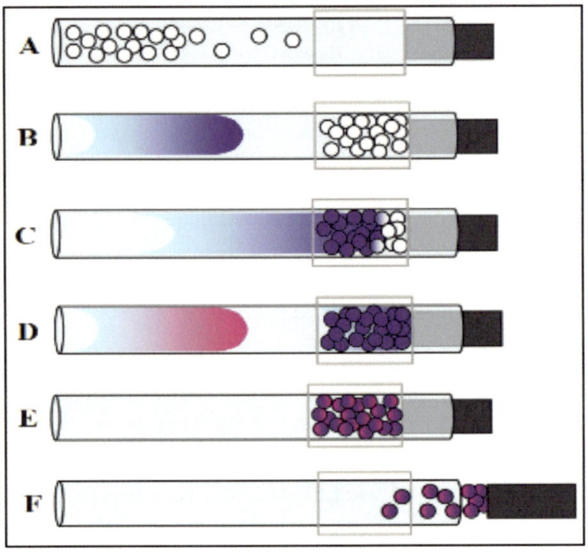

Figure 2.
Working principle of bead insertion analysis (BIA) . [44]: (A) injection of bio-activated granules into a flow channel to form a micro-column. (B) Sample injection (the sample binds to the activated granules). (C) Washing of the column to remove unbound sample components and matrix. (D) Labeled bio-component injection, incubation, and washing out of the unbound components. (E) Signal detection. (F) System regeneration.

reaction is detected spectrometrically or by measuring the fluorescence signal. The scheme of BIA operation is shown in **Figure 2** [44].

The amount of activated granules required to form a microcolumn is small (approx. 20 μg), which allows the assay to be a single-use one. This technique eliminates the need to regenerate the bio-recognition system, the risk of contamination, and the risk of denaturation of the bio-component bound to the granules. It allows operation in a continuous flow system. To ensure the reliability and accuracy of the results obtained, it is also essential to ensure a consistently high quality of the bio-activation of the granules [45].

2. Experimental part

2.1 Materials

Solutions of aptamer (5' GGC GCC AAA CAG GAC CAC CAT GAC AAT TAC CCA TAC CAC CTC ATT ATG CCC CAT CTC CGC 3′ (AN6, Integrated DNA Technologies)) and microcystin (Enzo Life Sciences, ALX-350-012-C100) were prepared in 0.01 M phosphate buffer saline (0.15 M NaCl, pH 7.2, PBS). Monoclonal MC-LR antibody (Enzo Life Sciences, MC10E7) solution was prepared in 0.5 M carbonate buffer (pH 9.5). The solutions were stored at 4°C. Epichlorohydrin was from Acros Organics (A0386058) and Coomassie brilliant blue R-250 was from Fluka AG (99%, CH-9470). Sephadex G-50 medium granules were from Pharmacia Fine Chemicals (FB-14567). All other chemicals used were at analytical grade. Buffer solutions were prepared using ultrapure MilliQ water (specific resistance 18.2 MΩ·cm).

2.2 Preparation of activated microgranules

Epichlorohydrin (Acros Organics, A0386058) was used to activate the Sephadex G-50 medium granules (Pharmacia Fine Chemicals, FB-14567). The antibody was covalently attached to the epoxy carbon of the epichlorohydrin via the amino group, using a previously published protocol with minor modifications was used [46]. First, 47 mg of granules were allowed to swell overnight at 4°C in 1 ml of water. After swelling, 400 μl of NaOH solution (concentration 0.1 M - 1 M) and 100 μl of epichlorohydrin to activate the granules were added and left on a shaker for 3 hours at room temperature. The beads were then washed twice with water and once with 0.5 M carbonate buffer (pH 9.5). The mixture was centrifuged at 2450 x g for 5 min after each washing step to separate beads. 0.5 to 2 ml of MC-LR antibody in 0.5 M carbonate buffer (pH 9.5) was added, with the antibody concentration varying from 10 to 350 μg/ml (the total amount). The mixture was incubated on a shaker for 24 h at room temperature. After incubation, the suspension was centrifuged (5 min at 2450 x g) and washed once with 0.5 M carbonate buffer (pH 9.5). To block free binding sites on the surface of the granules, ethanolamine solution (85 μl/ml in 0.5 M carbonate buffer) was added. The mixture was incubated on a shaker for 2 hours at room temperature. The suspension was centrifuged (5 min at 2450 x g), washed twice with water and several times with PBS buffer, and stored in PBS buffer at 4°C.

The yield of the attached antibody on the granules was evaluated with two different methods. First, it was visually inspected by adding 0.1% Coomassie brilliant blue R-250 (99% (Fluka AG, CH-9470)) to 30 μl of granules before adding ethanolamine. It was assessed by whether the granules turned blue, indicating the presence of bound protein on the granules. In addition, the protein content in the antibody solution was spectrophotometrically evaluated before and after the antibody attachment process. The protein content of the samples was determined at 280 nm, and the concentration was calculated using an absorption coefficient of 1.37 for IgG ($\varepsilon^{1\%}$).

2.3 Carrying out measurements with biosensor

The outflow channel of the BIA system was partially sealed with a moving cap, and 20 μl of bio-activated microgranules were injected into the measuring cell at a flow rate of 1 μl/sec to form a microcolumn. 30 μl of PBS buffer was added at a flow rate of 2 μl/sec to ensure the column's packing. A sample containing 150 μl of MC-LR was added at a flow rate of 1 μl/sec, the flow was stopped, and the system was incubated for 30 min. The measuring cell was washed with 150 μl of PBS at a flow rate of 2 μl/sec to remove the unbound toxin. Then 30 μl of MC-LR aptamer labeled with a fluorescence marker (Alexa Flour 647) was added at a flow rate of 1 μl/sec and incubated for 30 min. The concentration of the marker varied from 0.5 to 5 μg/ml. The unbound aptamer was removed from the microcolumn by adding 350 μl of PBS at a flow rate of 1 μl/sec. After each measurement, the cap was opened, and the system was washed at least four times with PBS buffer. PBS buffer with no added MC-LR was used for experimental determination of the system's background signal (all other measurement steps were left unchanged).

All measurements were performed in triplicate. Measurements were performed at room temperature. The fluorescence intensity was measured at 670 nm (excitation wavelength 650 nm) of the Alexa Flour 647 emission peak perpendicular to the excitation light. To calculate the signal change, the signal after washing off the unbound

MC-LR was subtracted from the final signal (signal after washing off unbound aptamer, recording started 5 min after completion of aptamer wash). The signal was recorded at 1-sec intervals. After stabilization, the mean signal was calculated as an average of 100 points to reduce experimental noise.

2.4 Cultivation and preparation of cyanobacteria sample

To cultivate *Microcystis aeruginosa* (Norwegian Culture Collection of Algae, K-0540) cells, 1 ml of culture was inoculated into 50 ml of liquid sterilized BG11+ medium, and grown under artificial light for 14 days at 16°C [47]. A LED lamp (16 h white/8 h dark, 6 W 3000 K) kept at a distance of 20 cm from the culture vessel was used as a light source. The cells were stored at -20°C.

After thawing, the samples were concentrated. Repeated centrifugation (5 min at 10000 x g) reduced the sample volume five times. An ultrasonic probe sonicator (Bandelin HD 2020 Sonopuls, horn ⌀ 3 mm) was used to disrupt the cyanobacterial cells for 1 min at a cycle intensity of 7/10 and a power of 75%.

2.5 The characterization of the formation of MC-LR complexes with size-exclusion chromatography (SEC)

The formation of MC-LR complexes with antibody/aptamer was studied with an ÄKTA Purifier 10 liquid chromatography system (GE Healthcare) equipped with a UPC detector (280 nm) and a conductivity detector. The column (height 29 cm and diameter 1 cm) was loaded with Sephacryl S-200 HR (GE Healthcare, product ID: 10090795) gel pre-expanded overnight at room temperature in PBS buffer (pH 7.2) and packed under pressure to ensure the high quality of packing. For analyses, we optimized the flow rate (0.18 to 0.39 ml/min), sample volume (50 and 100 µl), and sample concentration (0.05 to 1 mg/ml). The column was calibrated with different proteins with molar weights ranging from 20 to 240 kDa. Dextran blue (2000 kDa) and potassium dichromate (294 Da) were used to determine the column void volume and total volume. All optimizations, calibrations, and measurements were performed at 8°C.

The experiments were performed at an optimum flow rate of 0.18 ml/min, a sample volume of 50 µl, and the column was flushed with 70 ml of PBS buffer (pH 7.2). The aptamer/microcystin mixture was prepared, the aptamer was incubated with MC-LR for 30 min (1:1 molar ratio). To prepare the aptamer/microcystin/antibody mixture, the antibody was incubated with microcystin for 30 minutes and re-incubated for another 30 minutes with the aptamer (1:2:43 molar ratio).

3. Results and discussion

3.1 Aptasensor design

In the MC-LR aptasensor, a sandwich system consisting of an antibody, MC-LR, and an aptamer was used. MC-LR molecule is relatively small compared to the antibody and aptamer molecules (molecular weights 0.995, 150, and 18.17 kDa). To assure effective binding, the binding sites of the antibody and the aptamer to the MC-LR molecule should be different. According to the manufacturer, the MC-LR monoclonal antibody binds to the MC-LR molecule in position two [23]. The AN6 aptamer is used to detect the bound MC-LR molecules and was selected to bind to another

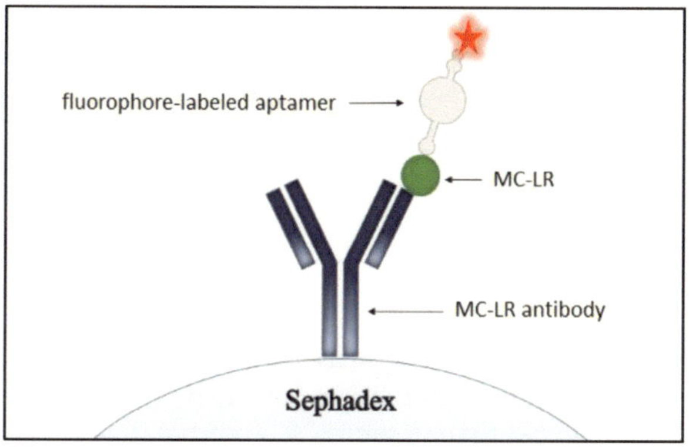

Figure 3.
MC-LR aptasensor.

characteristic amino acid residue in position four [41]. In addition to the amino acid residues in positions two and four in the MC-LR molecule, the ADDA residue in position five can be used for interactions. Still, its use would reduce the selectivity of the biosensor because ADDA is present in all microcystins and nodularin. The MC-LR determination scheme is shown in **Figure 3**.

The MC-LR Sephadex G-50 M granules were activated with MC10E7 monoclonal antibody for selective binding. The granules were chosen according to their size, to enable the collection of granules into a closed measuring channel, and the diameter of the granules must be >80 μm (Sephadex G-50 M is approximately 100–300 μm in a buffer solution) [45].

During immobilization, 44 ± 0.1% of the antibody in solution adhered to the surface of the microgranules. Visual inspection revealed that the granules turned light blue after mixing them with Coomassie brilliant blue, indicating the presence of bound antibodies on the granules. Coomassie brilliant blue stain is a widely used method for routine visualization of proteins because it makes complexes with essential amino acids, such as lysine, histidine, tyrosine, and arginine [48]. The efficiency of the immobilization process did not depend on the concentration of antibodies in the immobilization solution.

3.2 The formation of a detectable antibody/MC-LR/aptamer complex

The complex components have a significant molecular weight difference: MC-LR molecule 995 Da; the antibody and the aptamer of 150 kDa and 18.17 kDa, respectively. Considering these significant differences in molecular size, the potential formation of an antibody/MC-LR/aptamer triple complex was studied using SEC. The optimization was needed to achieve a sufficient resolution: flow rate, sample volume, and analyte concentration in a sample were modified. The best resolution over a significant range of molecular weights (100–200 kDa) was achieved at a flow rate of 0.18 ml/min, and a sample concentration of 0.5 mg/ml. Changing the sample volume did not significantly affect the resolution of the peaks, so 50 μl was chosen. These optimal conditions were used for all experiments. Dextran blue (M = 2000 kDa) was used to determine column void volume, and the total volume of the column was obtained with potassium dichromate. Individual compounds were analyzed and compared to investigate the formation of possible complexes. The chromatograms of

MC-LR monoclonal antibody, aptamer AN6, and microcystin alone were compared to chromatograms of component mixtures, which were incubated before analysis in different modes:

1. Aptamer solution was incubated with microcystin for 30 min before injection;

2. Antibody was incubated with microcystin for 30 min, and then the aptamer was added and incubated again for 30 min before injecting.

As SEC separates particles according to their size, several peaks were obtained from the various spatial structures of the aptamer AN6 [49], which moved through the column significantly faster than expected. The chromatogram of the aptamer showed two clear peaks at flow volumes of 8.10 ml and 11.50 ml (± 4%). For the antibody, it was also possible to identify two characteristic peaks at flow rates of 8.40 ml and 9.99 ml (± 4%).

Due to the aptamer AN6, it is impossible to characterize the chromatograms of mixtures by molecular weights; instead, the shape, intensity, and area of the peaks were compared. There were no differences in incubating the aptamer with MC-LR (mode 1) compared to the chromatogram of aptamer alone. The chromatogram of the mixture of the antibody, MC-LR, and the aptamer prepared according to mode 2 showed that there was no peak with an elution volume of 8.1 ml, and the intensity of the peak with an elution volume of 11.5 ml was increased (**Figure 4**), which may indicate interactions between different components. Comparing the areas under peaks for aptamer, antibody, and antibody/MC-LR/aptamer solutions, where the amount of material injected into the column was similar, the difference was less than

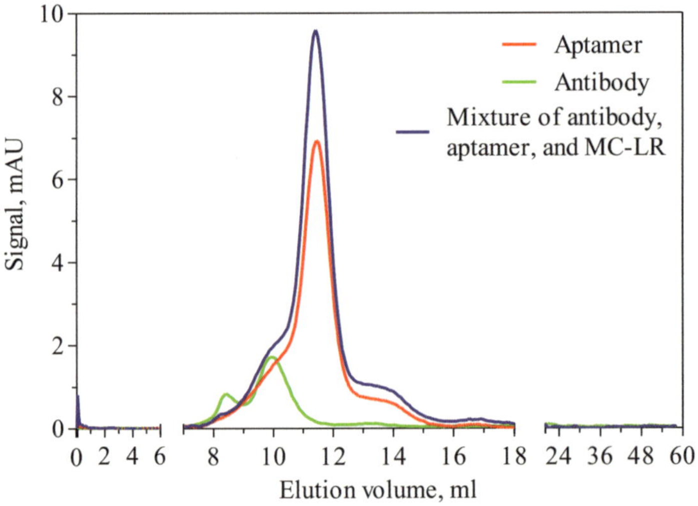

Figure 4.
Aptamer (red line), antibody (green line), and antibody, MC-LR and aptamer mixture (blue line) chromatograms (Sephacryl S-200 HR (1/29) column). Aptamer and antibody concentrations were 0.06 mg/ml and 0.24 mg/ml, respectively. In the mixture of antibody/MC-LR/aptamer the toxins concentration was 0.07 mg/ml (aptamers and antibody concentrations 0.06 mg/ml and 0.24 mg/ml, respectively). The flow rate was 0.18 ml/min, sample size of 50 μl.

one unit (approximately 6%), indicating that all substances injected into the column had passed through the column, and nothing was stuck into the column.

3.3 Determination of MC-LR

3.3.1 Optimization of the protocol

Measurements were performed using a protocol for detecting pathogens with a BIA-based sensor with some modifications [45]. To achieve a low limit of quantification, both antibody/MC-LR (MC-LR binding to activated microgranules) and MC-LR/aptamer (MC-LR binding to aptamer) incubation times were 30 minutes as the incubation at 15 minutes was not sufficient to obtain a reliable signal below the established WHO limit of 1 μg/l [6, 16]. The minimum volume of PBS for the efficient removal of unbound aptamer was 350 μl, as with lower PBS amounts, some of the aptamer remained in the flow channel and caused unstable signals (signal increased by hundreds of units in 5 minutes). We also optimized the aptamer concentration from 0.5 μg/ml to 5.0 μg/ml. With higher aptamer concentration, no stable end signal was achieved within 5 minutes. These results show aptamer concentration of 0.5 μg/ml was used in further experiments. The optimal protocol used for the determination of MC-LR was as follows:

- 20 μl of bio-activated gel was injected at a flow rate of 1 μl/s;
- 30 μl of PBS buffer was added at a flow rate of 2 μl/s;
- 150 μl of MC-LR sample was injected at a flow rate of 1 μl/s;
- MC-LR sample was incubated for 30 minutes to secure the attachment of MC-LR to the granules;
- The flow cell was washed with 150 μl PBS buffer at a flow rate of 2 μl/s;
- 30 μl of Alexa Flour 647-labeled aptamer AN6 was injected at a flow rate of 1 μl/s;
- The aptamer was incubated for 30 minutes to secure the attachment of the aptamer to the bound MC-LR;
- The flow cell was washed with 350 μl PBS buffer at a flow rate of 1 μl/s;
- The aptasensor signal was measured after its stabilization in 5 minutes.

3.3.2 The calibration of MC-LR biosensor

A calibration graph was plotted to characterize the sensitivity and operating range of the aptasensor (**Figure 5**).

The results showed that the signal of the aptasensor was linearly dependent on the concentration of MC-LR over a relatively wide concentration range from $1.3{\cdot}10^{-7}$ to $8.0{\cdot}10^{-4}$ mg/ml, and the experimental errors in this range were relatively minor from 0.8 to 4.3 AU. The coefficient of determination (R^2) of the graph was 0.97. The sensitivity of

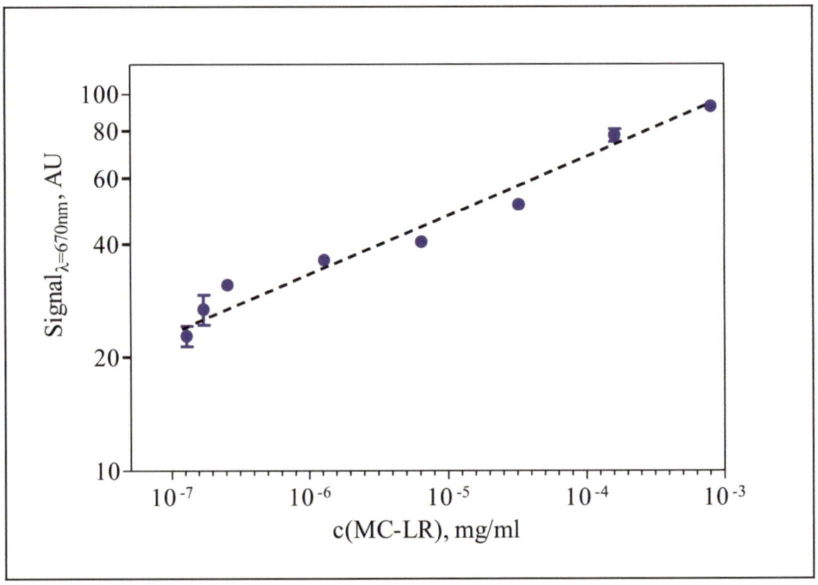

Figure 5.
The dependence of aptasensors signal on MC-LR concentration.

the aptasensor was characterized by the slope of the graph being 0.151 ± 0.006 logAU/log(mg/ml). The background signal was measured using PBS without added MC-LR, and the background value was 4.6 ± 0.9 AU. The theoretical detection limit (LOD) of the MC-LR aptasensor was calculated as the background signal + three standard deviations of the background signal, and the limit of quantification (LOQ) as the background signal +10 standard deviations of the background signal. The LOD and LOQ values for the aptasensor were $1.7 \cdot 10^{-8}$ mg/ml and $3.4 \cdot 10^{-8}$ mg/ml, respectively.

Comparing the LOD of the MC-LR aptasensor with the allowed limit of MC-LR in drinking water, established by the WHO (1 µg/l = 10^{-6} mg/ml) [50], the LOD value of the proposed aptasensor is significantly lower, serving as a good precondition for the application of this aptasensor for the determination of MC-LR content and monitoring quality of natural water bodies.

The analysis took approximately 75 minutes, of which 60 minutes were for the analyte binding and the formation of a detectable complex. Compared to laboratory-based methods for detecting MC-LR, typically taking 4–6 hours [18–20], this method allows the determination of cyanotoxins much faster. However, compared to other aptasensors, the results can be obtained within a longer time. The time of analysis can be reduced by reducing the incubation time. It is also interesting to mention that the average material cost for one measurement was estimated to be 4.4 €.

3.3.3 Testing of the aptasensor

A cyanobacterial culture was used to test the performance of the designed aptasensor. The sample of the cultured cyanobacteria treated with ultrasound to break up the bacterial cells was diluted 50 times as it contained broken blue-green algae cells, and the solution had green color. The MC-LR concentration in the diluted culture sample was $3.0 \cdot 10^{-7}$ mg/ml, indicating that the designed aptasensor was sensitive enough in the matrix, assumingly more complex than the one of natural water.

4. Conclusions

An aptasensor was designed and constructed to detect the cyanobacterial toxin MC-LR. The aptasensor was integrated with a bead injection system, where bio-activated micro-granules formed a disposable microcolumn in a partially closed flow channel. It took about 75 minutes to determine MC-LR. The aptasensor's detection limit (LOD) was $1.7 \cdot 10^{-8}$ mg/ml, and the limit of quantitation (LOQ) was $3.4 \cdot 10^{-8}$ mg/ml. The LOD and LOQ values of the aptasensor were below the allowed MC-LR limit of 1 μg/l in drinking water set by WHO. The MC-LR aptasensor was used for testing the MC-LR concentration in a cyanobacterial culture. The sensitivity of the aptasensor is sufficient to determine MC-LR in samples containing algae, which creates good conditions for using the constructed aptasensor in natural water bodies for water quality monitoring.

Conflict of interest

The authors have no conflicts of interest to declare. We certify that the submission is original work and is not under review at any other publication.

Author details

Rasmus Rohtla[1], Kairi Kivirand[1], Eerik Jõgi[1,2] and Toonika Rinken[1*]

1 Faculty of Science and Technology, Institute of Chemistry, University of Tartu, Estonia

2 Tartu Health Care College, Estonia

*Address all correspondence to: toonika.rinken@ut.ee

References

[1] U.S. Environmental Protection Agency. Health Effects from Cyanotoxins: Human Health Effects Caused by the Most Common Toxin-producing Cyanobacteria. https://www.epa.gov/cyanohabs/health-effects-cyanotoxins. [Accessed: May 27, 2022]

[2] Savadova K, Mazur-Marzec H, Karosienė J, et al. Effect of increased temperature on native and alien nuisance cyanobacteria from temperate lakes: An experimental approach. Toxins (Basel). 2018;**10**:445. DOI: 10.3390/toxins10110445

[3] Paerl HW, Huisman J. Climate change: A catalyst for global expansion of harmful cyanobacterial blooms. Environmental Microbiology Reports. 2019;**1**:27. DOI: 10.1111/j.1758-2229.2008.00004

[4] Wu X, Hou L, Lin X, et al. Application of novel nanomaterials for chemo- and biosensing of algal toxins in shellfish and water. Novel Nanomaterials for Biomedical, Environmental and Energy Applications. 2019. DOI: 10.1016/B978-0-12-814497-8.00012-6

[5] Campos A, Vasconcelos V. Molecular mechanisms of microcystin toxicity in animal cells. International Journal of Molecular Sciences. 2010;**11**:268. DOI: 10.3390/ijms11010268

[6] World Health Organization. Cyanobacterial toxins: microcystins. Background document for development of WHO Guidelines for drinking-water quality and Guidelines for safe recreational water environments. WHO/HEP/ECH/WSH/20206. https://apps.who.int/iris/bitstream/handle/10665/338066/WHO-HEP-ECH-WSH-2020.6-eng.pdf. [Accessed: May 27, 2022]

[7] Bouaïcha N, Miles CO, Beach DG, et al. Structural diversity, characterization and toxicology of microcystins. Toxins. 2019;**11**:714. DOI: 10.3390/toxins11120714

[8] Vesterkvist PSM, Misiorek JO, Spoof LEM, et al. Comparative cellular toxicity of hydrophilic and hydrophobic microcystins on Caco-2 cells. Toxins (Basel). 2012;**25**:1008. DOI: 10.3390/toxins4111008

[9] Rudolph-Böhner S, Mierke DF, Moroder L. Molecular structure of the cyanobacterial tumor-promoting microcystins. FEBS Letters. 1994;**349**:319. DOI: 10.1016/0014-5793(94)00680-6

[10] Toivola DM, Eriksson JE, Brautigan DL. Identification of protein phosphatase 2A as the primary target for microcystin-LR in rat liver homogenates. FEBS Letters. 1994;**344**:175. DOI: 10.1016/0014-5793(94)00382-3

[11] Sun Y, Zheng Q, Sun YT, et al. Microcystin-LR induces protein phosphatase 2A alteration in a human liver cell line. Environmental Toxicology. 2014;**29**:1236. DOI: 10.1002/tox.21854

[12] Pereira SR, Vasconcelos VM, Antunes A. Computational study of the covalent bonding of microcystins to cysteine residues - a reaction involved in the inhibition of the PPP family of protein phosphatases. The FEBS Journal. 2013;**280**:674. DOI: 10.1111/j.1742-4658.2011.08454.x

[13] Gagala I, Mankiewicz-Boczek J. The natural degradation of microcystins (cyanobacterial Hepatotoxins) in fresh water - the future of modern treatment systems and water quality improvement.

Polish Journal of Environment and Studies. 2012;**21**:1125

[14] Kabak B. The fate of mycotoxins during thermal food processing. Journal of the Science of Food and Agriculture. 2009;**89**:549. DOI: 10.1002/jsfa.3491

[15] National Center for Biotechnology Information. PubChem Compound Summary for CID 445434, Microcystin-LR. https://pubchem.ncbi.nlm.nih.gov/compound/445434. [Accessed: May 27, 2022]

[16] EPA (822-R-19-001): United States Environmental Protection Agency, Ross DP. Recommended Human Health Recreational Ambient Water Quality Criteria or Swimming Advisories for Microcystins and Cylindrospermopsin. Federal Register; 84. Available from: https://www.epa.gov/sites/default/files/2019-05/documents/hh-rec-criteria-habs-document-2019.pdf [Accessed: May 27, 2022]

[17] World Health Organization. Cyanobacterial toxins: Microcystin-LR in Drinking-water. WHO/SDE/WSH/0304/57. 1998 https://www.who.int/water_sanitation_health/dwq/chemicals/cyanobactoxins.pdf. [Accessed: May 27, 2022]

[18] Kaushik R, Balasubramanian R. Methods and approaches used for detection of cyanotoxins in environmental samples: A review. Critical Reviews in Environmental Science and Technology. 2013;**43**:1349. DOI: 10.1080/10643389.2011.644224

[19] Massey IY, Wu P, Wei J, et al. A mini-review on detection methods of microcystins. Toxins. 2020;**12**:641. DOI: 10.3390/toxins12100641

[20] Kumar P, Rautela A, Kesari V, et al. Recent developments in the methods of quantitative analysis of microcystins. Journal of Biochemical and Molecular Toxicology. 2020;**34**:e22582. DOI: 10.1002/jbt.22582

[21] Novus Biologicals. Microcystin-LR ELISA Kit (Colorimetric). https://www.novusbio.com/products/microcystin-lr-elisa-kit_ka1496. [Accessed: May 27, 2022]

[22] U.S. EPA. Method 546: Determination of Total Microcystins and Nodularins in Drinking Water and Ambient Water by Adda Enzyme-Linked Immunosorbent Assay. https://www.epa.gov/sites/default/files/2016-09/documents/, https://www.epa.gov/esam/method-546-determination-total-microcystins-and-nodularins-drinking-water-and-ambient-water. [Accessed: May 27, 2022]

[23] Enzo Life Sciences. Microcystin-LR monoclonal antibody (MC10E7). https://www.enzolifesciences.com/ALX-804-320/microcystin-lr-monoclonal-antibody-mc10e7/. [Accessed: May 27, 2022]

[24] Ward CJ, Beattie KA, Lee EYC, et al. Colorimetric protein phosphatase inhibition assay of laboratory strains and natural blooms of cyanobacteria: Comparisons with high-performance liquid chromatographic analysis for microcystins. FEMS Microbiology Letters. 1997;**153**:465. DOI: 10.1016/S0378-1097(97)00290-5

[25] Bouaïcha N, Maatouk I, Vincent G, et al. A colorimetric and fluorometric microplate assay for the detection of microcystin-LR in drinking water without pre-concentration. Food and Chemical Toxicology. 2002;**40**:1677. DOI: 10.1016/S0278-6915(02)00103-5

[26] Rapala J, Erkomaa K, Kukkonen J, et al. Detection of microcystins with

protein phosphatase inhibition assay, high-performance liquid chromatography-UV detection and enzyme-linked immunosorbent assay: Comparison of methods. Analytica Chimica Acta. 2002;**466**:213. DOI: 10.1016/S0003-2670(02)00588-3

[27] Ren L, Hu Z, Wang Q, et al. Regulation efficacy and mechanism of the toxicity of microcystin-LR targeting protein phosphatase 1 via the biodegradation pathway. Toxins (Basel). 2020;**12**:790. DOI: 10.3390/toxins12120790

[28] Dillon M, Zaczek-Moczydlowska MA, Edwards C, et al. Current trends and challenges for rapid smart diagnostics at point-of-site testing for marine toxins. Sensors. 2021;**21**:2499. DOI: 10.3390/s21072499

[29] Li X, Cheng R, Shi H, et al. A simple highly sensitive and selective aptamer-based colorimetric sensor for environmental toxins microcystin-LR in water samples. Journal of Hazardous Materials. 2016;**304**:474. DOI: 10.1016/j.jhazmat.2015.11.016

[30] Lv J, Zhao S, Wu S, et al. Upconversion nanoparticles grafted molybdenum disulfide nanosheets platform for microcystin-LR sensing. Biosensors & Bioelectronics. 2017;**90**:203. DOI: 10.1016/j.bios.2016.09.110

[31] He D, Wu Z, Cui B, et al. A novel SERS-based aptasensor for ultrasensitive sensing of microcystin-LR. Food Chemistry. 2019;**278**:197. DOI: 10.1016/j.foodchem.2018.11.071

[32] Eissa S, Ng A, Siaj M, et al. Label-free voltammetric aptasensor for the sensitive detection of microcystin-LR using graphene-modified electrodes. Analytical Chemistry. 2017;**86**:7551. DOI: 10.1021/ac501335k

[33] Tang L, Ouyang X, Peng B, et al. Highly sensitive detection of microcystin-LR under visible light using a self-powered photoelectrochemical aptasensor based on a CoO/Au/g-C3N4 Z-scheme heterojunction. Nanoscale. 2019;**11**:12198. DOI: 10.1039/c9nr03004b

[34] Pang P, Teng X, Chen M, et al. Ultrasensitive enzyme-free electrochemical immunosensor for microcystin-LR using molybdenum disulfide/gold nanoclusters nanocomposites as platform and Au@Pt core-shell nanoparticles as signal enhancer. Sensors Actuators, B Chemistry. 2018;**266**:400. DOI: 10.1016/j.snb.2018.03.154

[35] Nelson DL, Cox MM. Lehninger principles of biochemistry 7th edition. W.H. Freeman; 2017. IBAN-13: 978-1464126116

[36] Zeck A, Eikenberg A, Weller MG, et al. Highly sensitive immunoassay based on a monoclonal antibody specific for [4-arginine]microcystins. Analytica Chimica Acta. 2001;**441**:1. DOI: 10.1016/S0003-2670(01)01092-3

[37] Thiviyanathan V, Gorenstein DG. Aptamers and the next generation of diagnostic reagents. Proteomics - Clinical Applications. 2012;**6**:563. DOI: 10.1002/prca.201200042

[38] Sola M, Menon AP, Moreno B, et al. Aptamers against live targets: Is In vivo SELEX finally coming to the edge? Molecular Therapy - Nucleic Acids. 2020;**21**:192. DOI: 10.1016/j.omtn.2020.05.025

[39] Ye W, Liu T, Zhang W, et al. Marine toxins detection by biosensors

based on aptamers. Toxins. 2020;**12**:1. DOI: 10.3390/toxins12010001

[40] Cunha I, Biltes R, Sales MGF, et al. Aptamer-based biosensors to detect aquatic phycotoxins and cyanotoxins. Sensors (Switzerland). 2018;**18**:2367. DOI: 10.3390/s18072367

[41] Ng A, Chinnappan R, Eissa S, et al. Selection, characterization, and biosensing application of high affinity congener-specific microcystin-targeting aptamers. Environmental Science & Technology. 2012;**46**:10697. DOI: 10.1021/es301686k

[42] Ruzicka J, Scampavia L. From flow injection to bead injection. Analytical Chemistry. 1999;**71**:257A. DOI: 10.1021/ac990293i

[43] Idris AM. An overview of the generations and recent versions of flow injection techniques. Critical Reviews in Analytical Chemistry. 2010;**40**:150. DOI: 10.1080/10408340903103437

[44] Viirlaid E, Ilisson M, Kopanchuk S, et al. Immunoassay for rapid on-site detection of glyphosate herbicide. Environmental Monitoring and Assessment. 2019;**191**:507. DOI: 10.1007/s10661-019-7657-z

[45] Peedel D, Rinken T. Rapid biosensing of Staphylococcus aureus bacteria in milk. Anal. Methods. 2014;**6**:2642. DOI: 10.1039/c3ay42036a

[46] Juronen D, Kuusk A, Kivirand K, et al. Immunosensing system for rapid multiplex detection of mastitis-causing pathogens in milk. Talanta. 2018;**178**:949. DOI: 10.1016/j.talanta.2017.10.043

[47] www.dsmz.de. Cyanobacteria Medium BG11+. https://www.dsmz.de/microorganisms/medium/pdf/DSMZ_Medium1593.pdf. [Accessed: May 27, 2022]

[48] Coomassie Brilliant Blue Stain Protocol. https://conductscience.com/coomassie-brilliant-blue-stain/ [Accessed: May 27, 2022]

[49] Zhao L, Huang Y, Dong Y, et al. Aptamers and aptasensors for highly specific recognition and sensitive detection of marine biotoxins: Recent advances and perspectives. Toxins. 2018;**10**:427. DOI: 10.3390/toxins10110427

[50] Welker M, Steinberg C. Rates of humic substance photosensitized degradation of microcystin-LR in natural waters. Environmental Science & Technology. 2000;**34**:3415. DOI: 10.1021/es991274t